Student Solutions Manual

for

McKeague's

Elementary Algebra

Eighth Edition

Ross Rueger
College of the Sequoias

BROOKS/COLE

Australia • Brazil • Canada • Mexico • Singapore • Spain • United Kingdom • United States

Printed in the United States of America
1 2 3 4 5 6 7 11 10 09 08 07

Printer: Thomson/West
Cover Image: Glowimages/Getty Images

ISBN-13: 978-0-495-38313-0
ISBN-10: 0-495-38313-9

Thomson Higher Education
10 Davis Drive
Belmont, CA 94002-3098
USA

For more information about our products,
contact us at:
Thomson Learning Academic Resource Center
1-800-423-0563

For permission to use material from this text or
product, submit a request online at
http://www.thomsonrights.com.
Any additional questions about permissions can be
submitted by email to **thomsonrights@thomson.com.**

Contents

Preface

This *Student Solutions Manual* contains complete solutions to all odd-numbered exercises and all chapter review/test exercises of *Elementary Algebra* by Charles P. McKeague. I have attempted to format solutions for readability and accuracy, and apologize to you for any errors that you may encounter. If you have any comments, suggestions, error corrections, or alternative solutions please feel free to drop me a note or send an email (address below).

Please use this manual with some degree of caution. Be sure that you have attempted a solution, and re-attempted it, before you look it up in this manual. Algebra can only be learned by *doing*, and not by observing! As you use this manual, do not just read the solution but work it along with the manual, using my solution to check your work. If you use this manual in that fashion then it should be helpful to you as you do homework and study for tests.

I would like to thank a number of people for their assistance in preparing this manual. Thanks go to Laura Localio at Thomson Brooks/Cole Publishing for her valuable assistance and support. Special thanks go to Matt Bourez of College of the Sequoias for his meticulous error-checking of my solutions, and prompt return of my manuscript under tight deadlines. Also thanks go to Patrick McKeague for keeping me updated to manuscript corrections and prompt response to my emails.

I wish to express my appreciation to Pat McKeague for asking me to be involved with this textbook. This book provides a complete course in beginning algebra, and you will find the text very easy to read and understand. Good luck!

Ross Rueger
College of the Sequoias
matmanross@aol.com

January, 2007

Chapter 1
The Basics

1.1 Notation and Symbols

1. The equivalent expression is $x + 5 = 14$.

3. The equivalent expression is $5y < 30$.

5. The equivalent expression is $3y \leq y + 6$.

7. The equivalent expression is $\dfrac{x}{3} = x + 2$.

9. Expanding the expression: $3^2 = 3 \cdot 3 = 9$

11. Expanding the expression: $7^2 = 7 \cdot 7 = 49$

13. Expanding the expression: $2^3 = 2 \cdot 2 \cdot 2 = 8$

15. Expanding the expression: $4^3 = 4 \cdot 4 \cdot 4 = 64$

17. Expanding the expression: $2^4 = 2 \cdot 2 \cdot 2 \cdot 2 = 16$

19. Expanding the expression: $10^2 = 10 \cdot 10 = 100$

21. Expanding the expression: $11^2 = 11 \cdot 11 = 121$

23.   a. Using the order of operations: $2 \cdot 3 + 5 = 6 + 5 = 11$
     b. Using the order of operations: $2(3 + 5) = 2(8) = 16$

25.   a. Using the order of operations: $5 + 2 \cdot 6 = 5 + 12 = 17$
     b. Using the order of operations: $(5 + 2) \cdot 6 = 7 \cdot 6 = 42$

27.   a. Using the order of operations: $5 \cdot 4 + 5 \cdot 2 = 20 + 10 = 30$
     b. Using the order of operations: $5(4 + 2) = 5(6) = 30$

29.   a. Using the order of operations: $8 + 2(5 + 3) = 8 + 2(8) = 8 + 16 = 24$
     b. Using the order of operations: $(8 + 2)(5 + 3) = (10)(8) = 80$

31. Using the order of operations: $20 + 2(8 - 5) + 1 = 20 + 2(3) + 1 = 20 + 6 + 1 = 27$

33. Using the order of operations: $5 + 2(3 \cdot 4 - 1) + 8 = 5 + 2(12 - 1) + 8 = 5 + 2(11) + 8 = 5 + 22 + 8 = 35$

35. Using the order of operations: $4 + 8 \div 4 - 2 = 4 + 2 - 2 = 4$

37. Using the order of operations: $3 \cdot 8 + 10 \div 2 + 4 \cdot 2 = 24 + 5 + 8 = 37$

39.   a. Using the order of operations: $(5 + 3)(5 - 3) = (8)(2) = 16$
     b. Using the order of operations: $5^2 - 3^2 = 5 \cdot 5 - 3 \cdot 3 = 25 - 9 = 16$

41.   a. Using the order of operations: $(4 + 5)^2 = 9^2 = 9 \cdot 9 = 81$
     b. Using the order of operations: $4^2 + 5^2 = 4 \cdot 4 + 5 \cdot 5 = 16 + 25 = 41$

43. Using the order of operations: $2 \cdot 10^3 + 3 \cdot 10^2 + 4 \cdot 10 + 5 = 2000 + 300 + 40 + 5 = 2,345$

45. Using the order of operations: $10 - 2(4 \cdot 5 - 16) = 10 - 2(20 - 16) = 10 - 2(4) = 10 - 8 = 2$

47. Using the order of operations: $4[7 + 3(2 \cdot 9 - 8)] = 4[7 + 3(18 - 8)] = 4[7 + 3(10)] = 4(7 + 30) = 4(37) = 148$

49. Using the order of operations: $3(4 \cdot 5 - 12) + 6(7 \cdot 6 - 40) = 3(20 - 12) + 6(42 - 40) = 3(8) + 6(2) = 24 + 12 = 36$

51. Using the order of operations: $3^4 + 4^2 \div 2^3 - 5^2 = 81 + 16 \div 8 - 25 = 81 + 2 - 25 = 58$

53.   Using the order of operations: $5^2 + 3^4 \div 9^2 + 6^2 = 25 + 81 \div 81 + 36 = 25 + 1 + 36 = 62$
55.   Using the order of operations: $20 \div 2 \cdot 10 = 10 \cdot 10 = 100$
57.   Using the order of operations: $24 \div 8 \cdot 3 = 3 \cdot 3 = 9$
59.   Using the order of operations: $36 \div 6 \cdot 3 = 6 \cdot 3 = 18$
61.   Using the order of operations: $16 - 8 + 4 = 8 + 4 = 12$
63.   Using the order of operations: $24 - 14 + 8 = 10 + 8 = 18$
65.   Using the order of operations: $36 - 6 + 12 = 30 + 12 = 42$
67.   Computing the value: $0.08 + 0.09 = 0.17$ **69.**   Computing the value: $0.10 + 0.12 = 0.22$
71.   Computing the value: $4.8 - 2.5 = 2.3$ **73.**   Computing the value: $2.07 + 3.48 = 5.55$
75.   Computing the value: $0.12(2,000) = 240$ **77.**   Computing the value: $0.25(40) = 10$
79.   Computing the value: $510 \div 0.17 = 3,000$ **81.**   Computing the value: $240 \div 0.12 = 2,000$
83.   Using a calculator: $37.80 \div 1.07 \approx 35.33$ **85.**   Using a calculator: $555 \div 740 = 0.75$
87.   Using a calculator: $70 \div 210 \approx 0.33$ **89.**   Using a calculator: $6,000 \div 22 \approx 272.73$
91.   There are $5 \cdot 2 = 10$ cookies in the package.   **93.**   The total number of calories is $210 \cdot 2 = 420$ calories.
95.   **a.**   The amount of caffeine is: $6(100) = 600$ mg

     b.   The amount of caffeine is: $2(45) + 3(47) = 90 + 141 = 231$ mg

97.   Completing the table:

Activity	Calories burned in 1 hour
Bicycling	374
Bowling	265
Handball	680
Jogging	680
Skiing	544

99.   The next number is 5.   **101.**   The next number is 10.
103.   The next number is $5^2 = 25$.
105.   Since $2 + 2 = 4$ and $2 + 4 = 6$, the next number is $4 + 6 = 10$.

1.2 Real Numbers

1.   Labeling the point: **3.**   Labeling the point:

5.   Labeling the point: **7.**   Labeling the point:

9.   Building the fraction: $\frac{3}{4} = \frac{3}{4} \cdot \frac{6}{6} = \frac{18}{24}$ **11.**   Building the fraction: $\frac{1}{2} = \frac{1}{2} \cdot \frac{12}{12} = \frac{12}{24}$

13.   Building the fraction: $\frac{5}{8} = \frac{5}{8} \cdot \frac{3}{3} = \frac{15}{24}$ **15.**   Building the fraction: $\frac{3}{5} = \frac{3}{5} \cdot \frac{12}{12} = \frac{36}{60}$

17.   Building the fraction: $\frac{11}{30} = \frac{11}{30} \cdot \frac{2}{2} = \frac{22}{60}$

19.   Building the fraction: $\frac{1}{2} \cdot \frac{2}{2} = \frac{2}{4}$, so the missing numerator is 2.

21.   Building the fraction: $\frac{5}{9} \cdot \frac{5}{5} = \frac{25}{45}$, so the missing numerator is 25.

23.   Building the fraction: $\frac{3}{4} \cdot \frac{2}{2} = \frac{6}{8}$, so the missing numerator is 6.

25.   The opposite of 10 is –10, the reciprocal is $\frac{1}{10}$, and the absolute value is $|10| = 10$.

27.   The opposite of –3 is 3, the reciprocal is $-\frac{1}{3}$, and the absolute value is $|-3| = 3$.

29.   The opposite of x is $-x$, the reciprocal is $\frac{1}{x}$, and the absolute value is $|x|$.

31. The correct symbol is $<$: $-5 < -3$ **33.** The correct symbol is $>$: $-3 > -7$

35. Since $\left|-4\right| = 4$ and $-\left|-4\right| = -4$, the correct symbol is $>$: $\left|-4\right| > -\left|-4\right|$

37. Since $-\left|-7\right| = -7$, the correct symbol is $>$: $7 > -\left|-7\right|$

39. The correct symbol is $<$: $-\frac{3}{4} < -\frac{1}{4}$ **41.** The correct symbol is $<$: $-\frac{3}{2} < -\frac{3}{4}$

43. Simplifying the expression: $\left|8-2\right| = \left|6\right| = 6$

45. Simplifying the expression: $\left|5 \cdot 2^3 - 2 \cdot 3^2\right| = \left|5 \cdot 8 - 2 \cdot 9\right| = \left|40 - 18\right| = \left|22\right| = 22$

47. Simplifying the expression: $\left|7-2\right| - \left|4-2\right| = \left|5\right| - \left|2\right| = 5 - 2 = 3$

49. Simplifying the expression: $10 - \left|7 - 2(5-3)\right| = 10 - \left|7 - 2(2)\right| = 10 - \left|7 - 4\right| = 10 - \left|3\right| = 10 - 3 = 7$

51. Simplifying the expression:
$$\begin{aligned}
15 - \left|8 - 2(3 \cdot 4 - 9)\right| - 10 &= 15 - \left|8 - 2(12 - 9)\right| - 10 \\
&= 15 - \left|8 - 2(3)\right| - 10 \\
&= 15 - \left|8 - 6\right| - 10 \\
&= 15 - \left|2\right| - 10 \\
&= 15 - 2 - 10 \\
&= 3
\end{aligned}$$

53. Multiplying the fractions: $\frac{2}{3} \cdot \frac{4}{5} = \frac{8}{15}$ **55.** Multiplying the fractions: $\frac{1}{2}(3) = \frac{1}{2} \cdot \frac{3}{1} = \frac{3}{2}$

57. Multiplying the fractions: $\frac{4}{3} \cdot \frac{3}{4} = \frac{12}{12} = 1$ **59.** Multiplying the fractions: $3 \cdot \frac{1}{3} = \frac{3}{1} \cdot \frac{1}{3} = \frac{3}{3} = 1$

61.
  a. Multiplying the numbers: $\frac{1}{2}(4) = \frac{1}{2} \cdot \frac{4}{1} = \frac{4}{2} = 2$

  b. Multiplying the numbers: $\frac{1}{2}(8) = \frac{1}{2} \cdot \frac{8}{1} = \frac{8}{2} = 4$

  c. Multiplying the numbers: $\frac{1}{2}(16) = \frac{1}{2} \cdot \frac{16}{1} = \frac{16}{2} = 8$

  d. Multiplying the numbers: $\frac{1}{2}(0.06) = 0.03$

63.
  a. Multiplying the numbers: $\frac{3}{2}(4) = \frac{3}{2} \cdot \frac{4}{1} = \frac{12}{2} = 6$

  b. Multiplying the numbers: $\frac{3}{2}(8) = \frac{3}{2} \cdot \frac{8}{1} = \frac{24}{2} = 12$

  c. Multiplying the numbers: $\frac{3}{2}(16) = \frac{3}{2} \cdot \frac{16}{1} = \frac{48}{2} = 24$

  d. Multiplying the numbers: $\frac{3}{2}(0.06) = 0.09$

65. Expanding the exponent: $\left(\frac{3}{4}\right)^2 = \frac{3}{4} \cdot \frac{3}{4} = \frac{9}{16}$ **67.** Expanding the exponent: $\left(\frac{2}{3}\right)^3 = \frac{2}{3} \cdot \frac{2}{3} \cdot \frac{2}{3} = \frac{8}{27}$

69.
  a. Substituting $x = 5$: $2x - 6 = 2(5) - 6 = 10 - 6 = 4$

  b. Substituting $x = 10$: $2x - 6 = 2(10) - 6 = 20 - 6 = 14$

  c. Substituting $x = 15$: $2x - 6 = 2(15) - 6 = 30 - 6 = 24$

  d. Substituting $x = 20$: $2x - 6 = 2(20) - 6 = 40 - 6 = 34$

71.
  a. Substituting $x = 10$: $x + 2 = 10 + 2 = 12$ **b.** Substituting $x = 10$: $2x = 2(10) = 20$

  c. Substituting $x = 10$: $x^2 = (10)^2 = 100$ **d.** Substituting $x = 10$: $2^x = 2^{10} = 1{,}024$

73.
  a. Substituting $x = 4$: $x^2 + 1 = (4)^2 + 1 = 16 + 1 = 17$

  b. Substituting $x = 4$: $(x+1)^2 = (4+1)^2 = (5)^2 = 25$

  c. Substituting $x = 4$: $x^2 + 2x + 1 = (4)^2 + 2(4) + 1 = 16 + 8 + 1 = 25$

75. The perimeter is $4(1 \text{ in.}) = 4$ in., and the area is $(1 \text{ in.})^2 = 1 \text{ in.}^2$.

77. The perimeter is $2(1.5 \text{ in.}) + 2(0.75 \text{ in.}) = 3.0 \text{ in.} + 1.5 \text{ in.} = 4.5 \text{ in.}$, and the area is $(1.5 \text{ in.})(0.75 \text{ in.}) = 1.125 \text{ in.}^2$.

79. The perimeter is $2.75 \text{ cm} + 4 \text{ cm} + 3.5 \text{ cm} = 10.25 \text{ cm}$, and the area is $\frac{1}{2}(4 \text{ cm})(2.5 \text{ cm}) = 5.0 \text{ cm}^2$.

81. A loss of 8 yards corresponds to –8 on a number line. The total yards gained corresponds to –2 yards.

83. The temperature can be represented as –64°. The new (warmer) temperature corresponds to –54°.

85. His position corresponds to –100 feet. His new (deeper) position corresponds to –105 feet.

87. The area is given by: $8\frac{1}{2} \cdot 11 = \frac{17}{2} \cdot \frac{11}{1} = \frac{187}{2} = 93.5 \text{ in.}^2$

The perimeter is given by: $2\left(8\frac{1}{2}\right) + 2(11) = 17 + 22 = 39 \text{ in.}$

89. The calories consumed would be: $2(544) + 299 = 1,387 \text{ calories}$

91. The calories consumed by the 180 lb person would be $3(653) = 1,959 \text{ calories}$, while the calories consumed by the 120 lb person would be $3(435) = 1,305 \text{ calories}$. Thus the 180 lb person consumed $1959 - 1306 = 654$ more calories.

93. Answers will vary.

1.3 Addition of Real Numbers

1. Adding all positive and negative combinations of 3 and 5:
   $3 + 5 = 8$ $3 + (-5) = -2$
   $-3 + 5 = 2$ $(-3) + (-5) = -8$

3. Adding all positive and negative combinations of 15 and 20:
   $15 + 20 = 35$ $15 + (-20) = -5$
   $-15 + 20 = 5$ $(-15) + (-20) = -35$

5. Adding the numbers: $6 + (-3) = 3$

7. Adding the numbers: $13 + (-20) = -7$

9. Adding the numbers: $18 + (-32) = -14$

11. Adding the numbers: $-6 + 3 = -3$

13. Adding the numbers: $-30 + 5 = -25$

15. Adding the numbers: $-6 + (-6) = -12$

17. Adding the numbers: $-9 + (-10) = -19$

19. Adding the numbers: $-10 + (-15) = -25$

21. Performing the additions: $5 + (-6) + (-7) = 5 + (-13) = -8$

23. Performing the additions: $-7 + 8 + (-5) = -12 + 8 = -4$

25. Performing the additions: $5 + [6 + (-2)] + (-3) = 5 + 4 + (-3) = 9 + (-3) = 6$

27. Performing the additions: $[6 + (-2)] + [3 + (-1)] = 4 + 2 = 6$

29. Performing the additions: $20 + (-6) + [3 + (-9)] = 20 + (-6) + (-6) = 20 + (-12) = 8$

31. Performing the additions: $-3 + (-2) + [5 + (-4)] = -3 + (-2) + 1 = -5 + 1 = -4$

33. Performing the additions: $(-9 + 2) + [5 + (-8)] + (-4) = -7 + (-3) + (-4) = -14$

35. Performing the additions: $[-6 + (-4)] + [7 + (-5)] + (-9) = -10 + 2 + (-9) = -19 + 2 = -17$

37. Performing the additions: $(-6 + 9) + (-5) + (-4 + 3) + 7 = 3 + (-5) + (-1) + 7 = 10 + (-6) = 4$

39. Using order of operations: $-5 + 2(-3 + 7) = -5 + 2(4) = -5 + 8 = 3$

41. Using order of operations: $9 + 3(-8 + 10) = 9 + 3(2) = 9 + 6 = 15$

43. Using order of operations: $-10 + 2(-6 + 8) + (-2) = -10 + 2(2) + (-2) = -10 + 4 + (-2) = -12 + 4 = -8$

45. Using order of operations: $2(-4 + 7) + 3(-6 + 8) = 2(3) + 3(2) = 6 + 6 = 12$

47. Adding the decimals: $3.9 + 7.1 = 11.0 = 11$

49. Adding the decimals: $8.1 + 2.7 = 10.8$

51. Substituting $x = 7,000$ and $y = 8,000$: $0.06x + 0.07y = 0.06(7,000) + 0.07(8,000) = 420 + 560 = 980$

53. Substituting $x = 10$ and $y = 12$: $0.05x + 0.10y = 0.05(10) + 0.10(12) = 0.5 + 1.2 = 1.7$

55. The pattern is to add 5, so the next two terms are $18 + 5 = 23$ and $23 + 5 = 28$.

57. The pattern is to add 5, so the next two terms are $25 + 5 = 30$ and $30 + 5 = 35$.

59. The pattern is to add –5, so the next two terms are $5 + (-5) = 0$ and $0 + (-5) = -5$.

61. The pattern is to add –6, so the next two terms are $-6 + (-6) = -12$ and $-12 + (-6) = -18$.

63. The pattern is to add –4, so the next two terms are $0 + (-4) = -4$ and $-4 + (-4) = -8$.

65. Yes, since each successive odd number is 2 added to the previous one.

67. The expression is: $5 + 9 = 14$

69. The expression is: $[-7 + (-5)] + 4 = -12 + 4 = -8$

71. The expression is: $[-2 + (-3)] + 10 = -5 + 10 = 5$

73. The number is 3, since $-8 + 3 = -5$.

75. The number is -3, since $-6 + (-3) = -9$

77. The expression is $-12° + 4° = -8°$.

79. The expression is $\$10 + (-\$6) + (-\$8) = \$10 + (-\$14) = -\$4$.

81. The new balance is $-\$30 + \$40 = \$10$.

1.4 Subtraction of Real Numbers

1. Subtracting the numbers: $5 - 8 = 5 + (-8) = -3$

3. Subtracting the numbers: $3 - 9 = 3 + (-9) = -6$

5. Subtracting the numbers: $5 - 5 = 5 + (-5) = 0$

7. Subtracting the numbers: $-8 - 2 = -8 + (-2) = -10$

9. Subtracting the numbers: $-4 - 12 = -4 + (-12) = -16$

11. Subtracting the numbers: $-6 - 6 = -6 + (-6) = -12$

13. Subtracting the numbers: $-8 - (-1) = -8 + 1 = -7$

15. Subtracting the numbers: $15 - (-20) = 15 + 20 = 35$

17. Subtracting the numbers: $-4 - (-4) = -4 + 4 = 0$

19. Using order of operations: $3 - 2 - 5 = 3 + (-2) + (-5) = 3 + (-7) = -4$

21. Using order of operations: $9 - 2 - 3 = 9 + (-2) + (-3) = 9 + (-5) = 4$

23. Using order of operations: $-6 - 8 - 10 = -6 + (-8) + (-10) = -24$

25. Using order of operations: $-22 + 4 - 10 = -22 + 4 + (-10) = -32 + 4 = -28$

27. Using order of operations: $10 - (-20) - 5 = 10 + 20 + (-5) = 30 + (-5) = 25$

29. Using order of operations: $8 - (2 - 3) - 5 = 8 - (-1) - 5 = 8 + 1 + (-5) = 9 + (-5) = 4$

31. Using order of operations: $7 - (3 - 9) - 6 = 7 - (-6) - 6 = 7 + 6 + (-6) = 13 + (-6) = 7$

33. Using order of operations: $5 - (-8 - 6) - 2 = 5 - (-14) - 2 = 5 + 14 + (-2) = 19 + (-2) = 17$

35. Using order of operations: $-(5 - 7) - (2 - 8) = -(-2) - (-6) = 2 + 6 = 8$

37. Using order of operations: $-(3 - 10) - (6 - 3) = -(-7) - 3 = 7 + (-3) = 4$

39. Using order of operations: $16 - [(4 - 5) - 1] = 16 - (-1 - 1) = 16 - (-2) = 16 + 2 = 18$

41. Using order of operations: $5 - [(2 - 3) - 4] = 5 - (-1 - 4) = 5 - (-5) = 5 + 5 = 10$

43. Using order of operations:
$$21 - [-(3 - 4) - 2] - 5 = 21 - [-(-1) - 2] - 5 = 21 - (1 - 2) - 5 = 21 - (-1) - 5 = 21 + 1 + (-5) = 22 + (-5) = 17$$

45. Using order of operations: $2 \cdot 8 - 3 \cdot 5 = 16 - 15 = 16 + (-15) = 1$

47. Using order of operations: $3 \cdot 5 - 2 \cdot 7 = 15 - 14 = 15 + (-14) = 1$

49. Using order of operations: $5 \cdot 9 - 2 \cdot 3 - 6 \cdot 2 = 45 - 6 - 12 = 45 + (-6) + (-12) = 45 + (-18) = 27$

51. Using order of operations: $3 \cdot 8 - 2 \cdot 4 - 6 \cdot 7 = 24 - 8 - 42 = 24 + (-8) + (-42) = 24 + (-50) = -26$

53. Using order of operations: $2 \cdot 3^2 - 5 \cdot 2^2 = 2 \cdot 9 - 5 \cdot 4 = 18 - 20 = 18 + (-20) = -2$

55. Using order of operations: $4 \cdot 3^3 - 5 \cdot 2^3 = 4 \cdot 27 - 5 \cdot 8 = 108 - 40 = 108 + (-40) = 68$

57. Subtracting decimals: $-3.4 - 7.9 = -3.4 + (-7.9) = -11.3$

59. Subtracting decimals: $3.3 - 6.9 = 3.3 + (-6.9) = -3.6$

61. **a.** Substituting $x = -3$ and $y = -2$: $x + y - 4 = -3 + (-2) - 4 = -5 - 4 = -9$

   b. Substituting $x = -9$ and $y = 3$: $x + y - 4 = -9 + 3 - 4 = -6 - 4 = -10$

   c. Substituting $x = -\frac{3}{5}$ and $y = \frac{8}{5}$: $x + y - 4 = -\frac{3}{5} + \frac{8}{5} - 4 = \frac{5}{5} - 4 = 1 + (-4) = -3$

63. Writing the expression: $-7 - 4 = -7 + (-4) = -11$

65. Writing the expression: $12 - (-8) = 12 + 8 = 20$

67. Writing the expression: $-5 - (-7) = -5 + 7 = 2$

69. Writing the expression: $[4 + (-5)] - 17 = -1 - 17 = -1 + (-17) = -18$

71. Writing the expression: $8 - 5 = 8 + (-5) = 3$

73. Writing the expression: $-8 - 5 = -8 + (-5) = -13$

75. Writing the expression: $8 - (-5) = 8 + 5 = 13$

77. The number is 10, since $8 - 10 = 8 + (-10) = -2$.

79. The number is -2, since $8 - (-2) = 8 + 2 = 10$.

81. The expression is $\$1,500 - \$730 = \$770$.

83. The expression is $-\$35 + \$15 - \$20 = -\$35 + (-\$20) + \$15 = -\$55 + \$15 = -\$40$.

85. The expression is $\$98 - \$65 - \$53 = \$98 + (-\$65) + (-\$53) = \$98 + (-\$118) = -\$20$.

87. The sequence of values is $4500, $3950, $3400, $2850, and $2300. This is an arithmetic sequence, since –$550 is added to each value to obtain the new value.

89. The difference is $1000 \text{ feet} - 231 \text{ feet} = 769 \text{ feet}$.   **91.** He is 439 feet from the starting line.

93. 2 seconds have gone by.   **95.** The angles add to 90°, so $x = 90° - 55° = 35°$.

97. The angles add to 180°, so $x = 180° - 120° = 60°$.

99. **a.** Completing the table:

Day	Plant Height (inches)
0	0
1	0.5
2	1
3	1.5
4	3
5	4
6	6
7	9
8	13
9	18
10	23

b. Subtracting: $13 - 1.5 = 11.5$. The grass is 11.5 inches higher after 8 days than after 3 days.

1.5 Properties of Real Numbers

1. commutative property (of addition)   **3.** multiplicative inverse property

5. commutative property (of addition)   **7.** distributive property

9. commutative and associative properties (of addition)   **11.** commutative and associative properties (of addition)

13. commutative property (of addition)

15. commutative and associative properties (of multiplication)

17. commutative property (of multiplication)   **19.** additive inverse property

21. The expression should read $3(x + 2) = 3x + 6$.   **23.** The expression should read $9(a + b) = 9a + 9b$.

25. The expression should read $3(0) = 0$.   **27.** The expression should read $3 + (-3) = 0$.

29. The expression should read $10(1) = 10$.

31. Simplifying the expression: $4 + (2 + x) = (4 + 2) + x = 6 + x$

33. Simplifying the expression: $(x + 2) + 7 = x + (2 + 7) = x + 9$

35. Simplifying the expression: $3(5x) = (3 \bullet 5)x = 15x$ **37.** Simplifying the expression: $9(6y) = (9 \bullet 6)y = 54y$

39. Simplifying the expression: $\frac{1}{2}(3a) = \left(\frac{1}{2} \bullet 3\right)a = \frac{3}{2}a$ **41.** Simplifying the expression: $\frac{1}{3}(3x) = \left(\frac{1}{3} \bullet 3\right)x = 1x = x$

43. Simplifying the expression: $\frac{1}{2}(2y) = \left(\frac{1}{2} \bullet 2\right)y = 1y = y$ **45.** Simplifying the expression: $\frac{3}{4}\left(\frac{4}{3}x\right) = \left(\frac{3}{4} \bullet \frac{4}{3}\right)x = 1x = x$

47. Simplifying the expression: $\frac{6}{5}\left(\frac{5}{6}a\right) = \left(\frac{6}{5} \bullet \frac{5}{6}\right)a = 1a = a$

49. Applying the distributive property: $8(x + 2) = 8 \bullet x + 8 \bullet 2 = 8x + 16$

51. Applying the distributive property: $8(x - 2) = 8 \bullet x - 8 \bullet 2 = 8x - 16$

53. Applying the distributive property: $4(y + 1) = 4 \bullet y + 4 \bullet 1 = 4y + 4$

55. Applying the distributive property: $3(6x + 5) = 3 \bullet 6x + 3 \bullet 5 = 18x + 15$

57. Applying the distributive property: $2(3a + 7) = 2 \bullet 3a + 2 \bullet 7 = 6a + 14$

59. Applying the distributive property: $9(6y - 8) = 9 \bullet 6y - 9 \bullet 8 = 54y - 72$

61. Applying the distributive property: $\frac{1}{2}(3x - 6) = \frac{1}{2} \bullet 3x - \frac{1}{2} \bullet 6 = \frac{3}{2}x - 3$

63. Applying the distributive property: $\frac{1}{3}(3x + 6) = \frac{1}{3} \bullet 3x + \frac{1}{3} \bullet 6 = x + 2$

65. Applying the distributive property: $3(x+y) = 3x + 3y$

67. Applying the distributive property: $8(a-b) = 8a - 8b$

69. Applying the distributive property: $6(2x+3y) = 6 \cdot 2x + 6 \cdot 3y = 12x + 18y$

71. Applying the distributive property: $4(3a-2b) = 4 \cdot 3a - 4 \cdot 2b = 12a - 8b$

73. Applying the distributive property: $\frac{1}{2}(6x+4y) = \frac{1}{2} \cdot 6x + \frac{1}{2} \cdot 4y = 3x + 2y$

75. Applying the distributive property: $4(a+4)+9 = 4a+16+9 = 4a+25$

77. Applying the distributive property: $2(3x+5)+2 = 6x+10+2 = 6x+12$

79. Applying the distributive property: $7(2x+4)+10 = 14x+28+10 = 14x+38$

81. Applying the distributive property: $\frac{1}{2}(4x+2) = \frac{1}{2} \cdot 4x + \frac{1}{2} \cdot 2 = 2x+1$

83. Applying the distributive property: $\frac{3}{4}(8x-4) = \frac{3}{4} \cdot 8x - \frac{3}{4} \cdot 4 = 6x - 3$

85. Applying the distributive property: $\frac{5}{6}(6x+12) = \frac{5}{6} \cdot 6x + \frac{5}{6} \cdot 12 = 5x+10$

87. Applying the distributive property: $10\left(\frac{3}{5}x+\frac{1}{2}\right) = 10 \cdot \frac{3}{5}x + 10 \cdot \frac{1}{2} = 6x+5$

89. Applying the distributive property: $15\left(\frac{1}{3}x+\frac{2}{5}\right) = 15 \cdot \frac{1}{3}x + 15 \cdot \frac{2}{5} = 5x+6$

91. Applying the distributive property: $12\left(\frac{1}{2}m-\frac{5}{12}\right) = 12 \cdot \frac{1}{2}m - 12 \cdot \frac{5}{12} = 6m-5$

93. Applying the distributive property: $21\left(\frac{1}{3}+\frac{1}{7}x\right) = 21 \cdot \frac{1}{3} + 21 \cdot \frac{1}{7}x = 7+3x$

95. Applying the distributive property: $6\left(\frac{1}{2}x-\frac{1}{3}y\right) = 6 \cdot \frac{1}{2}x - 6 \cdot \frac{1}{3}y = 3x-2y$

97. Applying the distributive property: $0.09(x+2,000) = 0.09x + 180$

99. Applying the distributive property: $0.12(x+500) = 0.12x + 60$

101. Applying the distributive property: $a\left(1+\frac{1}{a}\right) = a \cdot 1 + a \cdot \frac{1}{a} = a+1$

103. Applying the distributive property: $a\left(\frac{1}{a}-1\right) = a \cdot \frac{1}{a} - a \cdot 1 = 1-a$

105. No. The man cannot reverse the order of putting on his socks and putting on his shoes.

107. No. The skydiver must jump out of the plane before pulling the rip cord.

109. Division is not a commutative operation. For example, $8 \div 4 = 2$ while $4 \div 8 = \frac{1}{2}$.

111. Computing the yearly take-home pay:

$$12(2400 - 480) = 12(1920) = \$23,040 \qquad 12 \cdot 2400 - 12 \cdot 480 = 28,800 - 5,760 = \$23,040$$

113. Computing her annual expenses:

$$3(650 + 225) = 3(875) = \$2,625 \qquad 3 \cdot 650 + 3 \cdot 225 = 1950 + 675 = \$2,625$$

1.6 Multiplication of Real Numbers

1. Finding the product: $7(-6) = -42$

3. Finding the product: $-8(2) = -16$

5. Finding the product: $-3(-1) = 3$

7. Finding the product: $-11(-11) = 121$

9. Using order of operations: $-3(2)(-1) = 6$

11. Using order of operations: $-3(-4)(-5) = -60$

13. Using order of operations: $-2(-4)(-3)(-1) = 24$

15. Using order of operations: $(-7)^2 = (-7)(-7) = 49$

17. Using order of operations: $(-3)^3 = (-3)(-3)(-3) = -27$

19. Using order of operations: $-2(2-5) = -2(-3) = 6$

21. Using order of operations: $-5(8-10) = -5(-2) = 10$

23. Using order of operations: $(4-7)(6-9) = (-3)(-3) = 9$

25. Using order of operations: $(-3-2)(-5-4) = (-5)(-9) = 45$

27. Using order of operations: $-3(-6)+4(-1) = 18+(-4) = 14$

29. Using order of operations: $2(3)-3(-4)+4(-5) = 6+12+(-20) = 18+(-20) = -2$

31. Using order of operations: $4(-3)^2 + 5(-6)^2 = 4(9) + 5(36) = 36+180 = 216$

33. Using order of operations: $7(-2)^3 - 2(-3)^3 = 7(-8) - 2(-27) = -56 + 54 = -2$
35. Using order of operations: $6 - 4(8 - 2) = 6 - 4(6) = 6 - 24 = 6 + (-24) = -18$
37. Using order of operations: $9 - 4(3 - 8) = 9 - 4(-5) = 9 + 20 = 29$
39. Using order of operations: $-4(3 - 8) - 6(2 - 5) = -4(-5) - 6(-3) = 20 + 18 = 38$
41. Using order of operations: $7 - 2\left[-6 - 4(-3)\right] = 7 - 2(-6 + 12) = 7 - 2(6) = 7 - 12 = 7 + (-12) = -5$
43. Using order of operations:

$$7 - 3\left[2(-4 - 4) - 3(-1 - 1)\right] = 7 - 3\left[2(-8) - 3(-2)\right] = 7 - 3(-16 + 6) = 7 - 3(-10) = 7 + 30 = 37$$

45. **a.** Simplifying: $5(-4)(-3) = (-20)(-3) = 60$

  b. Simplifying: $5(-4) - 3 = -20 - 3 = -23$

  c. Simplifying: $5 - 4(-3) = 5 - (-12) = 5 + 12 = 17$

  d. Simplifying: $5 - 4 - 3 = 1 - 3 = -2$

47. Multiplying the fractions: $-\frac{2}{3} \cdot \frac{5}{7} = -\frac{2 \cdot 5}{3 \cdot 7} = -\frac{10}{21}$ **49.** Multiplying the fractions: $-8\left(\frac{1}{2}\right) = -\frac{8}{1} \cdot \frac{1}{2} = -\frac{8}{2} = -4$

51. Multiplying the fractions: $\left(-\frac{3}{4}\right)^2 = \left(-\frac{3}{4}\right)\left(-\frac{3}{4}\right) = \frac{9}{16}$

53. **a.** Simplifying: $\frac{5}{8}(24) + \frac{3}{7}(28) = 15 + 12 = 27$ **b.** Simplifying: $\frac{5}{8}(24) - \frac{3}{7}(28) = 15 - 12 = 3$

  c. Simplifying: $\frac{5}{8}(-24) + \frac{3}{7}(-28) = -15 - 12 = -27$

  d. Simplifying: $-\frac{5}{8}(24) - \frac{3}{7}(28) = -15 - 12 = -27$

55. Simplifying: $\left(\frac{1}{2} \cdot 6\right)^2 = (3)^2 = (3)(3) = 9$ **57.** Simplifying: $\left(\frac{1}{2} \cdot 5\right)^2 = \left(\frac{5}{2}\right)^2 = \left(\frac{5}{2}\right)\left(\frac{5}{2}\right) = \frac{25}{4}$

59. Simplifying: $\left(\frac{1}{2}(-4)\right)^2 = (-2)^2 = (-2)(-2) = 4$ **61.** Simplifying: $\left(\frac{1}{2}(-3)\right)^2 = \left(-\frac{3}{2}\right)^2 = \left(-\frac{3}{2}\right)\left(-\frac{3}{2}\right) = \frac{9}{4}$

63. Multiplying the expressions: $-2(4x) = (-2 \cdot 4)x = -8x$
65. Multiplying the expressions: $-7(-6x) = \left[-7 \cdot (-6)\right]x = 42x$
67. Multiplying the expressions: $-\frac{1}{3}(-3x) = \left[-\frac{1}{3} \cdot (-3)\right]x = 1x = x$
69. Simplifying the expression: $-\frac{1}{2}(3x - 6) = -\frac{1}{2}(3x) - \frac{1}{2}(-6) = -\frac{3}{2}x + 3$
71. Simplifying the expression: $-3(2x - 5) - 7 = -6x + 15 - 7 = -6x + 8$
73. Simplifying the expression: $-5(3x + 4) - 10 = -15x - 20 - 10 = -15x - 30$
75. Simplifying the expression: $-4(3x + 5y) = -4(3x) - 4(5y) = -12x - 20y$
77. Simplifying the expression: $-2(3x + 5y) = -2(3x) - 2(5y) = -6x - 10y$
79. Simplifying the expression: $\frac{1}{2}(-3x + 6) = \frac{1}{2}(-3x) + \frac{1}{2}(6) = -\frac{3}{2}x + 3$
81. Simplifying the expression: $\frac{1}{3}(-2x + 6) = \frac{1}{3}(-2x) + \frac{1}{3}(6) = -\frac{2}{3}x + 2$
83. Simplifying the expression: $-\frac{1}{3}(-2x + 6) = -\frac{1}{3}(-2x) - \frac{1}{3}(6) = \frac{2}{3}x - 2$
85. Simplifying the expression: $8\left(-\frac{1}{4}x + \frac{1}{8}y\right) = 8\left(-\frac{1}{4}x\right) + 8\left(\frac{1}{8}y\right) = -2x + y$

87. **a.** Substituting $x = 0$: $-\frac{1}{3}x + 2 = -\frac{1}{3}(0) + 2 = 0 + 2 = 2$

  b. Substituting $x = 3$: $-\frac{1}{3}x + 2 = -\frac{1}{3}(3) + 2 = -1 + 2 = 1$

  c. Substituting $x = -3$: $-\frac{1}{3}x + 2 = -\frac{1}{3}(-3) + 2 = 1 + 2 = 3$

89.   **a.**   Substituting $x = 2$ and $y = -1$: $2x + y = 2(2) + (-1) = 4 - 1 = 3$

   b.   Substituting $x = 0$ and $y = 3$: $2x + y = 2(0) + (0) = 0 + 3 = 3$

   c.   Substituting $x = \frac{3}{2}$ and $y = -7$: $2x + y = 2\left(\frac{3}{2}\right) + (-7) = 3 + (-7) = -4$

91.   **a.**   Substituting $x = 4$: $2x^2 - 5x = 2(4)^2 - 5(4) = 2(16) - 5(4) = 32 - 20 = 12$

   b.   Substituting $x = -\frac{3}{2}$: $2x^2 - 5x = 2\left(-\frac{3}{2}\right)^2 - 5\left(-\frac{3}{2}\right) = 2\left(\frac{9}{4}\right) - 5\left(-\frac{3}{2}\right) = \frac{9}{2} + \frac{15}{2} = \frac{24}{2} = 12$

93.   **a.**   Substituting $y = 4$: $y(2y + 3) = 4\left[2(4) + 3\right] = 4(8 + 3) = 4(11) = 44$

   b.   Substituting $y = -\frac{11}{2}$: $y(2y + 3) = -\frac{11}{2}\left[2\left(-\frac{11}{2}\right) + 3\right] = -\frac{11}{2}(-11 + 3) = -\frac{11}{2}(-8) = \frac{88}{2} = 44$

95.   Writing the expression: $3(-10) + 5 = -30 + 5 = -25$ **97.**   Writing the expression: $2(-4x) = -8x$

99.   Writing the expression: $-9 \bullet 2 - 8 = -18 + (-8) = -26$

101.   The pattern is to multiply by 2, so the next number is $4 \bullet 2 = 8$.

103.   The pattern is to multiply by -2, so the next number is $40 \bullet (-2) = -80$.

105.   The pattern is to multiply by $\frac{1}{2}$, so the next number is $\frac{1}{4} \bullet \frac{1}{2} = \frac{1}{8}$.

107.   The pattern is to multiply by -2, so the next number is $12 \bullet (-2) = -24$.

109.   The amount lost is: $20(\$3) = \$60$ **111.**   The temperature is: $25° - 4(6°) = 25° - 24° = 1°\,F$

113.   **a.**   The difference is: $\$2,484 - \$1,940 = \$544$

   b.   The difference is: $\$1,115 - \$1,636 = -\$521$

   c.   The actual costs have decreased, due to increased tax credits and student grants.

1.7 Division of Real Numbers

1.   Finding the quotient: $\dfrac{8}{-4} = -2$ **3.**   Finding the quotient: $\dfrac{-48}{16} = -3$

5.   Finding the quotient: $\dfrac{-7}{21} = -\dfrac{1}{3}$ **7.**   Finding the quotient: $\dfrac{-39}{-13} = 3$

9.   Finding the quotient: $\dfrac{-6}{-42} = \dfrac{1}{7}$ **11.**   Finding the quotient: $\dfrac{0}{-32} = 0$

13.   Performing the operations: $-3 + 12 = 9$ **15.**   Performing the operations: $-3 - 12 = -3 + (-12) = -15$

17.   Performing the operations: $-3(12) = -36$ **19.**   Performing the operations: $-3 \div 12 = \dfrac{-3}{12} = -\dfrac{1}{4}$

21.   Dividing and reducing: $\dfrac{4}{5} \div \dfrac{3}{4} = \dfrac{4}{5} \bullet \dfrac{4}{3} = \dfrac{16}{15}$

23.   Dividing and reducing: $-\dfrac{5}{6} \div \left(-\dfrac{5}{8}\right) = -\dfrac{5}{6} \bullet \left(-\dfrac{8}{5}\right) = \dfrac{40}{30} = \dfrac{4}{3}$

25.   Dividing and reducing: $\dfrac{10}{13} \div \left(-\dfrac{5}{4}\right) = \dfrac{10}{13} \bullet \left(-\dfrac{4}{5}\right) = -\dfrac{40}{65} = -\dfrac{8}{13}$

27.   Dividing and reducing: $-\dfrac{5}{6} \div \dfrac{5}{6} = -\dfrac{5}{6} \bullet \dfrac{6}{5} = -\dfrac{30}{30} = -1$

29.   Dividing and reducing: $-\dfrac{3}{4} \div \left(-\dfrac{3}{4}\right) = -\dfrac{3}{4} \bullet \left(-\dfrac{4}{3}\right) = \dfrac{12}{12} = 1$

31.   Using order of operations: $\dfrac{3(-2)}{-10} = \dfrac{-6}{-10} = \dfrac{3}{5}$ **33.**   Using order of operations: $\dfrac{-5(-5)}{-15} = \dfrac{25}{-15} = -\dfrac{5}{3}$

35.   Using order of operations: $\dfrac{-8(-7)}{-28} = \dfrac{56}{-28} = -2$ **37.**   Using order of operations: $\dfrac{27}{4 - 13} = \dfrac{27}{-9} = -3$

39.   Using order of operations: $\dfrac{20 - 6}{5 - 5} = \dfrac{14}{0}$, which is undefined

41.   Using order of operations: $\dfrac{-3 + 9}{2 \bullet 5 - 10} = \dfrac{6}{10 - 10} = \dfrac{6}{0}$, which is undefined

43.   Using order of operations: $\dfrac{15(-5)-25}{2(-10)}=\dfrac{-75-25}{-20}=\dfrac{-100}{-20}=5$

45.   Using order of operations: $\dfrac{27-2(-4)}{-3(5)}=\dfrac{27+8}{-15}=\dfrac{35}{-15}=-\dfrac{7}{3}$

47.   Using order of operations: $\dfrac{12-6(-2)}{12(-2)}=\dfrac{12+12}{-24}=\dfrac{24}{-24}=-1$

49.   Using order of operations: $\dfrac{5^2-2^2}{-5+2}=\dfrac{25-4}{-3}=\dfrac{21}{-3}=-7$

51.   Using order of operations: $\dfrac{8^2-2^2}{8^2+2^2}=\dfrac{64-4}{64+4}=\dfrac{60}{68}=\dfrac{15}{17}$

53.   Using order of operations: $\dfrac{(5+3)^2}{-5^2-3^2}=\dfrac{8^2}{-25-9}=\dfrac{64}{-34}=-\dfrac{32}{17}$

55.   Using order of operations: $\dfrac{(8-4)^2}{8^2-4^2}=\dfrac{4^2}{64-16}=\dfrac{16}{48}=\dfrac{1}{3}$

57.   Using order of operations: $\dfrac{-4\cdot3^2-5\cdot2^2}{-8(7)}=\dfrac{-4\cdot9-5\cdot4}{-56}=\dfrac{-36-20}{-56}=\dfrac{-56}{-56}=1$

59.   Using order of operations: $\dfrac{3\cdot10^2+4\cdot10+5}{345}=\dfrac{300+40+5}{345}=\dfrac{345}{345}=1$

61.   Using order of operations: $\dfrac{7-[(2-3)-4]}{-1-2-3}=\dfrac{7-(-1-4)}{-6}=\dfrac{7-(-5)}{-6}=\dfrac{7+5}{-6}=\dfrac{12}{-6}=-2$

63.   Using order of operations: $\dfrac{6(-4)-2(5-8)}{-6-3-5}=\dfrac{-24-2(-3)}{-14}=\dfrac{-24+6}{-14}=\dfrac{-18}{-14}=\dfrac{9}{7}$

65.   Using order of operations: $\dfrac{3(-5-3)+4(7-9)}{5(-2)+3(-4)}=\dfrac{3(-8)+4(-2)}{-10+(-12)}=\dfrac{-24+(-8)}{-22}=\dfrac{-32}{-22}=\dfrac{16}{11}$

67.   Using order of operations: $\dfrac{|3-9|}{3-9}=\dfrac{|-6|}{-6}=\dfrac{6}{-6}=-1$

69.   Using order of operations: $\dfrac{2+0.15(10)}{10}=\dfrac{2+1.5}{10}=\dfrac{3.5}{10}=\dfrac{35}{100}=\dfrac{7}{20}=0.35$

71.   Using order of operations: $\dfrac{1-3}{3-1}=\dfrac{-2}{2}=-1$

73.   **a.**   Simplifying: $\dfrac{5-2}{3-1}=\dfrac{3}{2}$ **b.**   Simplifying: $\dfrac{2-5}{1-3}=\dfrac{-3}{-2}=\dfrac{3}{2}$

75.   **a.**   Simplifying: $\dfrac{-4-1}{5-(-2)}=\dfrac{-4+(-1)}{5+2}=\dfrac{-5}{7}=-\dfrac{5}{7}$ **b.**   Simplifying: $\dfrac{1-(-4)}{-2-5}=\dfrac{1+4}{-2+(-5)}=\dfrac{5}{-7}=-\dfrac{5}{7}$

77.   **a.**   Simplifying: $\dfrac{3+2.236}{2}=\dfrac{5.236}{2}=2.618$ **b.**   Simplifying: $\dfrac{3-2.236}{2}=\dfrac{0.764}{2}=0.382$

　　　　c.   Simplifying: $\dfrac{3+2.236}{2}+\dfrac{3-2.236}{2}=\dfrac{5.236}{2}+\dfrac{0.764}{2}=2.618+0.382=3$

79.   **a.**   Simplifying: $20\div4\cdot5=5\cdot5=25$ **b.**   Simplifying: $-20\div4\cdot5=-5\cdot5=-25$

　　　　c.   Simplifying: $20\div(-4)\cdot5=-5\cdot5=-25$ **d.**   Simplifying: $20\div4(-5)=5(-5)=-25$

　　　　e.   Simplifying: $-20\div4(-5)=-5(-5)=25$

81. **a.** Simplifying: $8 \div \frac{4}{5} = 8 \cdot \frac{5}{4} = 10$ **b.** Simplifying: $8 \div \frac{4}{5} - 10 = 8 \cdot \frac{5}{4} - 10 = 10 - 10 = 0$

c. Simplifying: $8 \div \frac{4}{5}(-10) = 8 \cdot \frac{5}{4}(-10) = 10(-10) = -100$

d. Simplifying: $8 \div \left(-\frac{4}{5}\right) - 10 = 8 \cdot \left(-\frac{5}{4}\right) - 10 = -10 - 10 = -20$

83. Applying the distributive property: $10\left(\frac{x}{2} + \frac{3}{5}\right) = 10\left(\frac{x}{2}\right) + 10\left(\frac{3}{5}\right) = 5x + 6$

85. Applying the distributive property: $15\left(\frac{x}{5} + \frac{4}{3}\right) = 15\left(\frac{x}{5}\right) + 15\left(\frac{4}{3}\right) = 3x + 20$

87. Applying the distributive property: $x\left(\frac{3}{x} + 1\right) = x\left(\frac{3}{x}\right) + x(1) = 3 + x$

89. Applying the distributive property: $21\left(\frac{x}{7} - \frac{y}{3}\right) = 21\left(\frac{x}{7}\right) - 21\left(\frac{y}{3}\right) = 3x - 7y$

91. Applying the distributive property: $a\left(\frac{3}{a} - \frac{2}{a}\right) = a\left(\frac{3}{a}\right) - a\left(\frac{2}{a}\right) = 3 - 2 = 1$

93. Applying the distributive property: $2y\left(\frac{1}{y} - \frac{1}{2}\right) = 2y\left(\frac{1}{y}\right) - 2y\left(\frac{1}{2}\right) = 2 - y$

95. The quotient is $\frac{-12}{-4} = 3$.   **97.** The number is -10, since $\frac{-10}{-5} = 2$.

99. The number is -3, since $\frac{27}{-3} = -9$.   **101.** The expression is: $\frac{-20}{4} - 3 = -5 - 3 = -8$

103. Each person would lose: $\frac{13600 - 15000}{4} = \frac{-1400}{4} = -350 = \$350$ loss

105. The change per hour is: $\frac{61° - 75°}{4} = \frac{-14°}{4} = -3.5°$ per hour

107. **a.** Since they predict \$50 revenue for every 25 people, their projected revenue is:

$\frac{\$50}{25 \text{ people}} \cdot 10,000 \text{ people} = \$20,000$

b. Since they predict \$50 revenue for every 25 people, their projected revenue is:

$\frac{\$50}{25 \text{ people}} \cdot 25,000 \text{ people} = \$50,000$

c. Since they predict \$50 revenue for every 25 people, their projected revenue is:

$\frac{\$50}{25 \text{ people}} \cdot 5,000 \text{ people} = \$10,000$

Since this projected revenue is more than the \$5,000 cost for the list, it is a wise purchase.

1.8 Subsets of the Real Numbers

1. The whole numbers are: 0, 1   **3.** The rational numbers are: $-3, -2.5, 0, 1, \frac{3}{2}$

5. The real numbers are: $-3, -2.5, 0, 1, \frac{3}{2}, \sqrt{15}$ **7.** The integers are: $-10, -8, -2, 9$

9. The irrational numbers are: π **11.** true

13. false   **15.** false

17. true

19. This number is composite: $48 = 6 \cdot 8 = (2 \cdot 3) \cdot (2 \cdot 2 \cdot 2) = 2^4 \cdot 3$

21. This number is prime.   **23.** This number is composite: $1023 = 3 \cdot 341 = 3 \cdot 11 \cdot 31$

25. Factoring the number: $144 = 12 \cdot 12 = (3 \cdot 4) \cdot (3 \cdot 4) = (3 \cdot 2 \cdot 2) \cdot (3 \cdot 2 \cdot 2) = 2^4 \cdot 3^2$

27. Factoring the number: $38 = 2 \cdot 19$

29. Factoring the number: $105 = 5 \cdot 21 = 5 \cdot (3 \cdot 7) = 3 \cdot 5 \cdot 7$

31. Factoring the number: $180 = 10 \cdot 18 = (2 \cdot 5) \cdot (3 \cdot 6) = (2 \cdot 5) \cdot (3 \cdot 2 \cdot 3) = 2^2 \cdot 3^2 \cdot 5$

33. Factoring the number: $385 = 5 \cdot 77 = 5 \cdot (7 \cdot 11) = 5 \cdot 7 \cdot 11$

35. Factoring the number: $121 = 11 \cdot 11 = 11^2$

37. Factoring the number: $420 = 10 \cdot 42 = (2 \cdot 5) \cdot (7 \cdot 6) = (2 \cdot 5) \cdot (7 \cdot 2 \cdot 3) = 2^2 \cdot 3 \cdot 5 \cdot 7$

39. Factoring the number: $620 = 10 \cdot 62 = (2 \cdot 5) \cdot (2 \cdot 31) = 2^2 \cdot 5 \cdot 31$

41. Reducing the fraction: $\dfrac{105}{165} = \dfrac{3 \cdot 5 \cdot 7}{3 \cdot 5 \cdot 11} = \dfrac{7}{11}$ **43.** Reducing the fraction: $\dfrac{525}{735} = \dfrac{3 \cdot 5 \cdot 5 \cdot 7}{3 \cdot 5 \cdot 7 \cdot 7} = \dfrac{5}{7}$

45. Reducing the fraction: $\dfrac{385}{455} = \dfrac{5 \cdot 7 \cdot 11}{5 \cdot 7 \cdot 13} = \dfrac{11}{13}$ **47.** Reducing the fraction: $\dfrac{322}{345} = \dfrac{2 \cdot 7 \cdot 23}{3 \cdot 5 \cdot 23} = \dfrac{2 \cdot 7}{3 \cdot 5} = \dfrac{14}{15}$

49. Reducing the fraction: $\dfrac{205}{369} = \dfrac{5 \cdot 41}{3 \cdot 3 \cdot 41} = \dfrac{5}{3 \cdot 3} = \dfrac{5}{9}$

51. Reducing the fraction: $\dfrac{215}{344} = \dfrac{5 \cdot 43}{2 \cdot 2 \cdot 2 \cdot 43} = \dfrac{5}{2 \cdot 2 \cdot 2} = \dfrac{5}{8}$

53. **a.** Adding: $50 + (-80) = -30$ **b.** Subtracting: $50 - (-80) = 50 + 80 = 130$

   c. Multiplying: $50(-80) = -4{,}000$ **d.** Dividing: $\dfrac{50}{-80} = -\dfrac{5}{8}$

55. Simplifying; $\dfrac{6.28}{9(3.14)} = \dfrac{6.28}{28.26} = \dfrac{628}{2826} = \dfrac{2 \cdot 314}{9 \cdot 314} = \dfrac{2}{9}$ **57.** Simplifying; $\dfrac{9.42}{2(3.14)} = \dfrac{9.42}{6.28} = \dfrac{942}{628} = \dfrac{3 \cdot 314}{2 \cdot 314} = \dfrac{3}{2}$

59. Simplifying; $\dfrac{32}{0.5} = \dfrac{320}{5} = 64$ **61.** Simplifying; $\dfrac{5{,}599}{11} = 509$

63. **a.** Substituting $x = 10$: $\dfrac{2 + 0.15x}{x} = \dfrac{2 + 0.15(10)}{10} = \dfrac{2 + 1.5}{10} = \dfrac{3.5}{10} = 0.35$

   b. Substituting $x = 15$: $\dfrac{2 + 0.15x}{x} = \dfrac{2 + 0.15(15)}{15} = \dfrac{2 + 2.25}{15} = \dfrac{4.25}{15} \approx 0.28$

   c. Substituting $x = 20$: $\dfrac{2 + 0.15x}{x} = \dfrac{2 + 0.15(20)}{20} = \dfrac{2 + 3}{20} = \dfrac{5}{20} = 0.25$

65. Factoring into prime numbers: $6^3 = (2 \cdot 3)^3 = 2^3 \cdot 3^3$

67. Factoring into prime numbers: $9^4 \cdot 16^2 = (3 \cdot 3)^4 \cdot (2 \cdot 2 \cdot 2 \cdot 2)^2 = 3^4 \cdot 3^4 \cdot 2^2 \cdot 2^2 \cdot 2^2 \cdot 2^2 = 2^8 \cdot 3^8$

69. Simplifying and factoring: $3 \cdot 8 + 3 \cdot 7 + 3 \cdot 5 = 24 + 21 + 15 = 60 = 6 \cdot 10 = (2 \cdot 3) \cdot (2 \cdot 5) = 2^2 \cdot 3 \cdot 5$

71. They are not a subset of the irrational numbers.

73. 8, 21, and 34 are Fibonacci numbers that are composite numbers.

1.9 Addition and Subtraction with Fractions

1. Combining the fractions: $\dfrac{3}{6} + \dfrac{1}{6} = \dfrac{4}{6} = \dfrac{2}{3}$ **3.** Combining the fractions: $\dfrac{3}{8} - \dfrac{5}{8} = -\dfrac{2}{8} = -\dfrac{1}{4}$

5. Combining the fractions: $-\dfrac{1}{4} + \dfrac{3}{4} = \dfrac{2}{4} = \dfrac{1}{2}$ **7.** Combining the fractions: $\dfrac{x}{3} - \dfrac{1}{3} = \dfrac{x-1}{3}$

9. Combining the fractions: $\dfrac{1}{4} + \dfrac{2}{4} + \dfrac{3}{4} = \dfrac{6}{4} = \dfrac{3}{2}$ **11.** Combining the fractions: $\dfrac{x+7}{2} - \dfrac{1}{2} = \dfrac{x+7-1}{2} = \dfrac{x+6}{2}$

13. Combining the fractions: $\dfrac{1}{10} - \dfrac{3}{10} - \dfrac{4}{10} = -\dfrac{6}{10} = -\dfrac{3}{5}$ **15.** Combining the fractions: $\dfrac{1}{a} + \dfrac{4}{a} + \dfrac{5}{a} = \dfrac{10}{a}$

17. Combining the fractions: $\dfrac{1}{8} + \dfrac{3}{4} = \dfrac{1}{8} + \dfrac{3 \cdot 2}{4 \cdot 2} = \dfrac{1}{8} + \dfrac{6}{8} = \dfrac{7}{8}$

19. Combining the fractions: $\frac{3}{10} - \frac{1}{5} = \frac{3}{10} - \frac{1 \cdot 2}{5 \cdot 2} = \frac{3}{10} - \frac{2}{10} = \frac{1}{10}$

21. Combining the fractions: $\frac{4}{9} + \frac{1}{3} = \frac{4}{9} + \frac{1 \cdot 3}{3 \cdot 3} = \frac{4}{9} + \frac{3}{9} = \frac{7}{9}$

23. Combining the fractions: $2 + \frac{1}{3} = \frac{2 \cdot 3}{1 \cdot 3} + \frac{1}{3} = \frac{6}{3} + \frac{1}{3} = \frac{7}{3}$

25. Combining the fractions: $-\frac{3}{4} + 1 = -\frac{3}{4} + \frac{1 \cdot 4}{1 \cdot 4} = -\frac{3}{4} + \frac{4}{4} = \frac{1}{4}$

27. Combining the fractions: $\frac{1}{2} + \frac{2}{3} = \frac{1 \cdot 3}{2 \cdot 3} + \frac{2 \cdot 2}{3 \cdot 2} = \frac{3}{6} + \frac{4}{6} = \frac{7}{6}$

29. Combining the fractions: $\frac{5}{12} - \left(-\frac{3}{8}\right) = \frac{5}{12} + \frac{3}{8} = \frac{5 \cdot 2}{12 \cdot 2} + \frac{3 \cdot 3}{8 \cdot 3} = \frac{10}{24} + \frac{9}{24} = \frac{19}{24}$

31. Combining the fractions: $-\frac{1}{20} + \frac{8}{30} = -\frac{1 \cdot 3}{20 \cdot 3} + \frac{8 \cdot 2}{30 \cdot 2} = -\frac{3}{60} + \frac{16}{60} = \frac{13}{60}$

33. First factor the denominators to find the LCM:

$30 = 2 \cdot 3 \cdot 5$

$42 = 2 \cdot 3 \cdot 7$

$LCM = 2 \cdot 3 \cdot 5 \cdot 7 = 210$

Combining the fractions: $\frac{17}{30} + \frac{11}{42} = \frac{17 \cdot 7}{30 \cdot 7} + \frac{11 \cdot 5}{42 \cdot 5} = \frac{119}{210} + \frac{55}{210} = \frac{174}{210} = \frac{2 \cdot 3 \cdot 29}{2 \cdot 3 \cdot 5 \cdot 7} = \frac{29}{5 \cdot 7} = \frac{29}{35}$

35. First factor the denominators to find the LCM:

$84 = 2 \cdot 2 \cdot 3 \cdot 7$

$90 = 2 \cdot 3 \cdot 3 \cdot 5$

$LCM = 2 \cdot 2 \cdot 3 \cdot 3 \cdot 5 \cdot 7 = 1260$

Combining the fractions: $\frac{25}{84} + \frac{41}{90} = \frac{25 \cdot 15}{84 \cdot 15} + \frac{41 \cdot 14}{90 \cdot 14} = \frac{375}{1260} + \frac{574}{1260} = \frac{949}{1260}$

37. First factor the denominators to find the LCM:

$126 = 2 \cdot 3 \cdot 3 \cdot 7$

$180 = 2 \cdot 2 \cdot 3 \cdot 3 \cdot 5$

$LCM = 2 \cdot 2 \cdot 3 \cdot 3 \cdot 5 \cdot 7 = 1260$

Combining the fractions:

$\frac{13}{126} - \frac{13}{180} = \frac{13 \cdot 10}{126 \cdot 10} - \frac{13 \cdot 7}{180 \cdot 7} = \frac{130}{1,260} - \frac{91}{1,260} = \frac{39}{1,260} = \frac{3 \cdot 13}{2 \cdot 2 \cdot 3 \cdot 3 \cdot 5 \cdot 7} = \frac{13}{2 \cdot 2 \cdot 3 \cdot 5 \cdot 7} = \frac{13}{420}$

39. Combining the fractions: $\frac{3}{4} + \frac{1}{8} + \frac{5}{6} = \frac{3 \cdot 6}{4 \cdot 6} + \frac{1 \cdot 3}{8 \cdot 3} + \frac{5 \cdot 4}{6 \cdot 4} = \frac{18}{24} + \frac{3}{24} + \frac{20}{24} = \frac{41}{24}$

41. Combining the fractions: $\frac{1}{2} + \frac{1}{3} + \frac{1}{4} + \frac{1}{6} = \frac{1 \cdot 6}{2 \cdot 6} + \frac{1 \cdot 4}{3 \cdot 4} + \frac{1 \cdot 3}{4 \cdot 3} + \frac{1 \cdot 2}{6 \cdot 2} = \frac{6}{12} + \frac{4}{12} + \frac{3}{12} + \frac{2}{12} = \frac{15}{12} = \frac{5}{4}$

43. Combining the fractions: $1 - \frac{5}{2} = 1 \cdot \frac{2}{2} - \frac{5}{2} = \frac{2}{2} - \frac{5}{2} = -\frac{3}{2}$

45. Combining the fractions: $1 + \frac{1}{2} = 1 \cdot \frac{2}{2} + \frac{1}{2} = \frac{2}{2} + \frac{1}{2} = \frac{3}{2}$

47. The pattern is to add $-\frac{1}{3}$, so the fourth term is: $-\frac{1}{3} + \left(-\frac{1}{3}\right) = -\frac{2}{3}$

49. The pattern is to add $\frac{2}{3}$, so the fourth term is: $\frac{5}{3} + \frac{2}{3} = \frac{7}{3}$

51. The pattern is to multiply by $\frac{1}{5}$, so the fourth term is: $\frac{1}{25} \cdot \frac{1}{5} = \frac{1}{125}$

53. Using order of operations: $9 - 3\left(\frac{5}{3}\right) = 9 - 5 = 4$

55. Using order of operations: $-\frac{1}{2} + 2\left(-\frac{3}{4}\right) = -\frac{1}{2} - \frac{3}{2} = -\frac{4}{2} = -2$

57. Using order of operations: $\frac{3}{5}(-10) + \frac{4}{7}(-21) = -6 - 12 = -18$

59. Using order of operations: $16\left(-\frac{1}{2}\right)^2 - 125\left(-\frac{2}{5}\right)^2 = 16\left(\frac{1}{4}\right) - 125\left(\frac{4}{25}\right) = 4 - 20 = -16$

61.   Using order of operations: $-\frac{4}{3} \div 2 \cdot 3 = -\frac{4}{3} \cdot \frac{1}{2} \cdot 3 = -\frac{2}{3} \cdot 3 = -2$

63.   Using order of operations: $-\frac{4}{3} \div 2(-3) = -\frac{4}{3} \cdot \frac{1}{2} \cdot (-3) = -\frac{2}{3} \cdot (-3) = 2$

65.   Using order of operations: $-6 \div \frac{1}{2} \cdot 12 = -6 \cdot 2 \cdot 12 = -144$

67.   Using order of operations: $-15 \div \frac{5}{3} \cdot 18 = -15 \cdot \frac{3}{5} \cdot 18 = -9 \cdot 18 = -162$

69.   Combining the fractions: $\frac{x}{4} + \frac{1}{5} = \frac{x}{4} \cdot \frac{5}{5} + \frac{1}{5} \cdot \frac{4}{4} = \frac{5x}{20} + \frac{4}{20} = \frac{5x+4}{20}$

71.   Combining the fractions: $\frac{1}{3} + \frac{a}{12} = \frac{1}{3} \cdot \frac{4}{4} + \frac{a}{12} = \frac{4}{12} + \frac{a}{12} = \frac{a+4}{12}$

73.   Combining the fractions: $\frac{x}{2} + \frac{1}{3} + \frac{x}{4} = \frac{x}{2} \cdot \frac{6}{6} + \frac{1}{3} \cdot \frac{4}{4} + \frac{x}{4} \cdot \frac{3}{3} = \frac{6x}{12} + \frac{4}{12} + \frac{3x}{12} = \frac{6x+4+3x}{12} = \frac{9x+4}{12}$

75.   Combining the fractions: $\frac{2}{x} + \frac{3}{5} = \frac{2}{x} \cdot \frac{5}{5} + \frac{3}{5} \cdot \frac{x}{x} = \frac{10}{5x} + \frac{3x}{5x} = \frac{3x+10}{5x}$

77.   Combining the fractions: $\frac{3}{7} + \frac{4}{y} = \frac{3}{7} \cdot \frac{y}{y} + \frac{4}{y} \cdot \frac{7}{7} = \frac{3y}{7y} + \frac{28}{7y} = \frac{3y+28}{7y}$

79.   Combining the fractions: $\frac{3}{a} + \frac{3}{4} + \frac{1}{5} = \frac{3}{a} \cdot \frac{20}{20} + \frac{3}{4} \cdot \frac{5a}{5a} + \frac{1}{5} \cdot \frac{4a}{4a} = \frac{60}{20a} + \frac{15a}{20a} + \frac{4a}{20a} = \frac{60+15a+4a}{20a} = \frac{19a+60}{20a}$

81.   Combining the fractions: $\frac{1}{2}x + \frac{1}{6}x = \left(\frac{1}{2} + \frac{1}{6}\right)x = \left(\frac{1}{2} \cdot \frac{3}{3} + \frac{1}{6}\right)x = \left(\frac{3}{6} + \frac{1}{6}\right)x = \frac{4}{6}x = \frac{2}{3}x$

83.   Combining the fractions: $\frac{1}{2}x - \frac{3}{4}x = \left(\frac{1}{2} - \frac{3}{4}\right)x = \left(\frac{1}{2} \cdot \frac{2}{2} - \frac{3}{4}\right)x = \left(\frac{2}{4} - \frac{3}{4}\right)x = -\frac{1}{4}x$

85.   Combining the fractions: $\frac{1}{3}x + \frac{3}{5}x = \left(\frac{1}{3} + \frac{3}{5}\right)x = \left(\frac{1}{3} \cdot \frac{5}{5} + \frac{3}{5} \cdot \frac{3}{3}\right)x = \left(\frac{5}{15} + \frac{9}{15}\right)x = \frac{14}{15}x$

87.   Combining the fractions: $\frac{3x}{4} + \frac{x}{6} = \frac{3x}{4} \cdot \frac{3}{3} + \frac{x}{6} \cdot \frac{2}{2} = \frac{9x}{12} + \frac{2x}{12} = \frac{11x}{12} = \frac{11}{12}x$

89.   Combining the fractions: $\frac{2x}{5} + \frac{5x}{8} = \frac{2x}{5} \cdot \frac{8}{8} + \frac{5x}{8} \cdot \frac{5}{5} = \frac{16x}{40} + \frac{25x}{40} = \frac{41x}{40} = \frac{41}{40}x$

91.   Combining the fractions: $1 - \frac{1}{x} = \frac{1}{1} \cdot \frac{x}{x} - \frac{1}{x} = \frac{x}{x} - \frac{1}{x} = \frac{x-1}{x}$

93.   **a.**   Adding the fractions: $\frac{3}{4} + \left(-\frac{1}{2}\right) = \frac{3}{4} - \frac{1}{2} \cdot \frac{2}{2} = \frac{3}{4} - \frac{2}{4} = \frac{1}{4}$

      b.   Subtracting the fractions: $\frac{3}{4} - \left(-\frac{1}{2}\right) = \frac{3}{4} + \frac{1}{2} \cdot \frac{2}{2} = \frac{3}{4} + \frac{2}{4} = \frac{5}{4}$

      c.   Multiplying the fractions: $\frac{3}{4}\left(-\frac{1}{2}\right) = -\frac{3}{8}$

      d.   Dividing the fractions: $\frac{3}{4} \div \left(-\frac{1}{2}\right) = \frac{3}{4} \cdot \left(-\frac{2}{1}\right) = -\frac{6}{4} = -\frac{3}{2}$

95.   Simplifying: $\left(1 - \frac{1}{2}\right)\left(1 - \frac{1}{3}\right) = \left(\frac{2}{2} - \frac{1}{2}\right)\left(\frac{3}{3} - \frac{1}{3}\right) = \left(\frac{1}{2}\right)\left(\frac{2}{3}\right) = \frac{2}{6} = \frac{1}{3}$

97.   Simplifying: $\left(1 + \frac{1}{2}\right)\left(1 - \frac{1}{2}\right) = \left(\frac{2}{2} + \frac{1}{2}\right)\left(\frac{2}{2} - \frac{1}{2}\right) = \left(\frac{3}{2}\right)\left(\frac{1}{2}\right) = \frac{3}{4}$

99.   **a.**   Substituting $x = 2$: $1 + \dfrac{1}{x} = 1 + \dfrac{1}{2} = \dfrac{2}{2} + \dfrac{1}{2} = \dfrac{3}{2}$

       b.   Substituting $x = 3$: $1 + \dfrac{1}{x} = 1 + \dfrac{1}{3} = \dfrac{3}{3} + \dfrac{1}{3} = \dfrac{4}{3}$

       c.   Substituting $x = 4$: $1 + \dfrac{1}{x} = 1 + \dfrac{1}{4} = \dfrac{4}{4} + \dfrac{1}{4} = \dfrac{5}{4}$

101.   **a.**   Substituting $x = 1$: $2x + \dfrac{6}{x} = 2(1) + \dfrac{6}{1} = 2 + 6 = 8$

       b.   Substituting $x = 2$: $2x + \dfrac{6}{x} = 2(2) + \dfrac{6}{2} = 4 + 3 = 7$

       c.   Substituting $x = 3$: $2x + \dfrac{6}{x} = 2(3) + \dfrac{6}{3} = 6 + 2 = 8$

Chapter 1 Review/Test

1. The expression is: $-7 + (-10) = -17$ **2.** The expression is: $(-7 + 4) + 5 = -3 + 5 = 2$

3. The expression is: $(-3 + 12) + 5 = 9 + 5 = 14$ **4.** The expression is: $4 - 9 = 4 + (-9) = -5$

5. The expression is: $9 - (-3) = 9 + 3 = 12$ **6.** The expression is: $-7 - (-9) = -7 + 9 = 2$

7. The expression is: $(-3)(-7) - 6 = 21 - 6 = 15$ **8.** The expression is: $5(-6) + 10 = -30 + 10 = -20$

9. The expression is: $2\left[-8(3x)\right] = 2(-24x) = -48x$ **10.** The expression is: $\dfrac{-25}{-5} = 5$

11. Simplifying: $\left|-1.8\right| = 1.8$ **12.** Simplifying: $-\left|-10\right| = -(10) = -10$

13. The opposite is -6, and the reciprocal is $\dfrac{1}{6}$.   **14.** The opposite is $\dfrac{12}{5}$, and the reciprocal is $-\dfrac{5}{12}$.

15. Multiplying the fractions: $\dfrac{1}{2}(-10) = \dfrac{1}{2} \bullet \left(-\dfrac{10}{1}\right) = -\dfrac{10}{2} = -5$

16. Multiplying the fractions: $\left(-\dfrac{4}{5}\right)\left(\dfrac{25}{16}\right) = -\dfrac{100}{80} = -\dfrac{5}{4}$

17. Adding: $-9 + 12 = 3$ **18.** Adding: $-18 + (-20) = -38$

19. Adding: $(-2) + (-8) + \left[-9 + (-6)\right] = -2 + (-8) + (-15) = -25$

20. Adding: $(-21) + 40 + (-23) + 5 = -44 + 45 = 1$ **21.** Subtracting: $6 - 9 = 6 + (-9) = -3$

22. Subtracting: $14 - (-8) = 14 + 8 = 22$ **23.** Subtracting: $-12 - (-8) = -12 + 8 = -4$

24. Subtracting: $4 - 9 - 15 = 4 + (-9) + (-15) = 4 + (-24) = -20$

25. Multiplying: $(-5)(6) = -30$ **26.** Multiplying: $4(-3) = -12$

27. Multiplying: $-2(3)(4) = -24$ **28.** Multiplying: $(-1)(-3)(-1)(-4) = 12$

29. Finding the quotient: $\dfrac{12}{-3} = -4$ **30.** Finding the quotient: $-\dfrac{8}{9} \div \dfrac{4}{3} = -\dfrac{8}{9} \bullet \dfrac{3}{4} = -\dfrac{24}{36} = -\dfrac{2}{3}$

31. Simplifying using order of operations: $4 \bullet 5 + 3 = 20 + 3 = 23$

32. Simplifying using order of operations: $9 \bullet 3 + 4 \bullet 5 = 27 + 20 = 47$

33. Simplifying using order of operations: $2^3 - 4 \bullet 3^2 + 5^2 = 8 - 4 \bullet 9 + 25 = 8 - 36 + 25 = 33 - 36 = -3$

34. Simplifying using order of operations:

$$12 - 3(2 \bullet 5 + 7) + 4 = 12 - 3(10 + 7) + 4 = 12 - 3(17) + 4 = 12 - 51 + 4 = -39 + 4 = -35$$

35. Simplifying using order of operations: $20 + 8 \div 4 + 2 \bullet 5 = 20 + 2 + 10 = 32$

36. Simplifying using order of operations: $2(3 - 5) - (2 - 8) = 2(-2) - (-6) = -4 + 6 = 2$

37. Simplifying using order of operations: $30 \div 3 \bullet 2 = 10 \bullet 2 = 20$

38. Simplifying using order of operations: $(-2)(3) - 4(-3) - 9 = -6 - (-12) - 9 = -6 + 12 + (-9) = -15 + 12 = -3$

39. Simplifying using order of operations:

$$3(4 - 7)^2 - 5(3 - 8)^2 = 3(-3)^2 - 5(-5)^2 = 3 \bullet 9 - 5 \bullet 25 = 27 - 125 = 27 + (-125) = -98$$

40. Simplifying using order of operations: $(-5 - 2)(-3 - 7) = (-7)(-10) = 70$

41. Simplifying using order of operations: $\dfrac{4(-3)}{-6} = \dfrac{-12}{-6} = 2$

42. Simplifying using order of operations: $\dfrac{3^2 + 5^2}{(3-5)^2} = \dfrac{9+25}{(-2)^2} = \dfrac{34}{4} = \dfrac{17}{2}$

43. Simplifying using order of operations: $\dfrac{15-10}{6-6} = \dfrac{5}{0}$, which is undefined

44. Simplifying using order of operations: $\dfrac{2(-7)+(-11)(-4)}{7-(-3)} = \dfrac{-14+44}{7+3} = \dfrac{30}{10} = 3$

45. associative property (of multiplication) 46. multiplicative identity property
47. commutative property (of addition) 48. additive inverse property
49. commutative and associative properties (of addition) 50. distributive property
51. Simplifying the expression: $7 + (5 + x) = (7 + 5) + x = 12 + x$
52. Simplifying the expression: $4(7a) = (4 \cdot 7)a = 28a$

53. Simplifying the expression: $\frac{1}{9}(9x) = \left(\frac{1}{9} \cdot 9\right)x = 1x = x$

54. Simplifying the expression: $\frac{4}{5}\left(\frac{5}{4}y\right) = \left(\frac{4}{5} \cdot \frac{5}{4}\right)y = 1y = y$

55. Applying the distributive property: $7(2x+3) = 7 \cdot 2x + 7 \cdot 3 = 14x + 21$
56. Applying the distributive property: $3(2a-4) = 3 \cdot 2a - 3 \cdot 4 = 6a - 12$

57. Applying the distributive property: $\frac{1}{2}(5x-6) = \frac{1}{2} \cdot 5x - \frac{1}{2} \cdot 6 = \frac{5}{2}x - 3$

58. Applying the distributive property: $-\frac{1}{2}(3x-6) = -\frac{1}{2} \cdot 3x - \left(-\frac{1}{2}\right) \cdot 6 = -\frac{3}{2}x + 3$

59. The rational numbers are: $-\frac{1}{3}, 0, 5, -4.5, \frac{2}{5}, -3$ 60. The whole numbers are: 0, 5

61. The irrational numbers are: $\sqrt{7}, \pi$ 62. The integers are: 0, 5, –3

63. Factoring the number: $90 = 9 \cdot 10 = (3 \cdot 3) \cdot (2 \cdot 5) = 2 \cdot 3^2 \cdot 5$

64. Factoring the number: $840 = 84 \cdot 10 = (21 \cdot 4) \cdot (2 \cdot 5) = (3 \cdot 7 \cdot 2 \cdot 2) \cdot (2 \cdot 5) = 2^3 \cdot 3 \cdot 5 \cdot 7$

65. First factor the denominators to find the LCM:
$$35 = 5 \cdot 7$$
$$42 = 2 \cdot 3 \cdot 7$$
$$\text{LCM} = 2 \cdot 3 \cdot 5 \cdot 7 = 210$$

Combining the fractions: $\dfrac{18}{35} + \dfrac{13}{42} = \dfrac{18 \cdot 6}{35 \cdot 6} + \dfrac{13 \cdot 5}{42 \cdot 5} = \dfrac{108}{210} + \dfrac{65}{210} = \dfrac{173}{210}$

66. Combining the fractions: $\dfrac{x}{6} + \dfrac{7}{12} = \dfrac{x}{6} \cdot \dfrac{2}{2} + \dfrac{7}{12} = \dfrac{2x}{12} + \dfrac{7}{12} = \dfrac{2x+7}{12}$

67. The pattern is to add –3, so the next number is: $1 + (-3) = -2$
68. The pattern is to multiply by –3, so the next number is: $-270 \cdot (-3) = 810$
69. The pattern is to add the previous two terms, so the next number is: $3 + 5 = 8$
70. The pattern is to add 2, so the next number is: $10 + 2 = 12$

71. The pattern is to add $-\frac{1}{2}$, so the next number is: $-\frac{1}{2} + \left(-\frac{1}{2}\right) = -\frac{2}{2} = -1$

72. The pattern is to multiply by $-\frac{1}{2}$, so the next number is: $-\frac{1}{8} \cdot \left(-\frac{1}{2}\right) = \frac{1}{16}$

Chapter 2
Linear Equations and Inequalities

2.1 Simplifying Expressions

1. Simplifying the expression: $3x - 6x = (3 - 6)x = -3x$
3. Simplifying the expression: $-2a + a = (-2 + 1)a = -a$
5. Simplifying the expression: $7x + 3x + 2x = (7 + 3 + 2)x = 12x$
7. Simplifying the expression: $3a - 2a + 5a = (3 - 2 + 5)a = 6a$
9. Simplifying the expression: $4x - 3 + 2x = 4x + 2x - 3 = 6x - 3$
11. Simplifying the expression: $3a + 4a + 5 = 7a + 5$
13. Simplifying the expression: $2x - 3 + 3x - 2 = 2x + 3x - 3 - 2 = 5x - 5$
15. Simplifying the expression: $3a - 1 + a + 3 = 3a + a - 1 + 3 = 4a + 2$
17. Simplifying the expression: $-4x + 8 - 5x - 10 = -4x - 5x + 8 - 10 = -9x - 2$
19. Simplifying the expression: $7a + 3 + 2a + 3a = 7a + 2a + 3a + 3 = 12a + 3$
21. Simplifying the expression: $5(2x - 1) + 4 = 10x - 5 + 4 = 10x - 1$
23. Simplifying the expression: $7(3y + 2) - 8 = 21y + 14 - 8 = 21y + 6$
25. Simplifying the expression: $-3(2x - 1) + 5 = -6x + 3 + 5 = -6x + 8$
27. Simplifying the expression: $5 - 2(a + 1) = 5 - 2a - 2 = -2a - 2 + 5 = -2a + 3$
29. Simplifying the expression: $6 - 4(x - 5) = 6 - 4x + 20 = -4x + 20 + 6 = -4x + 26$
31. Simplifying the expression: $-9 - 4(2 - y) + 1 = -9 - 8 + 4y + 1 = 4y + 1 - 9 - 8 = 4y - 16$
33. Simplifying the expression: $-6 + 2(2 - 3x) + 1 = -6 + 4 - 6x + 1 = -6x - 6 + 4 + 1 = -6x - 1$
35. Simplifying the expression: $(4x - 7) - (2x + 5) = 4x - 7 - 2x - 5 = 4x - 2x - 7 - 5 = 2x - 12$
37. Simplifying the expression: $8(2a + 4) - (6a - 1) = 16a + 32 - 6a + 1 = 16a - 6a + 32 + 1 = 10a + 33$
39. Simplifying the expression: $3(x - 2) + (x - 3) = 3x - 6 + x - 3 = 3x + x - 6 - 3 = 4x - 9$
41. Simplifying the expression: $4(2y - 8) - (y + 7) = 8y - 32 - y - 7 = 8y - y - 32 - 7 = 7y - 39$
43. Simplifying the expression: $-9(2x + 1) - (x + 5) = -18x - 9 - x - 5 = -18x - x - 9 - 5 = -19x - 14$
45. Evaluating when $x = 2$: $3x - 1 = 3(2) - 1 = 6 - 1 = 5$
47. Evaluating when $x = 2$: $-2x - 5 = -2(2) - 5 = -4 - 5 = -9$
49. Evaluating when $x = 2$: $x^2 - 8x + 16 = (2)^2 - 8(2) + 16 = 4 - 16 + 16 = 4$
51. Evaluating when $x = 2$: $(x - 4)^2 = (2 - 4)^2 = (-2)^2 = 4$
53. Evaluating when $x = -5$: $7x - 4 - x - 3 = 7(-5) - 4 - (-5) - 3 = -35 - 4 + 5 - 3 = -42 + 5 = -37$
    Now simplifying the expression: $7x - 4 - x - 3 = 7x - x - 4 - 3 = 6x - 7$
    Evaluating when $x = -5$: $6x - 7 = 6(-5) - 7 = -30 - 7 = -37$

Note that the two values are the same.

55. Evaluating when $x = -5$: $5(2x+1)+4 = 5[2(-5)+1]+4 = 5(-10+1)+4 = 5(-9)+4 = -45+4 = -41$

Now simplifying the expression: $5(2x+1)+4 = 10x+5+4 = 10x+9$

Evaluating when $x = -5$: $10x+9 = 10(-5)+9 = -50+9 = -41$

Note that the two values are the same.

57. Evaluating when $x = -3$ and $y = 5$: $x^2 - 2xy + y^2 = (-3)^2 - 2(-3)(5) + (5)^2 = 9 + 30 + 25 = 64$

59. Evaluating when $x = -3$ and $y = 5$: $(x-y)^2 = (-3-5)^2 = (-8)^2 = 64$

61. Evaluating when $x = -3$ and $y = 5$: $x^2 + 6xy + 9y^2 = (-3)^2 + 6(-3)(5) + 9(5)^2 = 9 - 90 + 225 = 144$

63. Evaluating when $x = -3$ and $y = 5$: $(x+3y)^2 = [-3 + 3(5)]^2 = (-3+15)^2 = (12)^2 = 144$

65. Evaluating when $x = \frac{1}{2}$: $12x - 3 = 12\left(\frac{1}{2}\right) - 3 = 6 - 3 = 3$

67. Evaluating when $x = \frac{1}{4}$: $12x - 3 = 12\left(\frac{1}{4}\right) - 3 = 3 - 3 = 0$

69. Evaluating when $x = \frac{3}{2}$: $12x - 3 = 12\left(\frac{3}{2}\right) - 3 = 18 - 3 = 15$

71. Evaluating when $x = \frac{3}{4}$: $12x - 3 = 12\left(\frac{3}{4}\right) - 3 = 9 - 3 = 6$

73. **a.** Substituting the values for n:

n	1	2	3	4
$3n$	3	6	9	12

b. Substituting the values for n:

n	1	2	3	4
n^3	1	8	27	64

75. Substituting $n = 1, 2, 3, 4$:

$n = 1$: $3(1) - 2 = 3 - 2 = 1$

$n = 2$: $3(2) - 2 = 6 - 2 = 4$

$n = 3$: $3(3) - 2 = 9 - 2 = 7$

$n = 4$: $3(4) - 2 = 12 - 2 = 10$

The sequence is $1, 4, 7, 10, \ldots$, which is an arithmetic sequence.

77. Substituting $n = 1, 2, 3, 4$:

$n = 1$: $(1)^2 - 2(1) + 1 = 1 - 2 + 1 = 0$

$n = 2$: $(2)^2 - 2(2) + 1 = 4 - 4 + 1 = 1$

$n = 3$: $(3)^2 - 2(3) + 1 = 9 - 6 + 1 = 4$

$n = 4$: $(4)^2 - 2(4) + 1 = 16 - 8 + 1 = 9$

The sequence is $0, 1, 4, 9, \ldots$, which is a sequence of squares.

79. Simplifying: $7 - 3(2y+1) = 7 - 6y - 3 = -6y + 4$

81. Simplifying: $0.08x + 0.09x = 0.17x$

83. Simplifying: $(x+y) + (x-y) = x + y + x - y = 2x$

85. Simplifying: $3x + 2(x-2) = 3x + 2x - 4 = 5x - 4$

87. Simplifying: $4(x+1) + 3(x-3) = 4x + 4 + 3x - 9 = 7x - 5$

89. Simplifying: $x + (x+3)(-3) = x - 3x - 9 = -2x - 9$

91. Simplifying: $3(4x-2) - (5x-8) = 12x - 6 - 5x + 8 = 7x + 2$

93. Simplifying: $-(3x+1) - (4x-7) = -3x - 1 - 4x + 7 = -7x + 6$

95. Simplifying: $(x+3y) + 3(2x-y) = x + 3y + 6x - 3y = 7x$

97. Simplifying: $3(2x+3y) - 2(3x+5y) = 6x + 9y - 6x - 10y = -y$

99. Simplifying: $-6\left(\frac{1}{2}x - \frac{1}{3}y\right) + 12\left(\frac{1}{4}x + \frac{2}{3}y\right) = -3x + 2y + 3x + 8y = 10y$

101. Simplifying: $0.08x + 0.09(x + 2,000) = 0.08x + 0.09x + 180 = 0.17x + 180$

103. Simplifying: $0.10x + 0.12(x + 500) = 0.10x + 0.12x + 60 = 0.22x + 60$

105. For the y terms to cancel, we must have $4y - ay = 0$, so $a = 4$.

Simplifying: $(5x + 4y) + 4(2x - y) = 5x + 4y + 8x - 4y = 13x$

107. Evaluating the expression: $b^2 - 4ac = (-5)^2 - 4(1)(-6) = 25 - (-24) = 25 + 24 = 49$

109. Evaluating the expression: $b^2 - 4ac = (4)^2 - 4(2)(-3) = 16 - (-24) = 16 + 24 = 40$

111. **a.** Substituting $x = 8,000$: $-0.0035(8000) + 70 = 42°F$

  b. Substituting $x = 12,000$: $-0.0035(12000) + 70 = 28°F$

  c. Substituting $x = 24,000$: $-0.0035(24000) + 70 = -14°F$

113. **a.** Substituting $t = 10$: $35 + 0.25(10) = \$37.50$ **b.** Substituting $t = 20$: $35 + 0.25(20) = \$40.00$

  c. Substituting $t = 30$: $35 + 0.25(30) = \$42.50$

115. Simplifying the expression: $G - 0.21G - 0.08G = 0.71G$. Substituting $G = \$1,250$: $0.71(\$1,250) = \$887.50$

117. Simplifying: $\frac{1}{8} - \frac{1}{6} = \frac{1}{8} \bullet \frac{3}{3} - \frac{1}{6} \bullet \frac{4}{4} = \frac{3}{24} - \frac{4}{24} = -\frac{1}{24}$ **119.** Simplifying: $\frac{5}{9} - \frac{4}{3} = \frac{5}{9} - \frac{4}{3} \bullet \frac{3}{3} = \frac{5}{9} - \frac{12}{9} = -\frac{7}{9}$

121. Simplifying: $-\frac{7}{30} + \frac{5}{28} = -\frac{7}{30} \bullet \frac{14}{14} + \frac{5}{28} \bullet \frac{15}{15} = -\frac{98}{420} + \frac{75}{420} = -\frac{23}{420}$

123. Simplifying: $17 - 5 = 12$ **125.** Simplifying: $2 - 5 = -3$

127. Simplifying: $-2.4 + (-7.3) = -9.7$ **129.** Simplifying: $-\frac{1}{2} + \left(-\frac{3}{4}\right) = -\frac{1}{2} \bullet \frac{2}{2} - \frac{3}{4} = -\frac{2}{4} - \frac{3}{4} = -\frac{5}{4}$

131. Simplifying: $4(2 \bullet 9 - 3) - 7 \bullet 9 = 4(18 - 3) - 63 = 4(15) - 63 = 60 - 63 = -3$

133. Simplifying: $4(2a - 3) - 7a = 8a - 12 - 7a = a - 12$

135. Subtracting: $-3 - \frac{1}{2} = \frac{-3}{1} - \frac{1}{2} = \frac{-3 \bullet 2}{1 \bullet 2} - \frac{1}{2} = \frac{-6}{2} - \frac{1}{2} = -\frac{7}{2}$

137. Adding: $\frac{4}{5} + \frac{1}{10} + \frac{3}{8} = \frac{4 \bullet 8}{5 \bullet 8} + \frac{1 \bullet 4}{10 \bullet 4} + \frac{3 \bullet 5}{8 \bullet 5} = \frac{32}{40} + \frac{4}{40} + \frac{15}{40} = \frac{51}{40}$

139. Evaluating when $x = 5$: $2(5) - 3 = 10 - 3 = 7$

2.2 Addition Property of Equality

1. Solving the equation:
$$x - 3 = 8$$
$$x - 3 + 3 = 8 + 3$$
$$x = 11$$

3. Solving the equation:
$$x + 2 = 6$$
$$x + 2 + (-2) = 6 + (-2)$$
$$x = 4$$

5. Solving the equation:
$$a + \frac{1}{2} = -\frac{1}{4}$$
$$a + \frac{1}{2} + \left(-\frac{1}{2}\right) = -\frac{1}{4} + \left(-\frac{1}{2}\right)$$
$$a = -\frac{1}{4} + \left(-\frac{2}{4}\right)$$
$$a = -\frac{3}{4}$$

7. Solving the equation:
$$x + 2.3 = -3.5$$
$$x + 2.3 + (-2.3) = -3.5 + (-2.3)$$
$$x = -5.8$$

9. Solving the equation:
$$y + 11 = -6$$
$$y + 11 + (-11) = -6 + (-11)$$
$$y = -17$$

11. Solving the equation:
$$x - \frac{5}{8} = -\frac{3}{4}$$
$$x - \frac{5}{8} + \frac{5}{8} = -\frac{3}{4} + \frac{5}{8}$$
$$x = -\frac{6}{8} + \frac{5}{8}$$
$$x = -\frac{1}{8}$$

13. Solving the equation:
$$m - 6 = 2m$$
$$m - 6 - m = 2m - m$$
$$m = -6$$

15. Solving the equation:
$$6.9 + x = 3.3$$
$$-6.9 + 6.9 + x = -6.9 + 3.3$$
$$x = -3.6$$

17. Solving the equation:
$$5a = 4a - 7$$
$$5a - 4a = 4a - 4a - 7$$
$$a = -7$$

19. Solving the equation:
$$-\frac{5}{9} = x - \frac{2}{5}$$
$$-\frac{5}{9} + \frac{2}{5} = x - \frac{2}{5} + \frac{2}{5}$$
$$-\frac{25}{45} + \frac{18}{45} = x$$
$$x = -\frac{7}{45}$$

21. Solving the equation:
$$4x + 2 - 3x = 4 + 1$$
$$x + 2 = 5$$
$$x + 2 + (-2) = 5 + (-2)$$
$$x = 3$$

23. Solving the equation:
$$8a - \frac{1}{2} - 7a = \frac{3}{4} + \frac{1}{8}$$
$$a - \frac{1}{2} = \frac{6}{8} + \frac{1}{8}$$
$$a - \frac{1}{2} = \frac{7}{8}$$
$$a - \frac{1}{2} + \frac{1}{2} = \frac{7}{8} + \frac{1}{2}$$
$$a = \frac{7}{8} + \frac{4}{8}$$
$$a = \frac{11}{8}$$

25. Solving the equation:
$$-3 - 4x + 5x = 18$$
$$-3 + x = 18$$
$$3 - 3 + x = 3 + 18$$
$$x = 21$$

27. Solving the equation:
$$-11x + 2 + 10x + 2x = 9$$
$$x + 2 = 9$$
$$x + 2 + (-2) = 9 + (-2)$$
$$x = 7$$

29. Solving the equation:
$$-2.5 + 4.8 = 8x - 1.2 - 7x$$
$$2.3 = x - 1.2$$
$$2.3 + 1.2 = x - 1.2 + 1.2$$
$$x = 3.5$$

31. Solving the equation:
$$2y - 10 + 3y - 4y = 18 - 6$$
$$y - 10 = 12$$
$$y - 10 + 10 = 12 + 10$$
$$y = 22$$

33. Solving the equation:
$$2(x + 3) - x = 4$$
$$2x + 6 - x = 4$$
$$x + 6 = 4$$
$$x + 6 + (-6) = 4 + (-6)$$
$$x = -2$$

35. Solving the equation:
$$-3(x - 4) + 4x = 3 - 7$$
$$-3x + 12 + 4x = -4$$
$$x + 12 = -4$$
$$x + 12 + (-12) = -4 + (-12)$$
$$x = -16$$

37. Solving the equation:
$$5(2a + 1) - 9a = 8 - 6$$
$$10a + 5 - 9a = 2$$
$$a + 5 = 2$$
$$a + 5 + (-5) = 2 + (-5)$$
$$a = -3$$

39. Solving the equation:
$$-(x + 3) + 2x - 1 = 6$$
$$-x - 3 + 2x - 1 = 6$$
$$x - 4 = 6$$
$$x - 4 + 4 = 6 + 4$$
$$x = 10$$

41. Solving the equation:
$$4y - 3(y - 6) + 2 = 8$$
$$4y - 3y + 18 + 2 = 8$$
$$y + 20 = 8$$
$$y + 20 + (-20) = 8 + (-20)$$
$$y = -12$$

43. Solving the equation:
$$-3(2m - 9) + 7(m - 4) = 12 - 9$$
$$-6m + 27 + 7m - 28 = 3$$
$$m - 1 = 3$$
$$m - 1 + 1 = 3 + 1$$
$$m = 4$$

45. Solving the equation:
$$4x = 3x + 2$$
$$4x + (-3x) = 3x + (-3x) + 2$$
$$x = 2$$

47. Solving the equation:
$$8a = 7a - 5$$
$$8a + (-7a) = 7a + (-7a) - 5$$
$$a = -5$$

49. Solving the equation:
$$2x = 3x + 1$$
$$(-2x) + 2x = (-2x) + 3x + 1$$
$$0 = x + 1$$
$$0 + (-1) = x + 1 + (-1)$$
$$x = -1$$

51. Solving the equation:
$$2y + 1 = 3y + 4$$
$$2y + (-2y) + 1 = 3y + (-2y) + 4$$
$$1 = y + 4$$
$$1 + (-4) = y + 4 + (-4)$$
$$y = -3$$

53. Solving the equation:
$$2m - 3 = m + 5$$
$$2m + (-m) - 3 = m + (-m) + 5$$
$$m - 3 = 5$$
$$m - 3 + 3 = 5 + 3$$
$$m = 8$$

55. Solving the equation:
$$4x - 7 = 5x + 1$$
$$4x + (-4x) - 7 = 5x + (-4x) + 1$$
$$-7 = x + 1$$
$$-7 + (-1) = x + 1 + (-1)$$
$$x = -8$$

57. Solving the equation:
$$4x + \frac{4}{3} = 5x - \frac{2}{3}$$
$$4x + (-4x) + \frac{4}{3} = 5x + (-4x) - \frac{2}{3}$$
$$\frac{4}{3} = x - \frac{2}{3}$$
$$\frac{4}{3} + \frac{2}{3} = x - \frac{2}{3} + \frac{2}{3}$$
$$x = \frac{6}{3} = 2$$

59. Solving the equation:
$$8a - 7.1 = 7a + 3.9$$
$$8a + (-7a) - 7.1 = 7a + (-7a) + 3.9$$
$$a - 7.1 = 3.9$$
$$a - 7.1 + 7.1 = 3.9 + 7.1$$
$$a = 11$$

61. **a.** Solving the equation:
$$2x = 3$$
$$x = \frac{3}{2}$$

b. Solving the equation:
$$2 + x = 3$$
$$2 + x - 2 = 3 - 2$$
$$x = 1$$

c. Solving the equation:
$$2x + 3 = 0$$
$$2x + 3 - 3 = 0 - 3$$
$$2x = -3$$
$$x = -\frac{3}{2}$$

d. Solving the equation:
$$2x + 3 = -5$$
$$2x + 3 - 3 = -5 - 3$$
$$2x = -8$$
$$x = -\frac{8}{2} = -4$$

e. Solving the equation:
$$2x + 3 = 7x - 5$$
$$2x + (-7x) + 3 = 7x + (-7x) - 5$$
$$-5x + 3 = -5$$
$$-5x + 3 + (-3) = -5 + (-3)$$
$$-5x = -8$$
$$x = \frac{8}{5}$$

63. **a.** Solving for R:
$$T + R + A = 100$$
$$88 + R + 6 = 100$$
$$94 + R = 100$$
$$R = 6\%$$

b. Solving for R:
$$T + R + A = 100$$
$$0 + R + 95 = 100$$
$$95 + R = 100$$
$$R = 5\%$$

c. Solving for A:
$$T + R + A = 100$$
$$0 + 98 + A = 100$$
$$98 + A = 100$$
$$A = 2\%$$

d. Solving for R:
$$T + R + A = 100$$
$$0 + R + 25 = 100$$
$$25 + R = 100$$
$$R = 75\%$$

65. Solving for x:
$$x + 55 + 55 = 180$$
$$x + 110 = 180$$
$$x = 70°$$

67. Applying the associative property: $3(6x) = (3 \cdot 6)x = 18x$

69. Applying the associative property: $\frac{1}{5}(5x) = \left(\frac{1}{5} \cdot 5\right)x = 1x = x$

71. Applying the associative property: $8\left(\frac{1}{8}y\right) = \left(8 \cdot \frac{1}{8}\right)y = 1y = y$

73. Applying the associative property: $-2\left(-\frac{1}{2}x\right) = \left[-2 \cdot \left(-\frac{1}{2}\right)\right]x = 1x = x$

75. Applying the associative property: $-\frac{4}{3}\left(-\frac{3}{4}a\right) = \left[-\frac{4}{3} \cdot \left(-\frac{3}{4}\right)\right]a = 1a = a$

77. Simplifying: $\frac{3}{2}\left(\frac{2}{3}y\right) = y$

79. Simplifying: $\frac{1}{5}(5x) = x$

81. Simplifying: $\frac{1}{5}(30) = 6$

83. Simplifying: $\frac{3}{2}(4) = \frac{12}{2} = 6$

85. Simplifying: $12\left(-\frac{3}{4}\right) = -\frac{36}{4} = -9$

87. Simplifying: $\frac{3}{2}\left(-\frac{5}{4}\right) = -\frac{15}{8}$

89. Simplifying: $-13 + (-5) = -18$

91. Simplifying: $-\frac{3}{4} + \left(-\frac{1}{2}\right) = -\frac{3}{4} - \frac{1}{2} \cdot \frac{2}{2} = -\frac{3}{4} - \frac{2}{4} = -\frac{5}{4}$

93. Simplifying: $7x + (-4x) = 3x$

2.3 Multiplication Property of Equality

1. Solving the equation:
$$5x = 10$$
$$\frac{1}{5}(5x) = \frac{1}{5}(10)$$
$$x = 2$$

3. Solving the equation:
$$7a = 28$$
$$\frac{1}{7}(7a) = \frac{1}{7}(28)$$
$$a = 4$$

5. Solving the equation:
$$-8x = 4$$
$$-\frac{1}{8}(-8x) = -\frac{1}{8}(4)$$
$$x = -\frac{1}{2}$$

7. Solving the equation:
$$8m = -16$$
$$\frac{1}{8}(8m) = \frac{1}{8}(-16)$$
$$m = -2$$

9. Solving the equation:
$$-3x = -9$$
$$-\frac{1}{3}(-3x) = -\frac{1}{3}(-9)$$
$$x = 3$$

11. Solving the equation:
$$-7y = -28$$
$$-\frac{1}{7}(-7y) = -\frac{1}{7}(-28)$$
$$y = 4$$

13. Solving the equation:
$$2x = 0$$
$$\frac{1}{2}(2x) = \frac{1}{2}(0)$$
$$x = 0$$

15. Solving the equation:
$$-5x = 0$$
$$-\frac{1}{5}(-5x) = -\frac{1}{5}(0)$$
$$x = 0$$

17. Solving the equation:
$$\frac{x}{3} = 2$$
$$3\left(\frac{x}{3}\right) = 3(2)$$
$$x = 6$$

19. Solving the equation:
$$-\frac{m}{5} = 10$$
$$-5\left(-\frac{m}{5}\right) = -5(10)$$
$$m = -50$$

21. Solving the equation:
$$-\frac{x}{2} = -\frac{3}{4}$$
$$-2\left(-\frac{x}{2}\right) = -2\left(-\frac{3}{4}\right)$$
$$x = \frac{3}{2}$$

23. Solving the equation:
$$\frac{2}{3}a = 8$$
$$\frac{3}{2}\left(\frac{2}{3}a\right) = \frac{3}{2}(8)$$
$$a = 12$$

25. Solving the equation:
$$-\tfrac{3}{5}x = \tfrac{9}{5}$$
$$-\tfrac{5}{3}\left(-\tfrac{3}{5}x\right) = -\tfrac{5}{3}\left(\tfrac{9}{5}\right)$$
$$x = -3$$

27. Solving the equation:
$$-\tfrac{5}{8}y = -20$$
$$-\tfrac{8}{5}\left(-\tfrac{5}{8}y\right) = -\tfrac{8}{5}(-20)$$
$$y = 32$$

29. Simplifying and then solving the equation:
$$-4x - 2x + 3x = 24$$
$$-3x = 24$$
$$-\tfrac{1}{3}(-3x) = -\tfrac{1}{3}(24)$$
$$x = -8$$

31. Simplifying and then solving the equation:
$$4x + 8x - 2x = 15 - 10$$
$$10x = 5$$
$$\tfrac{1}{10}(10x) = \tfrac{1}{10}(5)$$
$$x = \tfrac{1}{2}$$

33. Simplifying and then solving the equation:
$$-3 - 5 = 3x + 5x - 10x$$
$$-8 = -2x$$
$$-\tfrac{1}{2}(-8) = -\tfrac{1}{2}(-2x)$$
$$x = 4$$

35. Simplifying and then solving the equation:
$$18 - 13 = \tfrac{1}{2}a + \tfrac{3}{4}a - \tfrac{5}{8}a$$
$$8(5) = 8\left(\tfrac{1}{2}a + \tfrac{3}{4}a - \tfrac{5}{8}a\right)$$
$$40 = 4a + 6a - 5a$$
$$40 = 5a$$
$$\tfrac{1}{5}(40) = \tfrac{1}{5}(5a)$$
$$a = 8$$

37. Solving by multiplying both sides of the equation by −1:
$$-x = 4$$
$$-1(-x) = -1(4)$$
$$x = -4$$

39. Solving by multiplying both sides of the equation by −1:
$$-x = -4$$
$$-1(-x) = -1(-4)$$
$$x = 4$$

41. Solving by multiplying both sides of the equation by −1:
$$15 = -a$$
$$-1(15) = -1(-a)$$
$$a = -15$$

43. Solving by multiplying both sides of the equation by −1:
$$-y = \tfrac{1}{2}$$
$$-1(-y) = -1\left(\tfrac{1}{2}\right)$$
$$y = -\tfrac{1}{2}$$

45. Solving the equation:
$$3x - 2 = 7$$
$$3x - 2 + 2 = 7 + 2$$
$$3x = 9$$
$$\tfrac{1}{3}(3x) = \tfrac{1}{3}(9)$$
$$x = 3$$

47. Solving the equation:
$$2a + 1 = 3$$
$$2a + 1 + (-1) = 3 + (-1)$$
$$2a = 2$$
$$\tfrac{1}{2}(2a) = \tfrac{1}{2}(2)$$
$$a = 1$$

49. Using Method 2 to eliminate fractions:
$$\tfrac{1}{8} + \tfrac{1}{2}x = \tfrac{1}{4}$$
$$8\left(\tfrac{1}{8} + \tfrac{1}{2}x\right) = 8\left(\tfrac{1}{4}\right)$$
$$1 + 4x = 2$$
$$(-1) + 1 + 4x = (-1) + 2$$
$$4x = 1$$
$$\tfrac{1}{4}(4x) = \tfrac{1}{4}(1)$$
$$x = \tfrac{1}{4}$$

51. Solving the equation:
$$6x = 2x - 12$$
$$6x + (-2x) = 2x + (-2x) - 12$$
$$4x = -12$$
$$\tfrac{1}{4}(4x) = \tfrac{1}{4}(-12)$$
$$x = -3$$

53. Solving the equation:
$$2y = -4y + 18$$
$$2y + 4y = -4y + 4y + 18$$
$$6y = 18$$
$$\frac{1}{6}(6y) = \frac{1}{6}(18)$$
$$y = 3$$

55. Solving the equation:
$$-7x = -3x - 8$$
$$-7x + 3x = -3x + 3x - 8$$
$$-4x = -8$$
$$-\frac{1}{4}(-4x) = -\frac{1}{4}(-8)$$
$$x = 2$$

57. Solving the equation:
$$2x - 5 = 8x + 4$$
$$2x + (-8x) - 5 = 8x + (-8x) + 4$$
$$-6x - 5 = 4$$
$$-6x - 5 + 5 = 4 + 5$$
$$-6x = 9$$
$$-\frac{1}{6}(-6x) = -\frac{1}{6}(9)$$
$$x = -\frac{3}{2}$$

59. Using Method 2 to eliminate fractions:
$$x + \frac{1}{2} = \frac{1}{4}x - \frac{5}{8}$$
$$8\left(x + \frac{1}{2}\right) = 8\left(\frac{1}{4}x - \frac{5}{8}\right)$$
$$8x + 4 = 2x - 5$$
$$8x + (-2x) + 4 = 2x + (-2x) - 5$$
$$6x + 4 = -5$$
$$6x + 4 + (-4) = -5 + (-4)$$
$$6x = -9$$
$$\frac{1}{6}(6x) = \frac{1}{6}(-9)$$
$$x = -\frac{3}{2}$$

61. Solving the equation:
$$m + 2 = 6m - 3$$
$$m + (-6m) + 2 = 6m + (-6m) - 3$$
$$-5m + 2 = -3$$
$$-5m + 2 + (-2) = -3 + (-2)$$
$$-5m = -5$$
$$-\frac{1}{5}(-5m) = -\frac{1}{5}(-5)$$
$$m = 1$$

63. Using Method 2 to eliminate fractions:
$$\frac{1}{2}m - \frac{1}{4} = \frac{1}{12}m + \frac{1}{6}$$
$$12\left(\frac{1}{2}m - \frac{1}{4}\right) = 12\left(\frac{1}{12}m + \frac{1}{6}\right)$$
$$6m - 3 = m + 2$$
$$6m + (-m) - 3 = m + (-m) + 2$$
$$5m - 3 = 2$$
$$5m - 3 + 3 = 2 + 3$$
$$5m = 5$$
$$\frac{1}{5}(5m) = \frac{1}{5}(5)$$
$$m = 1$$

65. Solving the equation:
$$6y - 4 = 9y + 2$$
$$6y + (-9y) - 4 = 9y + (-9y) + 2$$
$$-3y - 4 = 2$$
$$-3y - 4 + 4 = 2 + 4$$
$$-3y = 6$$
$$-\frac{1}{3}(-3y) = -\frac{1}{3}(6)$$
$$y = -2$$

67. Using Method 2 to eliminate fractions:
$$\frac{3}{2}y + \frac{1}{3} = y - \frac{2}{3}$$
$$6\left(\frac{3}{2}y + \frac{1}{3}\right) = 6\left(y - \frac{2}{3}\right)$$
$$9y + 2 = 6y - 4$$
$$9y + (-6y) + 2 = 6y + (-6y) - 4$$
$$3y + 2 = -4$$
$$3y + 2 + (-2) = -4 + (-2)$$
$$3y = -6$$
$$\frac{1}{3}(3y) = \frac{1}{3}(-6)$$
$$y = -2$$

69. Solving the equation:
$$5x + 6 = 2$$
$$5x + 6 + (-6) = 2 + (-6)$$
$$5x = -4$$
$$x = -\frac{4}{5}$$

71. Solving the equation:
$$\frac{x}{2} = \frac{6}{12}$$
$$2 \bullet \frac{x}{2} = 2 \bullet \frac{6}{12}$$
$$x = \frac{12}{12} = 1$$

73. Solving the equation:

$$\frac{3}{x} = \frac{6}{7}$$
$$7x \bullet \frac{3}{x} = 7x \bullet \frac{6}{7}$$
$$21 = 6x$$
$$x = \frac{21}{6} = \frac{7}{2}$$

75. Solving the equation:

$$\frac{a}{3} = \frac{5}{12}$$
$$3 \bullet \frac{a}{3} = 3 \bullet \frac{5}{12}$$
$$a = \frac{15}{12} = \frac{5}{4}$$

77. Solving the equation:

$$\frac{10}{20} = \frac{20}{x}$$
$$20x \bullet \frac{10}{20} = 20x \bullet \frac{20}{x}$$
$$10x = 400$$
$$x = \frac{400}{10} = 40$$

79. Solving the equation:

$$\frac{2}{x} = \frac{6}{7}$$
$$7x \bullet \frac{2}{x} = 7x \bullet \frac{6}{7}$$
$$14 = 6x$$
$$x = \frac{14}{6} = \frac{7}{3}$$

81. Solving the equation:

$$7.5x = 1500$$
$$\frac{7.5x}{7.5} = \frac{1500}{7.5}$$
$$x = 200$$

The break-even point is 200 tickets.

83. Solving the equation:

$$G - 0.21G - 0.08G = 987.5$$
$$0.71G = 987.5$$
$$G \approx 1391$$

Your gross income is approximately $1,391.

85. Applying the distributive property: $2(3x - 5) = 2 \bullet 3x - 2 \bullet 5 = 6x - 10$

87. Applying the distributive property: $\frac{1}{2}(3x + 6) = \frac{1}{2} \bullet 3x + \frac{1}{2} \bullet 6 = \frac{3}{2}x + 3$

89. Applying the distributive property: $\frac{1}{3}(-3x + 6) = \frac{1}{3} \bullet (-3x) + \frac{1}{3} \bullet 6 = -x + 2$

91. Using the distributive property and combining like terms: $5(2x - 8) - 3 = 10x - 40 - 3 = 10x - 43$

93. Using the distributive property and combining like terms:
$$-2(3x + 5) + 3(x - 1) = -6x - 10 + 3x - 3 = -6x + 3x - 10 - 3 = -3x - 13$$

95. Using the distributive property and combining like terms: $7 - 3(2y + 1) = 7 - 6y - 3 = -6y + 7 - 3 = -6y + 4$

97. Solving the equation:

$$2x = 4$$
$$\frac{1}{2}(2x) = \frac{1}{2}(4)$$
$$x = 2$$

99. Solving the equation:

$$30 = 5x$$
$$5x = 30$$
$$\frac{1}{5}(5x) = \frac{1}{5}(30)$$
$$x = 6$$

101. Solving the equation:

$$0.17x = 510$$
$$x = \frac{510}{0.17} = 3,000$$

103. Simplifying: $3(x - 5) + 4 = 3x - 15 + 4 = 3x - 11$

105. Simplifying: $0.09(x + 2,000) = 0.09x + 180$

107. Simplifying: $7 - 3(2y + 1) = 7 - 6y - 3 = 4 - 6y = -6y + 4$

109. Simplifying: $3(2x - 5) - (2x - 4) = 6x - 15 - 2x + 4 = 4x - 11$

111. Simplifying: $10x + (-5x) = 5x$

113. Simplifying: $0.08x + 0.09x = 0.17x$

2.4 Solving Linear Equations

1.   Solving the equation:
$$2(x+3)=12$$
$$2x+6=12$$
$$2x+6+(-6)=12+(-6)$$
$$2x=6$$
$$\tfrac{1}{2}(2x)=\tfrac{1}{2}(6)$$
$$x=3$$

3.   Solving the equation:
$$6(x-1)=-18$$
$$6x-6=-18$$
$$6x-6+6=-18+6$$
$$6x=-12$$
$$\tfrac{1}{6}(6x)=\tfrac{1}{6}(-12)$$
$$x=-2$$

5.   Solving the equation:
$$2(4a+1)=-6$$
$$8a+2=-6$$
$$8a+2+(-2)=-6+(-2)$$
$$8a=-8$$
$$\tfrac{1}{8}(8a)=\tfrac{1}{8}(-8)$$
$$a=-1$$

7.   Solving the equation:
$$14=2(5x-3)$$
$$14=10x-6$$
$$14+6=10x-6+6$$
$$20=10x$$
$$\tfrac{1}{10}(20)=\tfrac{1}{10}(10x)$$
$$x=2$$

9.   Solving the equation:
$$-2(3y+5)=14$$
$$-6y-10=14$$
$$-6y-10+10=14+10$$
$$-6y=24$$
$$-\tfrac{1}{6}(-6y)=-\tfrac{1}{6}(24)$$
$$y=-4$$

11.   Solving the equation:
$$-5(2a+4)=0$$
$$-10a-20=0$$
$$-10a-20+20=0+20$$
$$-10a=20$$
$$-\tfrac{1}{10}(-10a)=-\tfrac{1}{10}(20)$$
$$a=-2$$

13.   Solving the equation:
$$1=\tfrac{1}{2}(4x+2)$$
$$1=2x+1$$
$$1+(-1)=2x+1+(-1)$$
$$0=2x$$
$$\tfrac{1}{2}(0)=\tfrac{1}{2}(2x)$$
$$x=0$$

15.   Solving the equation:
$$3(t-4)+5=-4$$
$$3t-12+5=-4$$
$$3t-7=-4$$
$$3t-7+7=-4+7$$
$$3t=3$$
$$\tfrac{1}{3}(3t)=\tfrac{1}{3}(3)$$
$$t=1$$

17.   Solving the equation:
$$4(2y+1)-7=1$$
$$8y+4-7=1$$
$$8y-3=1$$
$$8y-3+3=1+3$$
$$8y=4$$
$$\tfrac{1}{8}(8y)=\tfrac{1}{8}(4)$$
$$y=\tfrac{1}{2}$$

19.   Solving the equation:
$$\tfrac{1}{2}(x-3)=\tfrac{1}{4}(x+1)$$
$$\tfrac{1}{2}x-\tfrac{3}{2}=\tfrac{1}{4}x+\tfrac{1}{4}$$
$$4\left(\tfrac{1}{2}x-\tfrac{3}{2}\right)=4\left(\tfrac{1}{4}x+\tfrac{1}{4}\right)$$
$$2x-6=x+1$$
$$2x+(-x)-6=x+(-x)+1$$
$$x-6=1$$
$$x-6+6=1+6$$
$$x=7$$

21.   Solving the equation:
$$-0.7(2x-7)=0.3(11-4x)$$
$$-1.4x+4.9=3.3-1.2x$$
$$-1.4x+1.2x+4.9=3.3-1.2x+1.2x$$
$$-0.2x+4.9=3.3$$
$$-0.2x+4.9+(-4.9)=3.3+(-4.9)$$
$$-0.2x=-1.6$$
$$\frac{-0.2x}{-0.2}=\frac{-1.6}{-0.2}$$
$$x=8$$

23.   Solving the equation:
$$-2(3y+1)=3(1-6y)-9$$
$$-6y-2=3-18y-9$$
$$-6y-2=-18y-6$$
$$-6y+18y-2=-18y+18y-6$$
$$12y-2=-6$$
$$12y-2+2=-6+2$$
$$12y=-4$$
$$\tfrac{1}{12}(12y)=\tfrac{1}{12}(-4)$$
$$y=-\tfrac{1}{3}$$

25.   Solving the equation:

$$\frac{3}{4}(8x-4)+3=\frac{2}{5}(5x+10)-1$$
$$6x-3+3=2x+4-1$$
$$6x=2x+3$$
$$6x+(-2x)=2x+(-2x)+3$$
$$4x=3$$
$$\frac{1}{4}(4x)=\frac{1}{4}(3)$$
$$x=\frac{3}{4}$$

29.   Solving the equation:

$$6-5(2a-3)=1$$
$$6-10a+15=1$$
$$-10a+21=1$$
$$-10a+21+(-21)=1+(-21)$$
$$-10a=-20$$
$$-\frac{1}{10}(-10a)=-\frac{1}{10}(-20)$$
$$a=2$$

33.   Solving the equation:

$$2(t-3)+3(t-2)=28$$
$$2t-6+3t-6=28$$
$$5t-12=28$$
$$5t-12+12=28+12$$
$$5t=40$$
$$\frac{1}{5}(5t)=\frac{1}{5}(40)$$
$$t=8$$

37.   Solving the equation:

$$2(5x-3)-(2x-4)=5-(6x+1)$$
$$10x-6-2x+4=5-6x-1$$
$$8x-2=-6x+4$$
$$8x+6x-2=-6x+6x+4$$
$$14x-2=4$$
$$14x-2+2=4+2$$
$$14x=6$$
$$\frac{1}{14}(14x)=\frac{1}{14}(6)$$
$$x=\frac{3}{7}$$

41.   Solving the equation:

$$x+(2x-1)=2$$
$$3x-1=2$$
$$3x-1+1=2+1$$
$$3x=3$$
$$\frac{1}{3}(3x)=\frac{1}{3}(3)$$
$$x=1$$

27.   Solving the equation:

$$0.06x+0.08(100-x)=6.5$$
$$0.06x+8-0.08x=6.5$$
$$-0.02x+8=6.5$$
$$-0.02x+8+(-8)=6.5+(-8)$$
$$-0.02x=-1.5$$
$$\frac{-0.02x}{-0.02}=\frac{-1.5}{-0.02}$$
$$x=75$$

31.   Solving the equation:

$$0.2x-0.5=0.5-0.2(2x-13)$$
$$0.2x-0.5=0.5-0.4x+2.6$$
$$0.2x-0.5=-0.4x+3.1$$
$$0.2x+0.4x-0.5=-0.4x+0.4x+3.1$$
$$0.6x-0.5=3.1$$
$$0.6x-0.5+0.5=3.1+0.5$$
$$0.6x=3.6$$
$$\frac{0.6x}{0.6}=\frac{3.6}{0.6}$$
$$x=6$$

35.   Solving the equation:

$$5(x-2)-(3x+4)=3(6x-8)+10$$
$$5x-10-3x-4=18x-24+10$$
$$2x-14=18x-14$$
$$2x+(-18x)-14=18x+(-18x)-14$$
$$-16x-14=-14$$
$$-16x-14+14=-14+14$$
$$-16x=0$$
$$-\frac{1}{16}(-16x)=-\frac{1}{16}(0)$$
$$x=0$$

39.   Solving the equation:

$$-(3x+1)-(4x-7)=4-(3x+2)$$
$$-3x-1-4x+7=4-3x-2$$
$$-7x+6=-3x+2$$
$$-7x+3x+6=-3x+3x+2$$
$$-4x+6=2$$
$$-4x+6+(-6)=2+(-6)$$
$$-4x=-4$$
$$-\frac{1}{4}(-4x)=-\frac{1}{4}(-4)$$
$$x=1$$

43.   Solving the equation:

$$x-(3x+5)=-3$$
$$x-3x-5=-3$$
$$-2x-5=-3$$
$$-2x-5+5=-3+5$$
$$-2x=2$$
$$-\frac{1}{2}(-2x)=-\frac{1}{2}(2)$$
$$x=-1$$

45. Solving the equation:

$$15 = 3(x-1)$$
$$15 = 3x - 3$$
$$15 + 3 = 3x - 3 + 3$$
$$18 = 3x$$
$$\frac{1}{3}(18) = \frac{1}{3}(3x)$$
$$x = 6$$

47. Solving the equation:

$$4x - (-4x + 1) = 5$$
$$4x + 4x - 1 = 5$$
$$8x - 1 = 5$$
$$8x - 1 + 1 = 5 + 1$$
$$8x = 6$$
$$\frac{1}{8}(8x) = \frac{1}{8}(6)$$
$$x = \frac{3}{4}$$

49. Solving the equation:

$$5x - 8(2x - 5) = 7$$
$$5x - 16x + 40 = 7$$
$$-11x + 40 = 7$$
$$-11x + 40 - 40 = 7 - 40$$
$$-11x = -33$$
$$-\frac{1}{11}(-11x) = -\frac{1}{11}(-33)$$
$$x = 3$$

51. Solving the equation:

$$7(2y - 1) - 6y = -1$$
$$14y - 7 - 6y = -1$$
$$8y - 7 = -1$$
$$8y - 7 + 7 = -1 + 7$$
$$8y = 6$$
$$\frac{1}{8}(8y) = \frac{1}{8}(6)$$
$$y = \frac{3}{4}$$

53. Solving the equation:

$$0.2x + 0.5(12 - x) = 3.6$$
$$0.2x + 6 - 0.5x = 3.6$$
$$-0.3x + 6 = 3.6$$
$$-0.3x + 6 - 6 = 3.6 - 6$$
$$-0.3x = -2.4$$
$$x = \frac{-2.4}{-0.3} = 8$$

55. Solving the equation:

$$0.5x + 0.2(18 - x) = 5.4$$
$$0.5x + 3.6 - 0.2x = 5.4$$
$$0.3x + 3.6 = 5.4$$
$$0.3x + 3.6 - 3.6 = 5.4 - 3.6$$
$$0.3x = 1.8$$
$$x = \frac{1.8}{0.3} = 6$$

57. Solving the equation:

$$x + (x + 3)(-3) = x - 3$$
$$x - 3x - 9 = x - 3$$
$$-2x - 9 = x - 3$$
$$-2x - x - 9 = x - x - 3$$
$$-3x - 9 = -3$$
$$-3x - 9 + 9 = -3 + 9$$
$$-3x = 6$$
$$-\frac{1}{3}(-3x) = -\frac{1}{3}(6)$$
$$x = -2$$

59. Solving the equation:

$$5(x + 2) + 3(x - 1) = -9$$
$$5x + 10 + 3x - 3 = -9$$
$$8x + 7 = -9$$
$$8x + 7 - 7 = -9 - 7$$
$$8x = -16$$
$$\frac{1}{8}(8x) = \frac{1}{8}(-16)$$
$$x = -2$$

61. Solving the equation:

$$3(x - 3) + 2(2x) = 5$$
$$3x - 9 + 4x = 5$$
$$7x - 9 = 5$$
$$7x - 9 + 9 = 5 + 9$$
$$7x = 14$$
$$\frac{1}{7}(7x) = \frac{1}{7}(14)$$
$$x = 2$$

63. Solving the equation:

$$5(y + 2) = 4(y + 1)$$
$$5y + 10 = 4y + 4$$
$$5y - 4y + 10 = 4y - 4y + 4$$
$$y + 10 = 4$$
$$y + 10 - 10 = 4 - 10$$
$$y = -6$$

65. Solving the equation:

$$3x + 2(x - 2) = 6$$
$$3x + 2x - 4 = 6$$
$$5x - 4 = 6$$
$$5x - 4 + 4 = 6 + 4$$
$$5x = 10$$
$$\tfrac{1}{5}(5x) = \tfrac{1}{5}(10)$$
$$x = 2$$

67. Solving the equation:

$$50(x - 5) = 30(x + 5)$$
$$50x - 250 = 30x + 150$$
$$50x - 30x - 250 = 30x - 30x + 150$$
$$20x - 250 = 150$$
$$20x - 250 + 250 = 150 + 250$$
$$20x = 400$$
$$\tfrac{1}{20}(20x) = \tfrac{1}{20}(400)$$
$$x = 20$$

69. Solving the equation:

$$0.08x + 0.09(x + 2,000) = 860$$
$$0.08x + 0.09x + 180 = 860$$
$$0.17x + 180 = 860$$
$$0.17x + 180 - 180 = 860 - 180$$
$$0.17x = 680$$
$$x = \frac{680}{0.17} = 4,000$$

71. Solving the equation:

$$0.10x + 0.12(x + 500) = 214$$
$$0.10x + 0.12x + 60 = 214$$
$$0.22x + 60 = 214$$
$$0.22x + 60 - 60 = 214 - 60$$
$$0.22x = 154$$
$$x = \frac{154}{0.22} = 700$$

73. Solving the equation:

$$5x + 10(x + 8) = 245$$
$$5x + 10x + 80 = 245$$
$$15x + 80 = 245$$
$$15x + 80 - 80 = 245 - 80$$
$$15x = 165$$
$$x = \frac{165}{15} = 11$$

75. Solving the equation:

$$5x + 10(x + 3) + 25(x + 5) = 435$$
$$5x + 10x + 30 + 25x + 125 = 435$$
$$40x + 155 = 435$$
$$40x + 155 - 155 = 435 - 155$$
$$40x = 280$$
$$x = \frac{280}{40} = 7$$

77. a. Solving the equation:

$$4x - 5 = 0$$
$$4x - 5 + 5 = 0 + 5$$
$$4x = 5$$
$$\tfrac{1}{4}(4x) = \tfrac{1}{4}(5)$$
$$x = \tfrac{5}{4} = 1.25$$

b. Solving the equation:

$$4x - 5 = 25$$
$$4x - 5 + 5 = 25 + 5$$
$$4x = 30$$
$$\tfrac{1}{4}(4x) = \tfrac{1}{4}(30)$$
$$x = \tfrac{15}{2} = 7.5$$

c. Adding: $(4x - 5) + (2x + 25) = 6x + 20$

d. Solving the equation:

$$4x - 5 = 2x + 25$$
$$4x - 2x - 5 = 2x - 2x + 25$$
$$2x - 5 = 25$$
$$2x - 5 + 5 = 25 + 5$$
$$2x = 30$$
$$\tfrac{1}{2}(2x) = \tfrac{1}{2}(30)$$
$$x = 15$$

e. Multiplying: $4(x - 5) = 4x - 20$

f.   Solving the equation:
$$4(x-5)=2x+25$$
$$4x-20=2x+25$$
$$4x-2x-20=2x-2x+25$$
$$2x-20=25$$
$$2x-20+20=25+20$$
$$2x=45$$
$$\tfrac{1}{2}(2x)=\tfrac{1}{2}(45)$$
$$x=\tfrac{45}{2}=22.5$$

79.   Multiplying the fractions: $\tfrac{1}{2}(3)=\tfrac{1}{2}\cdot\tfrac{3}{1}=\tfrac{3}{2}$

81.   Multiplying the fractions: $\tfrac{2}{3}(6)=\tfrac{2}{3}\cdot\tfrac{6}{1}=\tfrac{12}{3}=4$

83.   Multiplying the fractions: $\tfrac{5}{9}\cdot\tfrac{9}{5}=\tfrac{45}{45}=1$

85.   Completing the table:

x	$3(x+2)$	$3x+2$	$3x+6$
0	6	2	6
1	9	5	9
2	12	8	12
3	15	11	15

87.   Completing the table:

a	$(2a+1)^2$	$4a^2+4a+1$
1	9	9
2	25	25
3	49	49

89.   Solving the equation:
$$40=2x+12$$
$$2x+12=40$$
$$2x=28$$
$$x=14$$

91.   Solving the equation:
$$12+2y=6$$
$$2y=-6$$
$$y=-3$$

93.   Solving the equation:
$$24x=6$$
$$x=\tfrac{6}{24}=\tfrac{1}{4}$$

95.   Solving the equation:
$$70=x\cdot210$$
$$x=\tfrac{70}{210}=\tfrac{1}{3}$$

97.   Simplifying: $\tfrac{1}{2}(-3x+6)=\tfrac{1}{2}(-3x)+\tfrac{1}{2}(6)=-\tfrac{3}{2}x+3$

2.5 Formulas

1.   Using the perimeter formula:
$$P=2l+2w$$
$$300=2l+2(50)$$
$$300=2l+100$$
$$200=2l$$
$$l=100$$
The length is 100 feet.

3.   **a.**   Substituting $x=3$:
$$2(0)+3y=6$$
$$0+3y=6$$
$$3y=6$$
$$y=2$$

b.   Substituting $y=1$:
$$2x+3(1)=6$$
$$2x+3=6$$
$$2x=3$$
$$x=\tfrac{3}{2}$$

c.   Substituting $x=3$:
$$2(3)+3y=6$$
$$6+3y=6$$
$$6+(-6)+3y=6+(-6)$$
$$3y=0$$
$$y=0$$

5. **a.** Substituting $x = 0$: $y = -\frac{1}{3}(0) + 2 = 0 + 2 = 2$

b. Substituting $y = 3$:

$$3 = -\frac{1}{3}x + 2$$
$$1 = -\frac{1}{3}x$$
$$3 = -x$$
$$x = -3$$

c. Substituting $x = 3$: $y = -\frac{1}{3}(3) + 2 = -1 + 2 = 1$

7. Substituting $x = -2$: $y = (-2 + 1)^2 - 3 = (-1)^2 - 3 = 1 - 3 = -2$

9. Substituting $x = 1$: $y = (1 + 1)^2 - 3 = (2)^2 - 3 = 4 - 3 = 1$

11. **a.** Substituting $x = 10$: $y = \frac{20}{10} = 2$ **b.** Substituting $x = 5$: $y = \frac{20}{5} = 4$

13. **a.** Substituting $y = 15$ and $x = 3$:

$$15 = K(3)$$
$$K = \frac{15}{3} = 5$$

b. Substituting $y = 72$ and $x = 4$:

$$72 = K(4)$$
$$K = \frac{72}{4} = 18$$

15. **a.** Substituting $x = 5$ and $y = 4$:

$$4 = \frac{K}{5}$$
$$K = 20$$

b. Substituting $x = 5$ and $y = 15$:

$$15 = \frac{K}{5}$$
$$K = 75$$

17. Solving for l:

$$lw = A$$
$$\frac{lw}{w} = \frac{A}{w}$$
$$l = \frac{A}{w}$$

19. Solving for h:

$$lwh = V$$
$$\frac{lwh}{lw} = \frac{V}{lw}$$
$$h = \frac{V}{lw}$$

21. Solving for a:

$$a + b + c = P$$
$$a + b + c - b - c = P - b - c$$
$$a = P - b - c$$

23. Solving for x:

$$x - 3y = -1$$
$$x - 3y + 3y = -1 + 3y$$
$$x = 3y - 1$$

25. Solving for y:

$$-3x + y = 6$$
$$-3x + 3x + y = 6 + 3x$$
$$y = 3x + 6$$

27. Solving for y:

$$2x + 3y = 6$$
$$-2x + 2x + 3y = -2x + 6$$
$$3y = -2x + 6$$
$$\frac{1}{3}(3y) = \frac{1}{3}(-2x + 6)$$
$$y = -\frac{2}{3}x + 2$$

29. Solving for w:

$$2l + 2w = P$$
$$2l - 2l + 2w = P - 2l$$
$$2w = P - 2l$$
$$\frac{2w}{2} = \frac{P - 2l}{2}$$
$$w = \frac{P - 2l}{2}$$

31. Solving for v:

$$vt + 16t^2 = h$$
$$vt + 16t^2 - 16t^2 = h - 16t^2$$
$$vt = h - 16t^2$$
$$\frac{vt}{t} = \frac{h - 16t^2}{t}$$
$$v = \frac{h - 16t^2}{t}$$

33. Solving for h:

$$\pi r^2 + 2\pi rh = A$$
$$\pi r^2 - \pi r^2 + 2\pi rh = A - \pi r^2$$
$$2\pi rh = A - \pi r^2$$
$$\frac{2\pi rh}{2\pi r} = \frac{A - \pi r^2}{2\pi r}$$
$$h = \frac{A - \pi r^2}{2\pi r}$$

35. **a.** Solving for y:

$$y - 3 = -2(x + 4)$$
$$y - 3 = -2x - 8$$
$$y = -2x - 5$$

b. Solving for y:

$$y - 5 = 4(x - 3)$$
$$y - 5 = 4x - 12$$
$$y = 4x - 7$$

37. **a.** Solving for y:

$$y - 1 = \frac{3}{4}(x - 1)$$
$$y - 1 = \frac{3}{4}x - \frac{3}{4}$$
$$y = \frac{3}{4}x + \frac{1}{4}$$

b. Solving for y:

$$y + 2 = \frac{3}{4}(x - 4)$$
$$y + 2 = \frac{3}{4}x - 3$$
$$y = \frac{3}{4}x - 5$$

39. **a.** Solving for y:

$$\frac{y - 1}{x} = \frac{3}{5}$$
$$5y - 5 = 3x$$
$$5y = 3x + 5$$
$$y = \frac{3}{5}x + 1$$

b. Solving for y:

$$\frac{y - 2}{x} = \frac{1}{2}$$
$$2y - 4 = x$$
$$2y = x + 4$$
$$y = \frac{1}{2}x + 2$$

c. Solving for y:

$$\frac{y - 3}{x} = 4$$
$$y - 3 = 4x$$
$$y = 4x + 3$$

41. Solving for y:

$$\frac{x}{7} - \frac{y}{3} = 1$$
$$-\frac{x}{7} + \frac{x}{7} - \frac{y}{3} = -\frac{x}{7} + 1$$
$$-\frac{y}{3} = -\frac{x}{7} + 1$$
$$-3\left(-\frac{y}{3}\right) = -3\left(-\frac{x}{7} + 1\right)$$
$$y = \frac{3}{7}x - 3$$

43. Solving for y:

$$-\frac{1}{4}x + \frac{1}{8}y = 1$$
$$-\frac{1}{4}x + \frac{1}{4}x + \frac{1}{8}y = 1 + \frac{1}{4}x$$
$$\frac{1}{8}y = \frac{1}{4}x + 1$$
$$8\left(\frac{1}{8}y\right) = 8\left(\frac{1}{4}x + 1\right)$$
$$y = 2x + 8$$

45. **a.** Solving the equation:
$$4x + 5 = 20$$
$$4x + 5 - 5 = 20 - 5$$
$$4x = 15$$
$$\tfrac{1}{4}(4x) = \tfrac{1}{4}(15)$$
$$x = \tfrac{15}{4} = 3.75$$

b. Evaluating when $x = 3$: $4x + 5 = 4(3) + 5 = 12 + 5 = 17$

c. Solving for y:
$$4x + 5y = 20$$
$$4x - 4x + 5y = 20 - 4x$$
$$5y = -4x + 20$$
$$\tfrac{1}{5}(5y) = \tfrac{1}{5}(-4x + 20)$$
$$y = -\tfrac{4}{5}x + 4$$

d. Solving for x:
$$4x + 5y = 20$$
$$4x + 5y - 5y = 20 - 5y$$
$$4x = -5y + 20$$
$$\tfrac{1}{4}(4x) = \tfrac{1}{4}(-5y + 20)$$
$$x = -\tfrac{5}{4}y + 5$$

47. The complement of $30°$ is $90° - 30° = 60°$, and the supplement is $180° - 30° = 150°$.

49. The complement of $45°$ is $90° - 45° = 45°$, and the supplement is $180° - 45° = 135°$.

51. Translating into an equation and solving:
$$x = 0.25 \cdot 40$$
$$x = 10$$
The number 10 is 25% of 40.

53. Translating into an equation and solving:
$$x = 0.12 \cdot 2000$$
$$x = 240$$
The number 240 is 12% of 2000.

55. Translating into an equation and solving:
$$x \cdot 28 = 7$$
$$28x = 7$$
$$\tfrac{1}{28}(28x) = \tfrac{1}{28}(7)$$
$$x = 0.25 = 25\%$$
The number 7 is 25% of 28.

57. Translating into an equation and solving:
$$x \cdot 40 = 14$$
$$40x = 14$$
$$\tfrac{1}{40}(40x) = \tfrac{1}{40}(14)$$
$$x = 0.35 = 35\%$$
The number 14 is 35% of 40.

59. Translating into an equation and solving:
$$0.50 \cdot x = 32$$
$$\frac{0.50x}{0.50} = \frac{32}{0.50}$$
$$x = 64$$
The number 32 is 50% of 64.

61. Translating into an equation and solving:
$$0.12 \cdot x = 240$$
$$\frac{0.12x}{0.12} = \frac{240}{0.12}$$
$$x = 2000$$
The number 240 is 12% of 2000.

63. Substituting $F = 212$: $C = \tfrac{5}{9}(212 - 32) = \tfrac{5}{9}(180) = 100°C$. This value agrees with the information in Table 1.

65. Substituting $F = 68$: $C = \tfrac{5}{9}(68 - 32) = \tfrac{5}{9}(36) = 20°C$. This value agrees with the information in Table 1.

67. Solving for C:
$$\tfrac{9}{5}C + 32 = F$$
$$\tfrac{9}{5}C + 32 - 32 = F - 32$$
$$\tfrac{9}{5}C = F - 32$$
$$\tfrac{5}{9}\left(\tfrac{9}{5}C\right) = \tfrac{5}{9}(F - 32)$$
$$C = \tfrac{5}{9}(F - 32)$$

69. We need to find what percent of 150 is 90:
$$x \cdot 150 = 90$$
$$\tfrac{1}{150}(150x) = \tfrac{1}{150}(90)$$
$$x = 0.60 = 60\%$$
So 60% of the calories in one serving of vanilla ice cream are fat calories.

71. We need to find what percent of 98 is 26:
$$x \cdot 98 = 26$$
$$\tfrac{1}{98}(98x) = \tfrac{1}{98}(26)$$
$$x \approx 0.265 = 26.5\%$$
So 26.5% of one serving of frozen yogurt are carbohydrates.

73. Solving for r:
$$2\pi r = C$$
$$2 \cdot \tfrac{22}{7} r = 44$$
$$\tfrac{44}{7} r = 44$$
$$\tfrac{7}{44}\left(\tfrac{44}{7} r\right) = \tfrac{7}{44}(44)$$
$$r = 7 \text{ meters}$$

75. Solving for r:
$$2\pi r = C$$
$$2 \cdot 3.14 r = 9.42$$
$$6.28 r = 9.42$$
$$\frac{6.28 r}{6.28} = \frac{9.42}{6.28}$$
$$r = 1.5 \text{ inches}$$

77. Solving for h:
$$\pi r^2 h = V$$
$$\tfrac{22}{7}\left(\tfrac{7}{22}\right)^2 h = 42$$
$$\tfrac{7}{22} h = 42$$
$$\tfrac{22}{7}\left(\tfrac{7}{22} h\right) = \tfrac{22}{7}(42)$$
$$h = 132 \text{ feet}$$

79. Solving for h:
$$\pi r^2 h = V$$
$$3.14(3)^2 h = 6.28$$
$$28.26 h = 6.28$$
$$\frac{28.26 h}{28.26} = \frac{6.28}{28.26}$$
$$h = \tfrac{2}{9} \text{ centimeters}$$

81. a. Simplifying: $27 - (-68) = 27 + 68 = 95$ b. Simplifying: $27 + (-68) = 27 - 68 = -41$

c. Simplifying: $-27 - 68 = -95$ d. Simplifying: $-27 + 68 = 41$

83. a. Simplifying: $-32 - (-41) = -32 + 41 = 9$ b. Simplifying: $-32 + (-41) = -32 - 41 = -73$

c. Simplifying: $-32 + 41 = 9$ d. Simplifying: $-32 - 41 = -73$

85. The sum of 4 and 1 is 5.   **87.** The difference of 6 and 2 is 4.

89. The difference of a number and 5 is -12.

91. The sum of a number and 3 is four times the difference of that number and 3.

93. An equivalent expression is: $2(6 + 3) = 2(9) = 18$ **95.** An equivalent expression is: $2(5) + 3 = 10 + 3 = 13$

97. An equivalent expression is: $x + 5 = 13$ **99.** An equivalent expression is: $5(x + 7) = 30$

2.6 Applications

1. Let x represent the number. The equation is:
$$x + 5 = 13$$
$$x = 8$$
The number is 8.

3. Let x represent the number. The equation is:
$$2x + 4 = 14$$
$$2x = 10$$
$$x = 5$$
The number is 5.

5. Let x represent the number. The equation is:
$$5(x + 7) = 30$$
$$5x + 35 = 30$$
$$5x = -5$$
$$x = -1$$
The number is -1.

7. Let x and $x + 2$ represent the two numbers. The equation is:
$$x + x + 2 = 8$$
$$2x + 2 = 8$$
$$2x = 6$$
$$x = 3$$
$$x + 2 = 5$$
The two numbers are 3 and 5.

9. Let x and $3x - 4$ represent the two numbers. The equation is:
$$\left(x + 3x - 4\right) + 5 = 25$$
$$4x + 1 = 25$$
$$4x = 24$$
$$x = 6$$
$$3x - 4 = 3(6) - 4 = 14$$
The two numbers are 6 and 14.

11. Completing the table:

	Four Years Ago	Now
Shelly	$x + 3 - 4 = x - 1$	$x + 3$
Michele	$x - 4$	x

The equation is:
$$x - 1 + x - 4 = 67$$
$$2x - 5 = 67$$
$$2x = 72$$
$$x = 36$$
$$x + 3 = 39$$
Shelly is 39 and Michele is 36.

13. Completing the table:

	Three Years Ago	Now
Cody	$2x - 3$	$2x$
Evan	$x - 3$	x

The equation is:
$$2x - 3 + x - 3 = 27$$
$$3x - 6 = 27$$
$$3x = 33$$
$$x = 11$$
$$2x = 22$$
Evan is 11 and Cody is 22.

15. Completing the table:

	Five Years Ago	Now
Fred	$x + 4 - 5 = x - 1$	$x + 4$
Barney	$x - 5$	x

The equation is:
$$x - 1 + x - 5 = 48$$
$$2x - 6 = 48$$
$$2x = 54$$
$$x = 27$$
$$x + 4 = 31$$
Barney is 27 and Fred is 31.

17. Completing the table:

	Now	Three Years from Now
Jack	$2x$	$2x + 3$
Lacy	x	$x + 3$

The equation is:
$$2x + 3 + x + 3 = 54$$
$$3x + 6 = 54$$
$$3x = 48$$
$$x = 16$$
$$2x = 32$$
Lacy is 16 and Jack is 32.

19. Completing the table:

	Now	Two Years from Now
Pat	$x + 20$	$x + 20 + 2 = x + 22$
Patrick	x	$x + 2$

The equation is:
$$x + 22 = 2(x + 2)$$
$$x + 22 = 2x + 4$$
$$22 = x + 4$$
$$x = 18$$
$$x + 20 = 38$$
Patrick is 18 and Pat is 38.

21. Using the formula $P = 4s$, the equation is:
$$4s = 36$$
$$s = 9$$
The length of each side is 9 inches.

23. Using the formula $P = 4s$, the equation is:
$$4s = 60$$
$$s = 15$$
The length of each side is 15 feet.

25. Let x, $3x$, and $x + 7$ represent the sides of the triangle. The equation is:
$$x + 3x + x + 7 = 62$$
$$5x + 7 = 62$$
$$5x = 55$$
$$x = 11$$
$$x + 7 = 18$$
$$3x = 33$$
The sides are 11 feet, 18 feet, and 33 feet.

27. Let x, $2x$, and $2x - 12$ represent the sides of the triangle. The equation is:
$$x + 2x + 2x - 12 = 53$$
$$5x - 12 = 53$$
$$5x = 65$$
$$x = 13$$
$$2x = 26$$
$$2x - 12 = 14$$
The sides are 13 feet, 14 feet, and 26 feet.

29. Let w represent the width and $w + 5$ represent the length. The equation is:
$$2w + 2(w + 5) = 34$$
$$2w + 2w + 10 = 34$$
$$4w + 10 = 34$$
$$4w = 24$$
$$w = 6$$
$$w + 5 = 11$$
The length is 11 inches and the width is 6 inches.

31. Let w represent the width and $2w + 7$ represent the length. The equation is:
$$2w + 2(2w + 7) = 68$$
$$2w + 4w + 14 = 68$$
$$6w + 14 = 68$$
$$6w = 54$$
$$w = 9$$
$$2w + 7 = 2(9) + 7 = 25$$
The length is 25 inches and the width is 9 inches.

33. Let w represent the width and $3w + 6$ represent the length. The equation is:
$$2w + 2(3w + 6) = 36$$
$$2w + 6w + 12 = 36$$
$$8w + 12 = 36$$
$$8w = 24$$
$$w = 3$$
$$3w + 6 = 3(3) + 6 = 15$$
The length is 15 feet and the width is 3 feet.

35. Completing the table:

	Dimes	Quarters
Number	x	$x + 5$
Value (cents)	$10(x)$	$25(x + 5)$

The equation is:
$$10(x) + 25(x + 5) = 440$$
$$10x + 25x + 125 = 440$$
$$35x + 125 = 440$$
$$35x = 315$$
$$x = 9$$
$$x + 5 = 14$$
Marissa has 9 dimes and 14 quarters.

37. Completing the table:

	Nickels	Quarters
Number	$x + 15$	x
Value (cents)	$5(x + 15)$	$25(x)$

The equation is:
$$5(x + 15) + 25(x) = 435$$
$$5x + 75 + 25x = 435$$
$$30x + 75 = 435$$
$$30x = 360$$
$$x = 12$$
$$x + 15 = 27$$
Tanner has 12 quarters and 27 nickels.

39. Completing the table:

	Nickels	Dimes
Number	x	$x+9$
Value (cents)	$5(x)$	$10(x+9)$

The equation is:

$$5(x)+10(x+9)=210$$
$$5x+10x+90=210$$
$$15x+90=210$$
$$15x=120$$
$$x=8$$
$$x+9=17$$

Sue has 8 nickels and 17 dimes.

41. Completing the table:

	Nickels	Dimes	Quarters
Number	x	$x+3$	$x+5$
Value (cents)	$5(x)$	$10(x+3)$	$25(x+5)$

The equation is:

$$5(x)+10(x+3)+25(x+5)=435$$
$$5x+10x+30+25x+125=435$$
$$40x+155=435$$
$$40x=280$$
$$x=7$$
$$x+3=10$$
$$x+5=12$$

Katie has 7 nickels, 10 dimes, and 12 quarters.

43. Completing the table:

	Nickels	Dimes	Quarters
Number	x	$x+6$	$2x$
Value (cents)	$5(x)$	$10(x+6)$	$25(2x)$

The equation is:

$$5(x)+10(x+6)+25(2x)=255$$
$$5x+10x+60+50x=255$$
$$65x+60=255$$
$$65x=195$$
$$x=3$$
$$x+6=9$$
$$2x=6$$

Cory has 3 nickels, 9 dimes, and 6 quarters.

45. The statement is: 4 is less than 10

47. The statement is: 9 is greater than or equal to -5

49. The correct statement is: $12 < 20$

51. The correct statement is: $-8 < -6$

53. Simplifying: $|8-3|-|5-2|=|5|-|3|=5-3=2$

55. Simplifying: $15-|9-3(7-5)|=15-|9-3(2)|=15-|9-6|=15-|3|=15-3=12$

57. Simplifying: $x+2x+2x=5x$

59. Simplifying: $x+0.075x=1.075x$

61. Simplifying: $0.09(x+2,000)=0.09x+180$

63. Solving the equation:

$$0.05x+0.06(x-1,500)=570$$
$$0.05x+0.06x-90=570$$
$$0.11x=660$$
$$x=6,000$$

65. Solving the equation:

$$x+2x+3x=180$$
$$6x=180$$
$$x=30$$

2.7 More Applications

1.   Let x and $x + 1$ represent the two numbers. The equation is:
$$x + x + 1 = 11$$
$$2x + 1 = 11$$
$$2x = 10$$
$$x = 5$$
$$x + 1 = 6$$
The numbers are 5 and 6.

3.   Let x and $x + 1$ represent the two numbers. The equation is:
$$x + x + 1 = -9$$
$$2x + 1 = -9$$
$$2x = -10$$
$$x = -5$$
$$x + 1 = -4$$
The numbers are -5 and -4.

5.   Let x and $x + 2$ represent the two numbers. The equation is:
$$x + x + 2 = 28$$
$$2x + 2 = 28$$
$$2x = 26$$
$$x = 13$$
$$x + 2 = 15$$
The numbers are 13 and 15.

7.   Let x and $x + 2$ represent the two numbers. The equation is:
$$x + x + 2 = 106$$
$$2x + 2 = 106$$
$$2x = 104$$
$$x = 52$$
$$x + 2 = 54$$
The numbers are 52 and 54.

9.   Let x and $x + 2$ represent the two numbers. The equation is:
$$x + x + 2 = -30$$
$$2x + 2 = -30$$
$$2x = -322$$
$$x = -16$$
$$x + 2 = -14$$
The numbers are -16 and -14.

11.   Let x, $x + 2$, and $x + 4$ represent the three numbers. The equation is:
$$x + x + 2 + x + 4 = 57$$
$$3x + 6 = 57$$
$$3x = 51$$
$$x = 17$$
$$x + 2 = 19$$
$$x + 4 = 21$$
The numbers are 17, 19, and 21.

13.   Let x, $x + 2$, and $x + 4$ represent the three numbers. The equation is:
$$x + x + 2 + x + 4 = 132$$
$$3x + 6 = 132$$
$$3x = 126$$
$$x = 42$$
$$x + 2 = 44$$
$$x + 4 = 46$$
The numbers are 42, 44, and 46.

15. Completing the table:

	Dollars Invested at 8%	Dollars Invested at 9%
Number of	x	$x + 2000$
Interest on	$0.08(x)$	$0.09(x + 2000)$

The equation is:
$$0.08(x) + 0.09(x + 2000) = 860$$
$$0.08x + 0.09x + 180 = 860$$
$$0.17x + 180 = 860$$
$$0.17x = 680$$
$$x = 4000$$
$$x + 2000 = 6000$$
You have $4,000 invested at 8% and $6,000 invested at 9%.

17. Completing the table:

	Dollars Invested at 10%	Dollars Invested at 12%
Number of	x	$x + 500$
Interest on	$0.10(x)$	$0.12(x + 500)$

The equation is:
$$0.10(x) + 0.12(x + 500) = 214$$
$$0.10x + 0.12x + 60 = 214$$
$$0.22x + 60 = 214$$
$$0.22x = 154$$
$$x = 700$$
$$x + 500 = 1200$$
Tyler has $700 invested at 10% and $1,200 invested at 12%.

19. Completing the table:

	Dollars Invested at 8%	Dollars Invested at 9%	Dollars Invested at 10%
Number of	x	$2x$	$3x$
Interest on	$0.08(x)$	$0.09(2x)$	$0.10(3x)$

The equation is:
$$0.08(x) + 0.09(2x) + 0.10(3x) = 280$$
$$0.08x + 0.18x + 0.30x = 280$$
$$0.56x = 280$$
$$x = 500$$
$$2x = 1000$$
$$3x = 1500$$
She has $500 invested at 8%, $1,000 invested at 9%, and $1,500 invested at 10%.

21. Let x represent the measure of the two equal angles, so $x + x = 2x$ represents the measure of the third angle. Since the sum of the three angles is 180°, the equation is:
$$x + x + 2x = 180°$$
$$4x = 180°$$
$$x = 45°$$
$$2x = 90°$$
The measures of the three angles are 45°, 45°, and 90°.

23.   Let x represent the measure of the largest angle. Then $\frac{1}{5}x$ represents the measure of the smallest angle, and

$2\left(\frac{1}{5}x\right) = \frac{2}{5}x$ represents the measure of the other angle. Since the sum of the three angles is 180°, the equation is:

$$x + \frac{1}{5}x + \frac{2}{5}x = 180°$$
$$\frac{5}{5}x + \frac{1}{5}x + \frac{2}{5}x = 180°$$
$$\frac{8}{5}x = 180°$$
$$x = 112.5°$$
$$\frac{1}{5}x = 22.5°$$
$$\frac{2}{5}x = 45°$$

The measures of the three angles are 22.5°, 45°, and 112.5°.

25.   Let x represent the measure of the other acute angle, and 90° is the measure of the right angle. Since the sum of the three angles is 180°, the equation is:

$$x + 37° + 90° = 180°$$
$$x + 127° = 180°$$
$$x = 53°$$

The other two angles are 53° and 90°.

27.   Let x represent the measure of the smallest angle, so $x + 20$ represents the measure of the second angle and $2x$ represents the measure of the third angle. Since the sum of the three angles is 180°, the equation is:

$$x + x + 20 + 2x = 180°$$
$$4x + 20 = 180°$$
$$4x = 160°$$
$$x = 40°$$
$$x + 20 = 60°$$
$$2x = 80°$$

The measures of the three angles are 40°, 60°, and 80°.

29.   Completing the table:

	Adult	Child
Number	x	$x + 6$
Income	$6(x)$	$4(x+6)$

The equation is:

$$6(x) + 4(x+6) = 184$$
$$6x + 4x + 24 = 184$$
$$10x + 24 = 184$$
$$10x = 160$$
$$x = 16$$
$$x + 6 = 22$$

Miguel sold 16 adult and 22 children's tickets.

31.   Let x represent the total minutes for the call. Then $0.41 is charged for the first minute, and $0.32 is charged for the additional $x - 1$ minutes. The equation is:

$$0.41(1) + 0.32(x-1) = 5.21$$
$$0.41 + 0.32x - 0.32 = 5.21$$
$$0.32x + 0.09 = 5.21$$
$$0.32x = 5.12$$
$$x = 16$$

The call was 16 minutes long.

33. Let x represent the hours Jo Ann worked that week. Then \$12/hour is paid for the first 35 hours and \$18/hour is paid for the additional $x-35$ hours. The equation is:

$$12(35)+18(x-35)=492$$
$$420+18x-630=492$$
$$18x-210=492$$
$$18x=702$$
$$x=39$$

Jo Ann worked 39 hours that week.

35. Let x and $x+2$ represent the two office numbers. The equation is:

$$x+x+2=14,660$$
$$2x+2=14,660$$
$$2x=14,658$$
$$x=7329$$
$$x+2=7331$$

They are in offices 7329 and 7331.

37. Let x represent Kendra's age and $x+2$ represent Marissa's age. The equation is:

$$x+2+2x=26$$
$$3x+2=26$$
$$3x=24$$
$$x=8$$
$$x+2=10$$

Kendra is 8 years old and Marissa is 10 years old.

39. For Jeff, the total time traveled is $\dfrac{425 \text{ miles}}{55 \text{ miles/hour}} \approx 7.72 \text{ hours} \approx 463 \text{ minutes}$. Since he left at 11:00 AM, he will arrive at 6:43 PM. For Carla, the total time traveled is $\dfrac{425 \text{ miles}}{65 \text{ miles/hour}} \approx 6.54 \text{ hours} \approx 392 \text{ minutes}$. Since she left at 1:00 PM, she will arrive at 7:32 PM. Thus Jeff will arrive in Lake Tahoe first.

41. Since $\frac{1}{5}$ mile $= 0.2$ mile, the taxi charge is \$1.25 for the first $\frac{1}{5}$ mile and \$0.25 per fifth mile for the remaining 7.3 miles. Since $7.3 \text{ miles} = \dfrac{7.3}{0.2} = 36.5 \text{ fifths}$, the total charge is: $\$1.25+\$0.25(36.5)\approx\$10.38$

43. Let w represent the width and $w+2$ represent the length. The equation is:

$$2w+2(w+2)=44$$
$$2w+2w+4=44$$
$$4w+4=44$$
$$4w=40$$
$$w=10$$
$$w+2=12$$

The length is 12 meters and the width is 10 meters.

45. Let x, $x+1$, and $x+2$ represent the measures of the three angles. Since the sum of the three angles is 180°, the equation is:

$$x+x+1+x+2=180°$$
$$3x+3=180°$$
$$3x=177°$$
$$x=59°$$
$$x+1=60°$$
$$x+2=61°$$

The measures of the three angles are 59°, 60°, and 61°.

47. If all 36 people are Elk's Lodge members (which would be the least amount), the cost of the lessons would be $\$3(36)=\$108$. Since half of the money is paid to Ike and Nancy, the least amount they could make is $\frac{1}{2}(\$108)=\$54$.

49. Yes. The total receipts were \$160, which is possible if there were 10 Elk's members and 26 nonmembers. Computing the total receipts: $10(\$3)+26(\$5)=\$30+\$130=\$160$

51. Substituting: $36x - 12 = 36\left(\frac{1}{4}\right) - 12 = 9 - 12 = -3$ **53.** Substituting: $36x - 12 = 36\left(\frac{1}{9}\right) - 12 = 4 - 12 = -8$

55. Substituting: $36x - 12 = 36\left(\frac{1}{3}\right) - 12 = 12 - 12 = 0$ **57.** Substituting: $36x - 12 = 36\left(\frac{5}{9}\right) - 12 = 20 - 12 = 8$

59. Evaluating when $x = -4$: $3(x - 4) = 3(-4 - 4) = 3(-8) = -24$

61. Evaluating when $x = -4$: $-5x + 8 = -5(-4) + 8 = 20 + 8 = 28$

63. Evaluating when $x = -4$: $\dfrac{x - 14}{36} = \dfrac{-4 - 14}{36} = -\dfrac{18}{36} = -\dfrac{1}{2}$

65. Evaluating when $x = -4$: $\dfrac{16}{x} + 3x = \dfrac{16}{-4} + 3(-4) = -4 - 12 = -16$

67. Evaluating when $x = -4$: $7x - \dfrac{12}{x} = 7(-4) - \dfrac{12}{-4} = -28 + 3 = -25$

69. Evaluating when $x = -4$: $8\left(\dfrac{x}{2} + 5\right) = 8\left(\dfrac{-4}{2} + 5\right) = 8(-2 + 5) = 8(3) = 24$

71. **a.**   Solving the equation: **b.**   Solving the equation:
$$x - 3 = 6$$
$$x = 9$$
$$x + 3 = 6$$
$$x = 3$$

c.   Solving the equation: **d.**   Solving the equation:
$$-x - 3 = 6$$
$$-x = 9$$
$$x = -9$$
$$-x + 3 = 6$$
$$-x = 3$$
$$x = -3$$

73. **a.**   Solving the equation: **b.**   Solving the equation:
$$\frac{x}{4} = -2$$
$$x = -2(4) = -8$$
$$-\frac{x}{4} = -2$$
$$x = -2(-4) = 8$$

c.   Solving the equation: **d.**   Solving the equation:
$$\frac{x}{4} = 2$$
$$x = 2(4) = 8$$
$$-\frac{x}{4} = 2$$
$$x = 2(-4) = -8$$

75.   Solving the equation: **77.**   Solving the equation:
$$2.5x - 3.48 = 4.9x + 2.07$$
$$-2.4x - 3.48 = 2.07$$
$$-2.4x = 5.55$$
$$x = -2.3125$$
$$3(x - 4) = -2$$
$$3x - 12 = -2$$
$$3x = 10$$
$$x = \frac{10}{3}$$

2.8 Linear Inequalities

1.   Solving the inequality: **3.**   Solving the inequality:
$$x - 5 < 7$$
$$x - 5 + 5 < 7 + 5$$
$$x < 12$$
$$a - 4 \le 8$$
$$a - 4 + 4 \le 8 + 4$$
$$a \le 12$$

Graphing the solution set: Graphing the solution set:

5.   Solving the inequality: **7.**   Solving the inequality:
$$x - 4.3 > 8.7$$
$$x - 4.3 + 4.3 > 8.7 + 4.3$$
$$x > 13$$
$$y + 6 \ge 10$$
$$y + 6 + (-6) \ge 10 + (-6)$$
$$y \ge 4$$

Graphing the solution set: Graphing the solution set:

9. Solving the inequality:
$$2 < x - 7$$
$$2 + 7 < x - 7 + 7$$
$$9 < x$$
$$x > 9$$
Graphing the solution set:

11. Solving the inequality:
$$3x < 6$$
$$\frac{1}{3}(3x) < \frac{1}{3}(6)$$
$$x < 2$$
Graphing the solution set:

13. Solving the inequality:
$$5a \le 25$$
$$\frac{1}{5}(5a) \le \frac{1}{5}(25)$$
$$a \le 5$$
Graphing the solution set:

15. Solving the inequality:
$$\frac{x}{3} > 5$$
$$3\left(\frac{x}{3}\right) > 3(5)$$
$$x > 15$$
Graphing the solution set:

17. Solving the inequality:
$$-2x > 6$$
$$-\frac{1}{2}(-2x) < -\frac{1}{2}(6)$$
$$x < -3$$
Graphing the solution set:

19. Solving the inequality:
$$-3x \ge -18$$
$$-\frac{1}{3}(-3x) \le -\frac{1}{3}(-18)$$
$$x \le 6$$
Graphing the solution set:

21. Solving the inequality:
$$-\frac{x}{5} \le 10$$
$$-5\left(-\frac{x}{5}\right) \ge -5(10)$$
$$x \ge -50$$
Graphing the solution set:

23. Solving the inequality:
$$-\frac{2}{3}y > 4$$
$$-\frac{3}{2}\left(-\frac{2}{3}y\right) < -\frac{3}{2}(4)$$
$$y < -6$$
Graphing the solution set:

25. Solving the inequality:
$$2x - 3 < 9$$
$$2x - 3 + 3 < 9 + 3$$
$$2x < 12$$
$$\frac{1}{2}(2x) < \frac{1}{2}(12)$$
$$x < 6$$
Graphing the solution set:

27. Solving the inequality:
$$-\frac{1}{5}y - \frac{1}{3} \le \frac{2}{3}$$
$$-\frac{1}{5}y - \frac{1}{3} + \frac{1}{3} \le \frac{2}{3} + \frac{1}{3}$$
$$-\frac{1}{5}y \le 1$$
$$-5\left(-\frac{1}{5}y\right) \ge -5(1)$$
$$y \ge -5$$
Graphing the solution set:

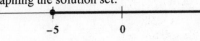

29. Solving the inequality:

$$-7.2x + 1.8 > -19.8$$
$$-7.2x + 1.8 - 1.8 > -19.8 - 1.8$$
$$-7.2x > -21.6$$
$$\frac{-7.2x}{-7.2} < \frac{-21.6}{-7.2}$$
$$x < 3$$

Graphing the solution set:

31. Solving the inequality:

$$\frac{2}{3}x - 5 \le 7$$
$$\frac{2}{3}x - 5 + 5 \le 7 + 5$$
$$\frac{2}{3}x \le 12$$
$$\frac{3}{2}\left(\frac{2}{3}x\right) \le \frac{3}{2}(12)$$
$$x \le 18$$

Graphing the solution set:

33. Solving the inequality:

$$-\frac{2}{5}a - 3 > 5$$
$$-\frac{2}{5}a - 3 + 3 > 5 + 3$$
$$-\frac{2}{5}a > 8$$
$$-\frac{5}{2}\left(-\frac{2}{5}a\right) < -\frac{5}{2}(8)$$
$$a < -20$$

Graphing the solution set:

35. Solving the inequality:

$$5 - \frac{3}{5}y > -10$$
$$-5 + 5 - \frac{3}{5}y > -5 + (-10)$$
$$-\frac{3}{5}y > -15$$
$$-\frac{5}{3}\left(-\frac{3}{5}y\right) < -\frac{5}{3}(-15)$$
$$y < 25$$

Graphing the solution set:

37. Solving the inequality:

$$0.3(a + 1) \le 1.2$$
$$0.3a + 0.3 \le 1.2$$
$$0.3a + 0.3 + (-0.3) \le 1.2 + (-0.3)$$
$$0.3a \le 0.9$$
$$\frac{0.3a}{0.3} \le \frac{0.9}{0.3}$$
$$a \le 3$$

Graphing the solution set:

39. Solving the inequality:

$$2(5 - 2x) \le -20$$
$$10 - 4x \le -20$$
$$-10 + 10 - 4x \le -10 + (-20)$$
$$-4x \le -30$$
$$-\frac{1}{4}(-4x) \ge -\frac{1}{4}(-30)$$
$$x \ge \frac{15}{2}$$

Graphing the solution set:

41. Solving the inequality:

$$3x - 5 > 8x$$
$$-3x + 3x - 5 > -3x + 8x$$
$$-5 > 5x$$
$$\frac{1}{5}(-5) > \frac{1}{5}(5x)$$
$$-1 > x$$
$$x < -1$$

Graphing the solution set:

43. First multiply by 6 to clear the inequality of fractions:

$$\frac{1}{3}y - \frac{1}{2} \le \frac{5}{6}y + \frac{1}{2}$$
$$6\left(\frac{1}{3}y - \frac{1}{2}\right) \le 6\left(\frac{5}{6}y + \frac{1}{2}\right)$$
$$2y - 3 \le 5y + 3$$
$$-5y + 2y - 3 \le -5y + 5y + 3$$
$$-3y - 3 \le 3$$
$$-3y - 3 + 3 \le 3 + 3$$
$$-3y \le 6$$
$$-\frac{1}{3}(-3y) \ge -\frac{1}{3}(6)$$
$$y \ge -2$$

Graphing the solution set:

45. Solving the inequality:

$$-2.8x + 8.4 < -14x - 2.8$$
$$-2.8x + 14x + 8.4 < -14x + 14x - 2.8$$
$$11.2x + 8.4 < -2.8$$
$$11.2x + 8.4 - 8.4 < -2.8 - 8.4$$
$$11.2x < -11.2$$
$$\frac{11.2x}{11.2} < \frac{-11.2}{11.2}$$
$$x < -1$$

Graphing the solution set:

$$-1 \qquad 0$$

47. Solving the inequality:

$$3(m - 2) - 4 \geq 7m + 14$$
$$3m - 6 - 4 \geq 7m + 14$$
$$3m - 10 \geq 7m + 14$$
$$-7m + 3m - 10 \geq -7m + 7m + 14$$
$$-4m - 10 \geq 14$$
$$-4m - 10 + 10 \geq 14 + 10$$
$$-4m \geq 24$$
$$-\tfrac{1}{4}(-4m) \leq -\tfrac{1}{4}(24)$$
$$m \leq -6$$

Graphing the solution set:

$$-6 \qquad 0$$

49. Solving the inequality:

$$3 - 4(x - 2) \leq -5x + 6$$
$$3 - 4x + 8 \leq -5x + 6$$
$$-4x + 11 \leq -5x + 6$$
$$-4x + 5x + 11 \leq -5x + 5x + 6$$
$$x + 11 \leq 6$$
$$x + 11 + (-11) \leq 6 + (-11)$$
$$x \leq -5$$

Graphing the solution set:

$$-5 \qquad 0$$

51. Solving for y:
$$3x + 2y < 6$$
$$2y < -3x + 6$$
$$y < -\tfrac{3}{2}x + 3$$

53. Solving for y:
$$2x - 5y > 10$$
$$-5y > -2x + 10$$
$$y < \tfrac{2}{5}x - 2$$

55. Solving for y:
$$-3x + 7y \leq 21$$
$$7y \leq 3x + 21$$
$$y \leq \tfrac{3}{7}x + 3$$

57. Solving for y:
$$2x - 4y \geq -4$$
$$-4y \geq -2x - 4$$
$$y \leq \tfrac{1}{2}x + 1$$

59. **a.** Evaluating when $x = 0$: $-5x + 3 = -5(0) + 3 = 0 + 3 = 3$

b. Solving the equation:
$$-5x + 3 = -7$$
$$-5x + 3 - 3 = -7 - 3$$
$$-5x = -10$$
$$-\tfrac{1}{5}(-5x) = -\tfrac{1}{5}(-10)$$
$$x = 2$$

c. Substituting $x = 0$:

$$-5(0) + 3 < -7$$
$$3 < -7 \quad \text{(false)}$$

No, $x = 0$ is not a solution to the inequality.

d. Solving the inequality:
$$-5x + 3 < -7$$
$$-5x + 3 - 3 < -7 - 3$$
$$-5x < -10$$
$$-\tfrac{1}{5}(-5x) > -\tfrac{1}{5}(-10)$$
$$x > 2$$

61. The inequality is $x < 3$.

63. The inequality is $x \geq 3$.

65.   Let x and $x + 1$ represent the integers. Solving the inequality:

$$x + x + 1 \geq 583$$
$$2x + 1 \geq 583$$
$$2x \geq 582$$
$$x \geq 291$$

The two numbers are at least 291.

67.   Let x represent the number. Solving the inequality:

$$2x + 6 < 10$$
$$2x < 4$$
$$x < 2$$

69.   Let x represent the number. Solving the inequality:

$$4x > x - 8$$
$$3x > -8$$
$$x > -\frac{8}{3}$$

71.   Let w represent the width, so $3w$ represents the length. Using the formula for perimeter:

$$2(w) + 2(3w) \geq 48$$
$$2w + 6w \geq 48$$
$$8w \geq 48$$
$$w \geq 6$$

The width is at least 6 meters.

73.   Let x, $x + 2$, and $x + 4$ represent the sides of the triangle. The inequality is:

$$x + (x + 2) + (x + 4) > 24$$
$$3x + 6 > 24$$
$$3x > 18$$
$$x > 6$$

The shortest side is an even number greater than 6 inches (greater than or equal to 8 inches).

75.   The inequality is $t \geq 100$.

77.   Let n represent the number of tickets they sell. The inequality is:

$$7.5n < 1500$$
$$n < 200$$

They will lose money if they sell less than 200 tickets. They will make a profit if they sell more than 200 tickets.

79.   Simplifying: $\frac{1}{6}(12x + 6) = \frac{1}{6} \cdot 12x + \frac{1}{6} \cdot 6 = 2x + 1$

81.   Simplifying: $\frac{2}{3}(-3x - 6) = \frac{2}{3} \cdot (-3x) - \frac{2}{3} \cdot 6 = -2x - 4$

83.   Simplifying: $3\left(\frac{5}{6}a + \frac{4}{9}\right) = 3 \cdot \frac{5}{6}a + 3 \cdot \frac{4}{9} = \frac{5}{2}a + \frac{4}{3}$

85.   Simplifying: $-3\left(\frac{2}{3}a + \frac{5}{6}\right) = -3 \cdot \frac{2}{3}a - 3 \cdot \frac{5}{6} = -2a - \frac{5}{2}$

87.   Simplifying: $\frac{1}{2}x + \frac{1}{6}x = \frac{3}{3} \cdot \frac{1}{2}x + \frac{1}{6}x = \frac{3}{6}x + \frac{1}{6}x = \frac{4}{6}x = \frac{2}{3}x$

89.   Simplifying: $\frac{2}{3}x - \frac{5}{6}x = \frac{2}{2} \cdot \frac{2}{3}x - \frac{5}{6}x = \frac{4}{6}x - \frac{5}{6}x = -\frac{1}{6}x$

91.   Simplifying: $\frac{3}{4}x + \frac{1}{6}x = \frac{3}{3} \cdot \frac{3}{4}x + \frac{2}{2} \cdot \frac{1}{6}x = \frac{9}{12}x + \frac{2}{12}x = \frac{11}{12}x$

93.   Simplifying: $\frac{2}{5}x + \frac{5}{8}x = \frac{8}{8} \cdot \frac{2}{5}x + \frac{5}{5} \cdot \frac{5}{8}x = \frac{16}{40}x + \frac{25}{40}x = \frac{41}{40}x$

95.   Solving the inequality:

$$2x - 1 \geq 3$$
$$2x \geq 4$$
$$x \geq 2$$

97.   Solving the inequality:

$$-2x > -8$$
$$x < 4$$

99.   Solving the inequality:

$$-3 \leq 4x + 1$$
$$4x \geq -4$$
$$x \geq -1$$

2.9 Compound Inequalities

1. Graphing the solution set:

-1 5

3. Graphing the solution set:

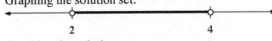

-3 0

5. Graphing the solution set:

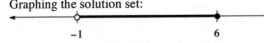

-1 6

7. Graphing the solution set:

2 4

9. Graphing the solution set:

-2 4

11. Graphing the solution set:

-1 5

13. Graphing the solution set:

-1 3

15. Graphing the solution set:

-3 -2

17. Solving the compound inequality:

$$3x - 1 < 5 \qquad \text{or} \qquad 5x - 5 > 10$$
$$3x < 6 \qquad\qquad\qquad 5x > 15$$
$$x < 2 \qquad\qquad\qquad x > 3$$

Graphing the solution set:

2 3

19. Solving the compound inequality:

$$x - 2 > -5 \qquad \text{and} \qquad x + 7 < 13$$
$$x > -3 \qquad\qquad\qquad x < 6$$

Graphing the solution set:

-3 6

21. Solving the compound inequality:

$$11x < 22 \qquad \text{or} \qquad 12x > 36$$
$$x < 2 \qquad\qquad\qquad x > 3$$

Graphing the solution set:

2 3

23. Solving the compound inequality:

$$3x - 5 < 10 \qquad \text{and} \qquad 2x + 1 > -5$$
$$3x < 15 \qquad\qquad\qquad 2x > -6$$
$$x < 5 \qquad\qquad\qquad x > -3$$

Graphing the solution set:

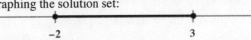

-3 5

25. Solving the compound inequality:

$$2x - 3 < 8 \qquad \text{and} \qquad 3x + 1 > -10$$
$$2x < 11 \qquad\qquad\qquad 3x > -11$$
$$x < \frac{11}{2} \qquad\qquad\qquad x > -\frac{11}{3}$$

Graphing the solution set:

$-11/3$ $11/2$

27. Solving the compound inequality:

$$2x - 1 < 3 \qquad \text{and} \qquad 3x - 2 > 1$$
$$2x < 4 \qquad\qquad\qquad 3x > 3$$
$$x < 2 \qquad\qquad\qquad x > 1$$

Graphing the solution set:

1 2

29. Solving the compound inequality:

$$-1 \le x - 5 \le 2$$
$$4 \le x \le 7$$

Graphing the solution set:

4 7

31. Solving the compound inequality:

$$-4 \le 2x \le 6$$
$$-2 \le x \le 3$$

Graphing the solution set:

-2 3

33. Solving the compound inequality:

$$-3 < 2x + 1 < 5$$
$$-4 < 2x < 4$$
$$-2 < x < 2$$

Graphing the solution set:

-2 2

35. Solving the compound inequality:

$$0 \le 3x + 2 \le 7$$
$$-2 \le 3x \le 5$$
$$-\frac{2}{3} \le x \le \frac{5}{3}$$

Graphing the solution set:

$-2/3$ $5/3$

37.   Solving the compound inequality:
$$-7 < 2x + 3 < 11$$
$$-10 < 2x < 8$$
$$-5 < x < 4$$

Graphing the solution set:

$$-5 \qquad\qquad 4$$

39.   Solving the compound inequality:
$$-1 \le 4x + 5 \le 9$$
$$-6 \le 4x \le 4$$
$$-\frac{3}{2} \le x \le 1$$

Graphing the solution set:

$$-3/2 \qquad\qquad 1$$

41.   The inequality is $-2 < x < 3$.

43.   The inequality is $-2 \le x \le 3$.

45.   **a.**   The three inequalities are $2x + x > 10$, $2x + 10 > x$, and $x + 10 > 2x$.

b.   For $x + 10 > 2x$ we have $x < 10$. For $3x > 10$ we have $x > \frac{10}{3}$. The compound inequality is $\frac{10}{3} < x < 10$.

47.   Graphing the inequality:

$$50 \qquad\qquad\qquad 266$$

49.   Let x represent the number. Solving the inequality:
$$5 < 2x - 3 < 7$$
$$8 < 2x < 10$$
$$4 < x < 5$$
The number is between 4 and 5.

51.   Let w represent the width and $w + 4$ represent the length. Using the perimeter formula:
$$20 < 2w + 2(w + 4) < 30$$
$$20 < 2w + 2w + 8 < 30$$
$$20 < 4w + 8 < 30$$
$$12 < 4w < 22$$
$$3 < w < \frac{11}{2}$$

The width is between 3 inches and $\frac{11}{2} = 5\frac{1}{2}$ inches.

53.   Finding the number: $0.25(32) = 8$

55.   Finding the number: $0.20(120) = 24$

57.   Finding the percent:
$$p \bullet 36 = 9$$
$$p = \frac{9}{36} = 0.25 = 25\%$$

59.   Finding the percent:
$$p \bullet 50 = 5$$
$$p = \frac{5}{50} = 0.10 = 10\%$$

61.   Finding the number:
$$0.20n = 16$$
$$n = \frac{16}{0.20} = 80$$

63.   Finding the number:
$$0.02n = 8$$
$$n = \frac{8}{0.02} = 400$$

65.   Simplifying the expression: $-\left|-5\right| = -(5) = -5$

67.   Simplifying the expression: $-3 - 4(-2) = -3 + 8 = 5$

69.   Simplifying the expression: $5\left|3 - 8\right| - 6\left|2 - 5\right| = 5\left|-5\right| - 6\left|-3\right| = 5(5) - 6(3) = 25 - 18 = 7$

71.   Simplifying the expression: $5 - 2\left[-3(5 - 7) - 8\right] = 5 - 2\left[-3(-2) - 8\right] = 5 - 2(6 - 8) = 5 - 2(-2) = 5 + 4 = 9$

73.   The expression is: $-3 - (-9) = -3 + 9 = 6$

75.   Applying the distributive property: $\frac{1}{2}(4x - 6) = \frac{1}{2} \bullet 4x - \frac{1}{2} \bullet 6 = 2x - 3$

77.   The integers are: $-3, 0, 2$

Chapter 2 Review/Test

1. Simplifying the expression: $5x - 8x = (5 - 8)x = -3x$
2. Simplifying the expression: $6x - 3 - 8x = 6x - 8x - 3 = -2x - 3$
3. Simplifying the expression: $-a + 2 + 5a - 9 = -a + 5a + 2 - 9 = 4a - 7$
4. Simplifying the expression: $5(2a - 1) - 4(3a - 2) = 10a - 5 - 12a + 8 = 10a - 12a - 5 + 8 = -2a + 3$
5. Simplifying the expression: $6 - 2(3y + 1) - 4 = 6 - 6y - 2 - 4 = -6y + 6 - 2 - 4 = -6y$
6. Simplifying the expression: $4 - 2(3x - 1) - 5 = 4 - 6x + 2 - 5 = -6x + 4 + 2 - 5 = -6x + 1$
7. Evaluating when $x = 3$: $7x - 2 = 7(3) - 2 = 21 - 2 = 19$
8. Evaluating when $x = 3$: $-4x - 5 + 2x = -2x - 5 = -2(3) - 5 = -6 - 5 = -11$
9. Evaluating when $x = 3$: $-x - 2x - 3x = -6x = -6(3) = -18$
10. Evaluating when $x = -2$: $5x - 3 = 5(-2) - 3 = -10 - 3 = -13$
11. Evaluating when $x = -2$: $-3x + 2 = -3(-2) + 2 = 6 + 2 = 8$
12. Evaluating when $x = -2$: $7 - x - 3 = 4 - x = 4 - (-2) = 4 + 2 = 6$

13. Solving the equation:

$$x + 2 = -6$$
$$x + 2 + (-2) = -6 + (-2)$$
$$x = -8$$

14. Solving the equation:

$$x - \tfrac{1}{2} = \tfrac{4}{7}$$
$$x - \tfrac{1}{2} + \tfrac{1}{2} = \tfrac{4}{7} + \tfrac{1}{2}$$
$$x = \tfrac{8}{14} + \tfrac{7}{14}$$
$$x = \tfrac{15}{14}$$

15. Solving the equation:

$$10 - 3y + 4y = 12$$
$$10 + y = 12$$
$$-10 + 10 + y = -10 + 12$$
$$y = 2$$

16. Solving the equation:

$$-3 - 4 = -y - 2 + 2y$$
$$-7 = y - 2$$
$$-7 + 2 = y - 2 + 2$$
$$y = -5$$

17. Solving the equation:

$$2x = -10$$
$$\tfrac{1}{2}(2x) = \tfrac{1}{2}(-10)$$
$$x = -5$$

18. Solving the equation:

$$3x = 0$$
$$\tfrac{1}{3}(3x) = \tfrac{1}{3}(0)$$
$$x = 0$$

19. Solving the equation:

$$\frac{x}{3} = 4$$
$$3\left(\frac{x}{3}\right) = 3(4)$$
$$x = 12$$

20. Solving the equation:

$$-\frac{x}{4} = 2$$
$$-4\left(-\frac{x}{4}\right) = -4(2)$$
$$x = -8$$

21. Solving the equation:

$$3a - 2 = 5a$$
$$-3a + 3a - 2 = -3a + 5a$$
$$-2 = 2a$$
$$\tfrac{1}{2}(-2) = \tfrac{1}{2}(2a)$$
$$a = -1$$

22. First multiply by 10 to clear the equation of fractions:

$$\tfrac{7}{10}a = \tfrac{1}{5}a + \tfrac{1}{2}$$
$$10\left(\tfrac{7}{10}a\right) = 10\left(\tfrac{1}{5}a + \tfrac{1}{2}\right)$$
$$7a = 2a + 5$$
$$-2a + 7a = -2a + 2a + 5$$
$$5a = 5$$
$$\tfrac{1}{5}(5a) = \tfrac{1}{5}(5)$$
$$a = 1$$

23. Solving the equation:
$$3x + 2 = 5x - 8$$
$$3x + (-5x) + 2 = 5x + (-5x) - 8$$
$$-2x + 2 = -8$$
$$-2x + 2 + (-2) = -8 + (-2)$$
$$-2x = -10$$
$$-\tfrac{1}{2}(-2x) = -\tfrac{1}{2}(-10)$$
$$x = 5$$

24. Solving the equation:
$$6x - 3 = x + 7$$
$$6x + (-x) - 3 = x + (-x) + 7$$
$$5x - 3 = 7$$
$$5x - 3 + 3 = 7 + 3$$
$$5x = 10$$
$$\tfrac{1}{5}(5x) = \tfrac{1}{5}(10)$$
$$x = 2$$

25. Solving the equation:
$$0.7x - 0.1 = 0.5x - 0.1$$
$$0.7x + (-0.5x) - 0.1 = 0.5x + (-0.5x) - 0.1$$
$$0.2x - 0.1 = -0.1$$
$$0.2x - 0.1 + 0.1 = -0.1 + 0.1$$
$$0.2x = 0$$
$$\frac{0.2x}{0.2} = \frac{0}{0.2}$$
$$x = 0$$

26. Solving the equation:
$$0.2x - 0.3 = 0.8x - 0.3$$
$$0.2x + (-0.8x) - 0.3 = 0.8x + (-0.8x) - 0.3$$
$$-0.6x - 0.3 = -0.3$$
$$-0.6x - 0.3 + 0.3 = -0.3 + 0.3$$
$$-0.6x = 0$$
$$\frac{-0.6x}{-0.6} = \frac{0}{-0.6}$$
$$x = 0$$

27. Simplifying and then solving the equation:
$$2(x - 5) = 10$$
$$2x - 10 = 10$$
$$2x - 10 + 10 = 10 + 10$$
$$2x = 20$$
$$\tfrac{1}{2}(2x) = \tfrac{1}{2}(20)$$
$$x = 10$$

28. Simplifying and then solving the equation:
$$12 = 2(5x - 4)$$
$$12 = 10x - 8$$
$$12 + 8 = 10x - 8 + 8$$
$$20 = 10x$$
$$\tfrac{1}{10}(20) = \tfrac{1}{10}(10x)$$
$$x = 2$$

29. Simplifying and then solving the equation:
$$\tfrac{1}{2}(3t - 2) + \tfrac{1}{2} = \tfrac{5}{2}$$
$$\tfrac{3}{2}t - 1 + \tfrac{1}{2} = \tfrac{5}{2}$$
$$\tfrac{3}{2}t - \tfrac{1}{2} = \tfrac{5}{2}$$
$$\tfrac{3}{2}t - \tfrac{1}{2} + \tfrac{1}{2} = \tfrac{5}{2} + \tfrac{1}{2}$$
$$\tfrac{3}{2}t = 3$$
$$\tfrac{2}{3}\left(\tfrac{3}{2}t\right) = \tfrac{2}{3}(3)$$
$$t = 2$$

30. Simplifying and then solving the equation:
$$\tfrac{3}{5}(5x - 10) = \tfrac{2}{3}(9x + 3)$$
$$3x - 6 = 6x + 2$$
$$3x + (-6x) - 6 = 6x + (-6x) + 2$$
$$-3x - 6 = 2$$
$$-3x - 6 + 6 = 2 + 6$$
$$-3x = 8$$
$$-\tfrac{1}{3}(-3x) = -\tfrac{1}{3}(8)$$
$$x = -\tfrac{8}{3}$$

31. Simplifying and then solving the equation:
$$2(3x + 7) = 4(5x - 1) + 18$$
$$6x + 14 = 20x - 4 + 18$$
$$6x + 14 = 20x + 14$$
$$6x + (-20x) + 14 = 20x + (-20x) + 14$$
$$-14x + 14 = 14$$
$$-14x + 14 + (-14) = 14 + (-14)$$
$$-14x = 0$$
$$-\tfrac{1}{14}(-14x) = -\tfrac{1}{14}(0)$$
$$x = 0$$

32. Simplifying and then solving the equation:
$$7 - 3(y + 4) = 10$$
$$7 - 3y - 12 = 10$$
$$-3y - 5 = 10$$
$$-3y - 5 + 5 = 10 + 5$$
$$-3y = 15$$
$$-\tfrac{1}{3}(-3y) = -\tfrac{1}{3}(15)$$
$$y = -5$$

33. Substituting $x = 5$:
$$4(5) - 5y = 20$$
$$20 - 5y = 20$$
$$-5y = 0$$
$$y = 0$$

34. Substituting $x = 0$:
$$4(0) - 5y = 20$$
$$-5y = 20$$
$$y = -4$$

35. Substituting $x = -5$:
$$4(-5) - 5y = 20$$
$$-20 - 5y = 20$$
$$-5y = 40$$
$$y = -8$$

36. Substituting $x = 10$:
$$4(10) - 5y = 20$$
$$40 - 5y = 20$$
$$-5y = -20$$
$$y = 4$$

37. Solving for y:
$$2x - 5y = 10$$
$$-5y = -2x + 10$$
$$y = \frac{2}{5}x - 2$$

38. Solving for y:
$$5x - 2y = 10$$
$$-2y = -5x + 10$$
$$y = \frac{5}{2}x - 5$$

39. Solving for h:
$$\pi r^2 h = V$$
$$\frac{\pi r^2 h}{\pi r^2} = \frac{V}{\pi r^2}$$
$$h = \frac{V}{\pi r^2}$$

40. Solving for w:
$$2l + 2w = P$$
$$2w = P - 2l$$
$$w = \frac{P - 2l}{2}$$

41. Computing the amount:
$$0.86(240) = x$$
$$x = 206.4$$

So 86% of 240 is 206.4

42. Finding the percent:
$$p \cdot 2000 = 180$$
$$p = \frac{180}{2000}$$
$$p = 0.09 = 9\%$$

So 9% of 2000 is 180.

43. Let x represent the number. The equation is:
$$2x + 6 = 28$$
$$2x = 22$$
$$x = 11$$
The number is 11.

44. Let w represent the width and $5w$ represent the length. Using the perimeter formula:
$$2w + 2(5w) = 60$$
$$2w + 10w = 60$$
$$12w = 60$$
$$w = 5$$
$$5w = 25$$
The width is 5 meters and the length is 25 meters.

45. Completing the table:

	Dollars Invested at 9%	Dollars Invested at 10%
Number of	x	$x + 300$
Interest on	$0.09(x)$	$0.10(x + 300)$

The equation is:
$$0.09(x) + 0.10(x + 300) = 125$$
$$0.09x + 0.10x + 30 = 125$$
$$0.19x + 30 = 125$$
$$0.19x = 95$$
$$x = 500$$
$$x + 300 = 800$$
The man invested $500 at 9% and $800 at 10%.

46. Completing the table:

	Nickels	Dimes
Number	x	$15 - x$
Value (cents)	$5(x)$	$10(15 - x)$

The equation is:
$$5(x) + 10(15 - x) = 100$$
$$5x + 150 - 10x = 100$$
$$-5x + 150 = 100$$
$$-5x = -150$$
$$x = 10$$
$$15 - x = 5$$
The collection has 10 nickels and 5 dimes.

47. Solving the inequality:
$$-2x < 4$$
$$-\frac{1}{2}(-2x) > -\frac{1}{2}(4)$$
$$x > -2$$

48. Solving the inequality:
$$-5x > -10$$
$$-\frac{1}{5}(-5x) < -\frac{1}{5}(-10)$$
$$x < 2$$

49. Solving the inequality:

$$-\frac{a}{2} \le -3$$

$$-2\left(-\frac{a}{2}\right) \ge -2(-3)$$

$$a \ge 6$$

50. Solving the inequality:

$$-\frac{a}{3} > 5$$

$$-3\left(-\frac{a}{3}\right) < -3(5)$$

$$a < -15$$

51. Solving the inequality:

$$-4x + 5 > 37$$

$$-4x > 32$$

$$x < -8$$

Graphing the solution set:

$$-8 \qquad 0$$

52. Solving the inequality:

$$2x + 10 < 5x - 11$$

$$-3x + 10 < -11$$

$$-3x < -21$$

$$x > 7$$

Graphing the solution set:

$$0 \qquad 7$$

53. Solving the inequality:

$$2(3t + 1) + 6 \ge 5(2t + 4)$$

$$6t + 2 + 6 \ge 10t + 20$$

$$6t + 8 \ge 10t + 20$$

$$-4t + 8 \ge 20$$

$$-4t \ge 12$$

$$t \le -3$$

Graphing the solution set:

$$-3 \qquad 0$$

54. Graphing the solution set:

$$-2 \qquad 5$$

55. Solving the compound inequality:

$$-5x \ge 25 \qquad \text{or} \qquad 2x - 3 \ge 9$$
$$x \le -5 \qquad\qquad\qquad 2x \ge 12$$
$$x \le -5 \qquad\qquad\qquad x \ge 6$$

Graphing the solution set:

$$-5 \qquad\qquad 6$$

56. Solving the compound inequality:

$$-1 < 2x + 1 < 9$$

$$-2 < 2x < 8$$

$$-1 < x < 4$$

Graphing the solution set:

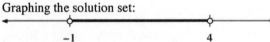

$$-1 \qquad\qquad 4$$

Chapter 3
Linear Equations and Inequalities in Two Variables

3.1 Paired Data and Graphing Ordered Pairs

1. Graphing the ordered pair:

3. Graphing the ordered pair:

5. Graphing the ordered pair:

7. Graphing the ordered pair:

9.   Graphing the ordered pair:

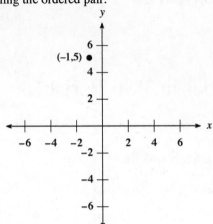

11.   Graphing the ordered pair:

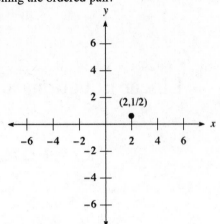

13.   Graphing the ordered pair:

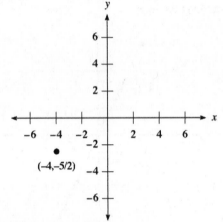

15.   Graphing the ordered pair:

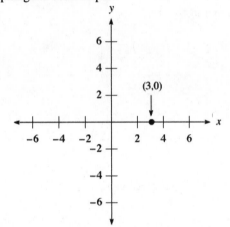

17.   Graphing the ordered pair:

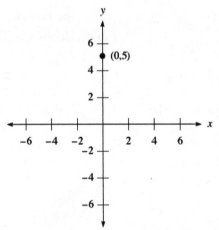

19.   The coordinates are $(-4,4)$.

21.   The coordinates are $(-4,2)$.

23.   The coordinates are $(-3,0)$.

25.   The coordinates are $(2,-2)$.

27.   The coordinates are $(-5,-5)$.

29. Graphing the line:

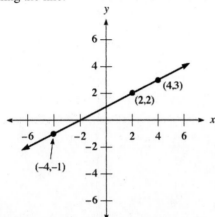

Yes, the point $(2,2)$ lies on the line.

31. Graphing the line:

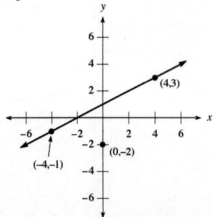

No, the point $(0,-2)$ does not lie on the line.

33. Graphing the line:

Yes, the point $(0,0)$ lies on the line.

35. Graphing the line:

No, the point $(2,-1)$ does not lie on the line.

37. Graphing the line:

Yes, the point $(3,0)$ lies on the line.

39. No, the x-coordinate of every point on this line is 3.

41.   Graphing the line:

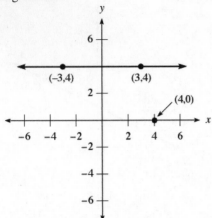

No, the point $(4,0)$ does not lie on the line.

43.   No, the *y*-coordinate of every point on this line is 4.

45.   **a.**   Three ordered pairs on the graph are (5,40), (10,80), and (20,160).
   b.   She will earn $320 for working 40 hours.
   c.   If her check is $240, she worked 30 hours that week.
   d.   No. She should be paid $280 for working 35 hours, not $260.

47.   Constructing a line graph:

49.   Constructing a line graph:

51.   Six ordered pairs are (1981,5), (1985,20.2), (1990,34.4), (1995,44.8), (2000,65.4), and (2004,99.4).

53.   Point A is $(6-5,2) = (1,2)$, and point B is $(6,2+5) = (6,7)$.

55.   Point A is $(7-5,2) = (2,2)$, point B is $(2,2+3) = (2,5)$, and point C is $(2+5,5) = (7,5)$.

57.   Simplifying: $\dfrac{x}{5} + \dfrac{3}{5} = \dfrac{x+3}{5}$

59.   Simplifying: $\dfrac{2}{7} - \dfrac{a}{7} = \dfrac{2-a}{7}$

61.   Simplifying: $\dfrac{1}{14} - \dfrac{y}{7} = \dfrac{1}{14} - \dfrac{y}{7} \bullet \dfrac{2}{2} = \dfrac{1-2y}{14}$

63.   Simplifying: $\dfrac{1}{2} + \dfrac{3}{x} = \dfrac{1}{2} \bullet \dfrac{x}{x} + \dfrac{3}{x} \bullet \dfrac{2}{2} = \dfrac{x+6}{2x}$

65.   Simplifying: $\dfrac{5+x}{6} - \dfrac{5}{6} = \dfrac{5+x-5}{6} = \dfrac{x}{6}$

67.   Simplifying: $\dfrac{4}{x} + \dfrac{1}{2} = \dfrac{4}{x} \bullet \dfrac{2}{2} + \dfrac{1}{2} \bullet \dfrac{x}{x} = \dfrac{8+x}{2x}$

69. **a.** Substituting $y = 4$:

$$2x + 3(4) = 6$$
$$2x + 12 = 6$$
$$2x = -6$$
$$x = -3$$

b. Substituting $y = -2$:

$$2x + 3(-2) = 6$$
$$2x - 6 = 6$$
$$2x = 12$$
$$x = 6$$

c. Substituting $x = 3$:

$$2(3) + 3y = 6$$
$$6 + 3y = 6$$
$$3y = 0$$
$$y = 0$$

d. Substituting $x = 9$:

$$2(9) + 3y = 6$$
$$18 + 3y = 6$$
$$3y = -12$$
$$y = -4$$

71. **a.** Substituting $y = 7$:

$$2x - 1 = 7$$
$$2x = 8$$
$$x = 4$$

b. Substituting $y = 3$:

$$2x - 1 = 3$$
$$2x = 4$$
$$x = 2$$

c. Substituting $x = 0$:

$$y = 2(0) - 1 = 0 - 1 = -1$$

d. Substituting $x = 5$:

$$y = 2(5) - 1 = 10 - 1 = 9$$

3.2 Solutions to Linear Equations in Two Variables

1. Substituting $x = 0$, $y = 0$, and $y = -6$:

$$2(0) + y = 6 \qquad 2x + 0 = 6 \qquad 2x + (-6) = 6$$
$$0 + y = 6 \qquad\quad 2x = 6 \qquad\qquad 2x = 12$$
$$y = 6 \qquad\qquad x = 3 \qquad\qquad x = 6$$

The ordered pairs are $(0, 6)$, $(3, 0)$, and $(6, -6)$.

3. Substituting $x = 0$, $y = 0$, and $x = -4$:

$$3(0) + 4y = 12 \qquad 3x + 4(0) = 12 \qquad 3(-4) + 4y = 12$$
$$0 + 4y = 12 \qquad\quad 3x + 0 = 12 \qquad\quad -12 + 4y = 12$$
$$4y = 12 \qquad\qquad 3x = 12 \qquad\qquad 4y = 24$$
$$y = 3 \qquad\qquad x = 4 \qquad\qquad y = 6$$

The ordered pairs are $(0, 3)$, $(4, 0)$, and $(-4, 6)$.

5. Substituting $x = 1$, $y = 0$, and $x = 5$:

$$y = 4(1) - 3 \qquad 0 = 4x - 3 \qquad y = 4(5) - 3$$
$$y = 4 - 3 \qquad\quad 3 = 4x \qquad\qquad y = 20 - 3$$
$$y = 1 \qquad\qquad x = \tfrac{3}{4} \qquad\qquad y = 17$$

The ordered pairs are $(1, 1)$, $\left(\tfrac{3}{4}, 0\right)$, and $(5, 17)$.

7. Substituting $x = 2$, $y = 6$, and $x = 0$:

$$y = 7(2) - 1 \qquad 6 = 7x - 1 \qquad y = 7(0) - 1$$
$$y = 14 - 1 \qquad\quad 7 = 7x \qquad\qquad y = 0 - 1$$
$$y = 13 \qquad\qquad x = 1 \qquad\qquad y = -1$$

The ordered pairs are $(2, 13)$, $(1, 6)$, and $(0, -1)$.

9. Substituting $y = 4$, $y = -3$, and $y = 0$ results (in each case) in $x = -5$. The ordered pairs are $(-5, 4)$, $(-5, -3)$, and $(-5, 0)$.

11. Completing the table:

x	y
1	3
−3	−9
4	12
6	18

13. Completing the table:

x	y
0	0
−1/2	−2
−3	−12
3	12

15.   Completing the table:

x	y
2	3
3	2
5	0
9	−4

17.   Completing the table:

x	y
2	0
3	2
1	−2
−3	−10

19.   Completing the table:

x	y
0	−1
−1	−7
−3	−19
3/2	8

21.   Substituting each ordered pair into the equation:

$$(2,3): \ 2(2)-5(3)=4-15=-11 \neq 10$$

$$(0,-2): \ 2(0)-5(-2)=0+10=10$$

$$\left(\tfrac{5}{2},1\right): \ 2\left(\tfrac{5}{2}\right)-5(1)=5-5=0 \neq 10$$

Only the ordered pair $(0,-2)$ is a solution.

23.   Substituting each ordered pair into the equation:

$$(1,5): \ 7(1)-2=7-2=5$$

$$(0,-2): \ 7(0)-2=0-2=-2$$

$$(-2,-16): \ 7(-2)-2=-14-2=-16$$

All the ordered pairs $(1,5)$, $(0,-2)$ and $(-2,-16)$ are solutions.

25.   Substituting each ordered pair into the equation:

$$(1,6): \ 6(1)=6$$

$$(-2,-12): \ 6(-2)=-12$$

$$(0,0): \ 6(0)=0$$

All the ordered pairs $(1,6)$, $(-2,-12)$ and $(0,0)$ are solutions.

27.   Substituting each ordered pair into the equation:

$$(1,1): \ 1+1=2 \neq 0$$

$$(2,-2): \ 2+(-2)=0$$

$$(3,3): \ 3+3=6 \neq 0$$

Only the ordered pair $(2,-2)$ is a solution.

29.   Since $x=3$, the ordered pair $(5,3)$ cannot be a solution. The ordered pairs $(3,0)$ and $(3,-3)$ are solutions.

31.   Substituting $w=3$:

$$2l+2(3)=30$$
$$2l+6=30$$
$$2l=24$$
$$l=12$$

The length is 12 inches.

33.   **a.**   This is correct, since; $y = 12(5) = \$60$

   b.   This is not correct, since: $y = 12(9) = \$108$. Her check should be for \$108.

   c.   This is not correct, since: $y = 12(7) = \$84$. Her check should be for \$84.

   d.   This is correct, since: $y = 12(14) = \$168$

35.   **a.**   Substituting $t = 5$: $V = -45,000(5) + 600,000 = -225,000 + 600,000 = \$375,000$

   b.   Solving when $V = 330,000$:
$$-45,000t + 600,000 = 330,000$$
$$-45,000t = -270,000$$
$$t = 6$$
   The crane will be worth \$330,000 at the end of 6 years.

   c.   Substituting $t = 9$: $V = -45,000(9) + 600,000 = -405,000 + 600,000 = \$195,000$

   No, the crane will be worth \$195,000 after 9 years.

   d.   The crane cost \$600,000 (the value when $t = 0$).

37.   Simplifying: $\dfrac{11(-5) - 17}{2(-6)} = \dfrac{-55 - 17}{-12} = \dfrac{-72}{-12} = 6$ **39.**   Simplifying: $\dfrac{13(-6) + 18}{4(-5)} = \dfrac{-78 + 18}{-20} = \dfrac{-60}{-20} = 3$

41.   Simplifying: $\dfrac{7^2 - 5^2}{(7-5)^2} = \dfrac{49 - 25}{2^2} = \dfrac{24}{4} = 6$

43.   Simplifying: $\dfrac{-3 \bullet 4^2 - 3 \bullet 2^4}{-3(8)} = \dfrac{-3 \bullet 16 - 3 \bullet 16}{-3(8)} = \dfrac{-48 - 48}{-24} = \dfrac{-96}{-24} = 4$

45.   Substituting $x = 4$:
$$3(4) + 2y = 6$$
$$12 + 2y = 6$$
$$2y = -6$$
$$y = -3$$

47.   Substituting $x = 0$: $y = -\frac{1}{3}(0) + 2 = 0 + 2 = 2$ **49.**   Substituting $x = 2$: $y = \frac{3}{2}(2) - 3 = 3 - 3 = 0$

51.   Solving for y: **53.**   Solving for y:
$$3x - 2y = 6$$
$$5x + y = 4 \qquad\qquad -2y = -3x + 6$$
$$y = -5x + 4 \qquad\qquad y = \tfrac{3}{2}x - 3$$

3.3 Graphing Linear Equations in Two Variables

1.   The ordered pairs are $(0,4)$, $(2,2)$, and $(4,0)$:   **3.**   The ordered pairs are $(0,3)$, $(2,1)$, and $(4,-1)$:

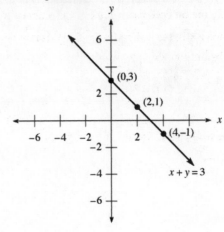

5.   The ordered pairs are $(0,0)$, $(-2,-4)$, and $(2,4)$:

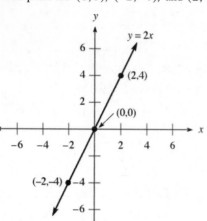

7.   The ordered pairs are $(-3,-1)$, $(0,0)$, and $(3,1)$:

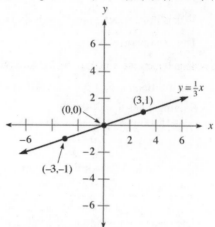

9.   The ordered pairs are $(0,1)$, $(-1,-1)$, and $(1,3)$:

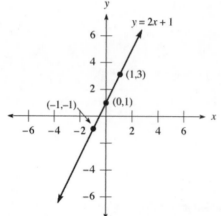

11.   The ordered pairs are $(0,4)$, $(-1,4)$, and $(2,4)$:

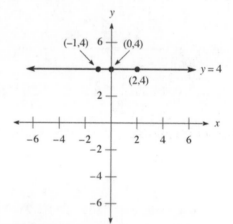

13.   The ordered pairs are $(-2,2)$, $(0,3)$, and $(2,4)$:

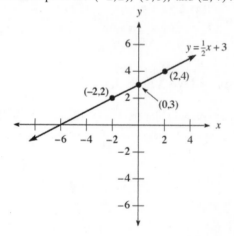

15.   The ordered pairs are $(-3,3)$, $(0,1)$, and $(3,-1)$:

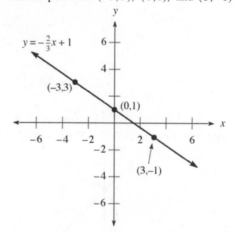

17. Solving for y:

$$2x + y = 3$$
$$y = -2x + 3$$

The ordered pairs are $(-1, 5)$, $(0, 3)$, and $(1, 1)$:

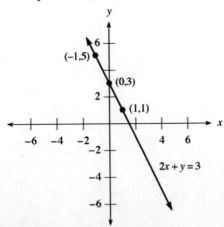

19. Solving for y:

$$3x + 2y = 6$$
$$2y = -3x + 6$$
$$y = -\frac{3}{2}x + 3$$

The ordered pairs are $(0, 3)$, $(2, 0)$, and $(4, -3)$:

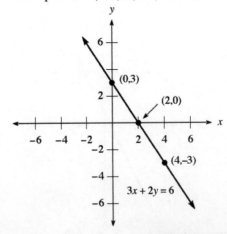

21. Solving for y:

$$-x + 2y = 6$$
$$2y = x + 6$$
$$y = \frac{1}{2}x + 3$$

The ordered pairs are $(-2, 2)$, $(0, 3)$, and $(2, 4)$:

23. Three solutions are $(-4, 2)$, $(0, 0)$, and $(4, -2)$:

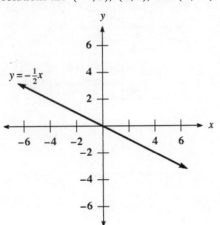

25. Three solutions are $(-1, -4)$, $(0, -1)$, and $(1, 2)$:

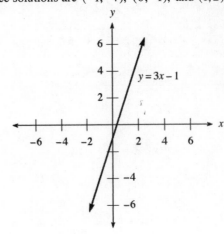

27. Solving for y:

$$-2x + y = 1$$
$$y = 2x + 1$$

Three solutions are $(-2,-3)$, $(0,1)$, and $(2,5)$:

29. Solving for y:

$$3x + 4y = 8$$
$$4y = -3x + 8$$
$$y = -\frac{3}{4}x + 2$$

Three solutions are $(-4,5)$, $(0,2)$, and $(4,-1)$:

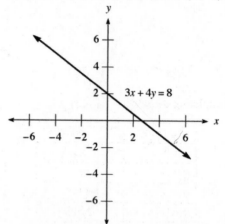

31. Three solutions are $(-2,-4)$, $(-2,0)$, and $(-2,4)$:

33. Three solutions are $(-4,2)$, $(0,2)$, and $(4,2)$:

35. Graphing the equation:

37. Graphing the equation:

39.   Graphing the equation:

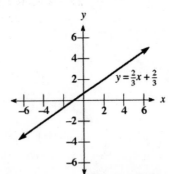

41.   Completing the table:

Equation	H, V, and/or O
$x = 3$	V
$y = 3$	H
$y = 3x$	O
$y = 0$	O,H

43.   Completing the table:

Equation	H, V, and/or O
$x = -\frac{3}{5}$	V
$y = -\frac{3}{5}$	H
$y = -\frac{3}{5}x$	O
$x = 0$	O,V

45.   **a.**   Solving the equation:
$$2x + 5 = 10$$
$$2x = 5$$
$$x = \frac{5}{2}$$

   b.   Substituting $y = 0$:
$$2x + 5(0) = 10$$
$$2x = 10$$
$$x = 5$$

   c.   Substituting $x = 0$:
$$2(0) + 5y = 10$$
$$5y = 10$$
$$y = 2$$

   d.   Graphing the line:

   e.   Solving for y:
$$2x + 5y = 10$$
$$5y = -2x + 10$$
$$y = -\frac{2}{5}x + 2$$

47.   Simplifying: $\frac{1}{2}(4x + 10) = \frac{1}{2} \cdot 4x + \frac{1}{2} \cdot 10 = 2x + 5$

49.   Simplifying: $\frac{2}{3}(3x - 9) = \frac{2}{3} \cdot 3x - \frac{2}{3} \cdot 9 = 2x - 6$

51.   Simplifying: $\frac{3}{4}(4x + 10) = \frac{3}{4} \cdot 4x + \frac{3}{4} \cdot 10 = 3x + \frac{15}{2}$

53.   Simplifying: $\frac{3}{5}(10x + 15) = \frac{3}{5} \cdot 10x + \frac{3}{5} \cdot 15 = 6x + 9$

55.   Simplifying: $5\left(\frac{2}{5}x + 10\right) = 5 \cdot \frac{2}{5}x + 5 \cdot 10 = 2x + 50$

57.   Simplifying: $4\left(\frac{3}{2}x - 7\right) = 4 \cdot \frac{3}{2}x - 4 \cdot 7 = 6x - 28$

59. Simplifying: $\frac{3}{4}(2x+12y) = \frac{3}{4} \bullet 2x + \frac{3}{4} \bullet 12y = \frac{3}{2}x + 9y$

61. Simplifying: $\frac{1}{2}(5x-10y)+6 = \frac{1}{2} \bullet 5x - \frac{1}{2} \bullet 10y + 6 = \frac{5}{2}x - 5y + 6$

63.   **a.**   Substituting $y = 0$:   **b.**   Substituting $x = 0$:

$$3x - 2(0) = 6 \qquad\qquad\qquad 3(0) - 2y = 6$$
$$3x - 0 = 6 \qquad\qquad\qquad\qquad 0 - 2y = 6$$
$$3x = 6 \qquad\qquad\qquad\qquad\qquad -2y = 6$$
$$x = 2 \qquad\qquad\qquad\qquad\qquad y = -3$$

65.   **a.**   Substituting $y = 0$:   **b.**   Substituting $x = 0$:

$$-x + 2(0) = 4 \qquad\qquad\qquad -(0) + 2y = 4$$
$$-x + 0 = 4 \qquad\qquad\qquad\qquad 0 + 2y = 4$$
$$-x = 4 \qquad\qquad\qquad\qquad\qquad 2y = 4$$
$$x = -4 \qquad\qquad\qquad\qquad\qquad y = 2$$

67.   **a.**   Substituting $y = 0$:   **b.**   Substituting $x = 0$:

$$0 = -\frac{1}{3}x + 2$$
$$-\frac{1}{3}x = -2 \qquad\qquad\qquad\qquad y = -\frac{1}{3}(0) + 2 = 2$$
$$x = 6$$

3.4 More on Graphing: Intercepts

1.   To find the x-intercept, let $y = 0$:   **3.**   To find the x-intercept, let $y = 0$:

$$2x + 0 = 4 \qquad\qquad\qquad\qquad\qquad -x + 0 = 3$$
$$2x = 4 \qquad\qquad\qquad\qquad\qquad\qquad -x = 3$$
$$x = 2 \qquad\qquad\qquad\qquad\qquad\qquad x = -3$$

To find the y-intercept, let $x = 0$:   To find the y-intercept, let $x = 0$:

$$2(0) + y = 4 \qquad\qquad\qquad\qquad\qquad -0 + y = 3$$
$$0 + y = 4 \qquad\qquad\qquad\qquad\qquad\qquad y = 3$$
$$y = 4$$

Graphing the line: Graphing the line:

 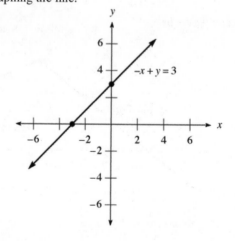

5. To find the *x*-intercept, let $y = 0$:
$$-x + 2(0) = 2$$
$$-x = 2$$
$$x = -2$$
To find the *y*-intercept, let $x = 0$:
$$-0 + 2y = 2$$
$$2y = 2$$
$$y = 1$$
Graphing the line:

7. To find the *x*-intercept, let $y = 0$:
$$5x + 2(0) = 10$$
$$5x = 10$$
$$x = 2$$
To find the *y*-intercept, let $x = 0$:
$$5(0) + 2y = 10$$
$$2y = 10$$
$$y = 5$$
Graphing the line:

9. To find the *x*-intercept, let $y = 0$:
$$-4x + 5(0) = 20$$
$$-4x = 20$$
$$x = -5$$
To find the *y*-intercept, let $x = 0$:
$$-4(0) + 5y = 20$$
$$5y = 20$$
$$y = 4$$
Graphing the line:

11. To find the *x*-intercept, let $y = 0$:
$$3x - 4(0) = -4$$
$$3x = -4$$
$$x = -\frac{4}{3}$$
To find the *y*-intercept, let $x = 0$:
$$3(0) - 4y = -4$$
$$-4y = -4$$
$$y = 1$$
Graphing the line:

13.   To find the x-intercept, let $y = 0$:

$$x - 3(0) = 2$$
$$x = 2$$

To find the y-intercept, let $x = 0$:

$$0 - 3y = 2$$
$$-3y = 2$$
$$y = -\frac{2}{3}$$

Graphing the line:

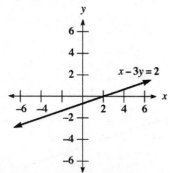

15.   To find the x-intercept, let $y = 0$:

$$2x - 3(0) = -2$$
$$2x = -2$$
$$x = -1$$

To find the y-intercept, let $x = 0$:

$$2(0) - 3y = -2$$
$$-3y = -2$$
$$y = \frac{2}{3}$$

Graphing the line:

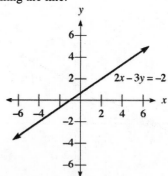

17.   To find the x-intercept, let $y = 0$:

$$2x - 6 = 0$$
$$2x = 6$$
$$x = 3$$

To find the y-intercept, let $x = 0$:

$$y = 2(0) - 6$$
$$y = -6$$

Graphing the line:

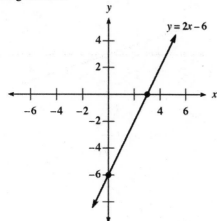

19.   To find the x-intercept, let $y = 0$:

$$2x - 1 = 0$$
$$2x = 1$$
$$x = \frac{1}{2}$$

To find the y-intercept, let $x = 0$:

$$y = 2(0) - 1$$
$$y = -1$$

Graphing the line:

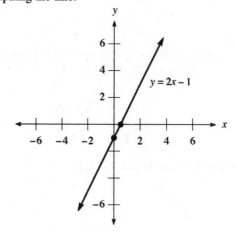

21. To find the *x*-intercept, let $y = 0$:

$$\tfrac{1}{2}x + 3 = 0$$
$$\tfrac{1}{2}x = -3$$
$$x = -6$$

To find the *y*-intercept, let $x = 0$:

$$y = \tfrac{1}{2}(0) + 3$$
$$y = 3$$

Graphing the line:

23. To find the *x*-intercept, let $y = 0$:

$$-\tfrac{1}{3}x - 2 = 0$$
$$-\tfrac{1}{3}x = 2$$
$$x = -6$$

To find the *y*-intercept, let $x = 0$:

$$y = -\tfrac{1}{3}(0) - 2$$
$$y = -2$$

Graphing the line:

25. Another point on the line is $(2, -4)$:

27. Another point on the line is $(3, -1)$:

29. Another point on the line is $(3, 2)$:

31. Completing the table:

Equation	x-intercept	y-intercept
$3x + 4y = 12$	4	3
$3x + 4y = 4$	$\frac{4}{3}$	1
$3x + 4y = 3$	1	$\frac{3}{4}$
$3x + 4y = 2$	$\frac{2}{3}$	$\frac{1}{2}$

33. Completing the table:

Equation	x-intercept	y-intercept
$x - 3y = 2$	2	$-\frac{2}{3}$
$y = \frac{1}{3}x - \frac{2}{3}$	2	$-\frac{2}{3}$
$x - 3y = 0$	0	0
$y = \frac{1}{3}x$	0	0

35. a. Solving the equation:

$$2x - 3 = -3$$
$$2x = 0$$
$$x = 0$$

b. Substituting $y = 0$:

$$2x - 3(0) = -3$$
$$2x - 0 = -3$$
$$2x = -3$$
$$x = -\frac{3}{2}$$

c. Substituting $x = 0$:

$$2(0) - 3y = -3$$
$$0 - 3y = -3$$
$$-3y = -3$$
$$y = 1$$

d. Graphing the equation:

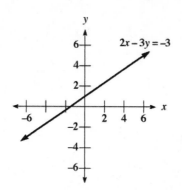

e. Solving for y:

$$2x - 3y = -3$$
$$-3y = -2x - 3$$
$$y = \frac{2}{3}x + 1$$

37. Graphing the line:

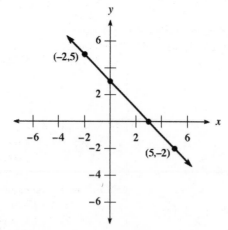

The x- and y-intercepts are both 3.

39. The x-intercept is -1 and the y-intercept is -3.

x	y
-2	1
0	-1
-1	0
1	-2

41. Completing the table:

43. Graphing the line:

The x-intercept is 3.

45. Graphing the line:

The y-intercept is 4.

47. **a.** The equation is $10x + 12y = 480$.

b. Let $y = 0$:
$$10x + 12(0) = 480$$
$$10x = 480$$
$$x = 48$$

Let $x = 0$:
$$10(0) + 12y = 480$$
$$12y = 480$$
$$y = 40$$

The x-intercept is 48 and the y-intercept is 40.

c. Graphing the equation in the first quadrant:

d. When $x = 36$, $y = 10$. She worked 10 hours at $12 per hour.

e. When $y = 25$, $x = 18$. She worked 18 hours at $10 per hour.

49. Solving the equation:
$$-12y - 4 = -148$$
$$-12y = -144$$
$$y = 12$$

51. Solving the equation:
$$-5y - 4 = 51$$
$$-5y = 55$$
$$y = -11$$

53. Solving the equation:
$$11x - 12 = -78$$
$$11x = -66$$
$$x = -6$$

55. Solving the equation:
$$9x + 3 = 66$$
$$9x = 63$$
$$x = 7$$

57. Solving the equation:
$$-9c - 6 = 12$$
$$-9c = 18$$
$$c = -2$$

59. Solving the equation:
$$4 + 13c = -9$$
$$13c = -13$$
$$c = -1$$

61. Solving the equation:
$$3y - 12 = 30$$
$$3y = 42$$
$$y = 14$$

63. Solving the equation:
$$-11y + 9 = 75$$
$$-11y = 66$$
$$y = -6$$

65. **a.** Evaluating: $\dfrac{5-2}{3-1} = \dfrac{3}{2}$

b. Evaluating: $\dfrac{2-5}{1-3} = \dfrac{-3}{-2} = \dfrac{3}{2}$

67. **a.** Evaluating when $x = 3$ and $y = 5$: $\dfrac{y-2}{x-1} = \dfrac{5-2}{3-1} = \dfrac{3}{2}$

b. Evaluating when $x = 3$ and $y = 5$: $\dfrac{2-y}{1-x} = \dfrac{2-5}{1-3} = \dfrac{-3}{-2} = \dfrac{3}{2}$

3.5 The Slope of a Line

1. The slope is given by: $m = \dfrac{4-1}{4-2} = \dfrac{3}{2}$

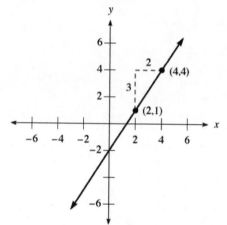

3. The slope is given by: $m = \dfrac{2-4}{5-1} = \dfrac{-2}{4} = -\dfrac{1}{2}$

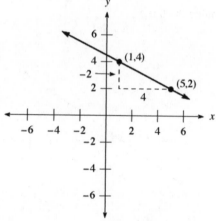

5. The slope is given by: $m = \dfrac{2-(-3)}{4-1} = \dfrac{5}{3}$

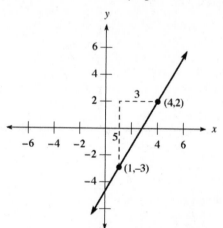

7. The slope is given by: $m = \dfrac{3-(-2)}{1-(-3)} = \dfrac{5}{4}$

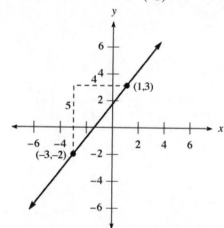

9. The slope is given by: $m = \dfrac{-2-2}{3-(-3)} = \dfrac{-4}{6} = -\dfrac{2}{3}$

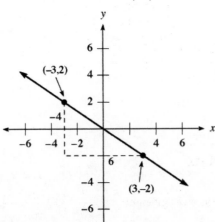

11. The slope is given by: $m = \dfrac{-2-(-5)}{3-2} = \dfrac{3}{1} = 3$

13. Graphing the line:

$y = \dfrac{2}{3}x + 1$

15. Graphing the line:

$y = \dfrac{3}{2}x - 3$

17. Graphing the line:

$y = -\dfrac{4}{3}x + 5$

19. Graphing the line:

$y = 2x + 1$

21.   Graphing the line:

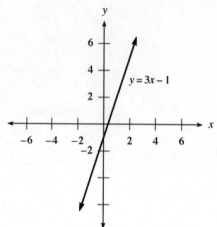

23.   The y-intercept is 2, and the slope is given by: $m = \dfrac{5-(-1)}{1-(-1)} = \dfrac{6}{2} = 3$

25.   The y-intercept is -2, and the slope is given by: $m = \dfrac{2-0}{2-1} = \dfrac{2}{1} = 2$

27.   The slope is given by: $m = \dfrac{0-(-2)}{3-0} = \dfrac{2}{3}$ **29.**   The slope is given by: $m = \dfrac{0-2}{4-0} = \dfrac{-2}{4} = -\dfrac{1}{2}$

          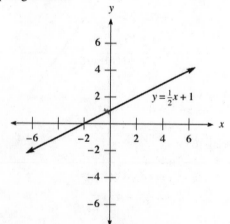

31.   Graphing the line: **33.**   Graphing the line:

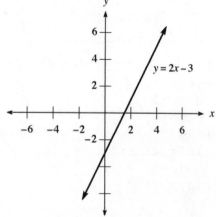

The slope is 2 and the y-intercept is -3.   The slope is $\dfrac{1}{2}$ and the y-intercept is 1.

35. Using the slope formula:

$$\frac{y-2}{6-4} = 2$$

$$\frac{y-2}{2} = 2$$

$$y - 2 = 4$$

$$y = 6$$

37. The slopes are given in the table:

Equation	Slope
$x = 3$	undefined
$y = 3$	0
$y = 3x$	3

39. The slopes are given in the table:

Equation	Slope
$y = -\frac{2}{3}$	0
$x = -\frac{2}{3}$	undefined
$y = -\frac{2}{3}x$	$-\frac{2}{3}$

41. Finding the slopes:

A: $\frac{1-0.5}{2-1} = \frac{0.5}{1} = 0.5$ B: $\frac{6-4}{6-5} = \frac{2}{1} = 2$ C: $\frac{23-18}{10-9} = \frac{5}{1} = 5$

43. Solving the equation:

$$\frac{1}{2}(4x+10) = 11$$

$$2x + 5 = 11$$

$$2x = 6$$

$$x = 3$$

45. Solving the equation:

$$\frac{2}{3}(3x-9) = 24$$

$$2x - 6 = 24$$

$$2x = 30$$

$$x = 15$$

47. Solving the equation:

$$\frac{3}{4}(4x+8) = -12$$

$$3x + 6 = -12$$

$$3x = -18$$

$$x = -6$$

49. Solving the equation:

$$\frac{3}{5}(10x+15) = 45$$

$$6x + 9 = 45$$

$$6x = 36$$

$$x = 6$$

51. Solving the equation:

$$5\left(\frac{2}{5}x+10\right) = -28$$

$$2x + 50 = -28$$

$$2x = -78$$

$$x = -39$$

53. Solving the equation:

$$4\left(\frac{3}{2}x-7\right) = -4$$

$$6x - 28 = -4$$

$$6x = 24$$

$$x = 4$$

55. Solving the equation:

$$\frac{3}{4}(2x+12) = 24$$

$$\frac{3}{2}x + 9 = 24$$

$$\frac{3}{2}x = 15$$

$$x = \frac{2}{3}(15) = 10$$

57. Solving the equation:

$$\frac{1}{2}(5x-10)+6 = -49$$

$$\frac{5}{2}x - 5 + 6 = -49$$

$$\frac{5}{2}x + 1 = -49$$

$$\frac{5}{2}x = -50$$

$$x = \frac{2}{5}(-50) = -20$$

59. Solving for y:

$$-2x + y = 4$$

$$y = 2x + 4$$

61. Solving for y:

$$3x + y = 3$$

$$y = -3x + 3$$

63. Solving for y:

$$4x - 5y = 20$$

$$-5y = -4x + 20$$

$$y = \frac{4}{5}x - 4$$

65. Solving for y:

$$y - 3 = -2(x+4)$$

$$y - 3 = -2x - 8$$

$$y = -2x - 5$$

67. Solving for y:

$$y - 3 = -\frac{2}{3}(x+3)$$
$$y - 3 = -\frac{2}{3}x - 2$$
$$y = -\frac{2}{3}x + 1$$

69. Solving for y:

$$\frac{y-1}{x} = \frac{3}{2}$$
$$2y - 2 = 3x$$
$$2y = 3x + 2$$
$$y = \frac{3}{2}x + 1$$

3.6 Finding the Equation of a Line

1. The slope-intercept form is $y = \frac{2}{3}x + 1$.

3. The slope-intercept form is $y = \frac{3}{2}x - 1$.

5. The slope-intercept form is $y = -\frac{2}{5}x + 3$.

7. The slope-intercept form is $y = 2x - 4$.

9. Solving for y:

$$-2x + y = 4$$
$$y = 2x + 4$$

The slope is 2 and the y-intercept is 4:

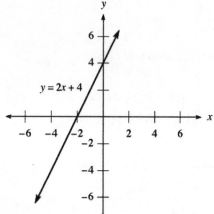

11. Solving for y:

$$3x + y = 3$$
$$y = -3x + 3$$

The slope is -3 and the y-intercept is 3:

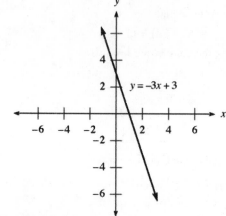

13. Solving for y:

$$3x + 2y = 6$$
$$2y = -3x + 6$$
$$y = -\frac{3}{2}x + 3$$

The slope is $-\frac{3}{2}$ and the y-intercept is 3:

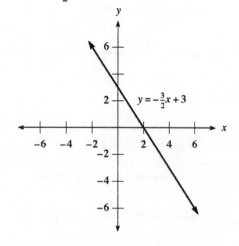

15. Solving for y:

$$4x - 5y = 20$$
$$-5y = -4x + 20$$
$$y = \frac{4}{5}x - 4$$

The slope is $\frac{4}{5}$ and the y-intercept is -4:

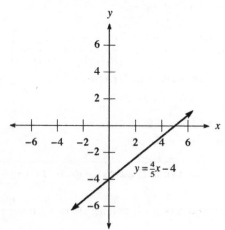

17.   Solving for y:
$$-2x - 5y = 10$$
$$-5y = 2x + 10$$
$$y = -\tfrac{2}{5}x - 2$$

The slope is $-\tfrac{2}{5}$ and the y-intercept is -2:

19.   Using the point-slope formula:
$$y - (-5) = 2(x - (-2))$$
$$y + 5 = 2(x + 2)$$
$$y + 5 = 2x + 4$$
$$y = 2x - 1$$

21.   Using the point-slope formula:
$$y - 1 = -\tfrac{1}{2}(x - (-4))$$
$$y - 1 = -\tfrac{1}{2}(x + 4)$$
$$y - 1 = -\tfrac{1}{2}x - 2$$
$$y = -\tfrac{1}{2}x - 1$$

23.   Using the point-slope formula:
$$y - (-3) = \tfrac{3}{2}(x - 2)$$
$$y + 3 = \tfrac{3}{2}x - 3$$
$$y = \tfrac{3}{2}x - 6$$

25.   Using the point-slope formula:
$$y - 4 = -3(x - (-1))$$
$$y - 4 = -3(x + 1)$$
$$y - 4 = -3x - 3$$
$$y = -3x + 1$$

27.   Finding the slope: $m = \dfrac{-1 - (-4)}{1 - (-2)} = \dfrac{-1 + 4}{1 + 2} = \dfrac{3}{3} = 1$

Using the point-slope formula:
$$y - (-4) = 1(x - (-2))$$
$$y + 4 = x + 2$$
$$y = x - 2$$

29.   Finding the slope: $m = \dfrac{1 - (-5)}{2 - (-1)} = \dfrac{1 + 5}{2 + 1} = \dfrac{6}{3} = 2$

Using the point-slope formula:
$$y - 1 = 2(x - 2)$$
$$y - 1 = 2x - 4$$
$$y = 2x - 3$$

31.   Finding the slope: $m = \dfrac{6 - (-2)}{3 - (-3)} = \dfrac{6 + 2}{3 + 3} = \dfrac{8}{6} = \dfrac{4}{3}$

Using the point-slope formula:
$$y - 6 = \tfrac{4}{3}(x - 3)$$
$$y - 6 = \tfrac{4}{3}x - 4$$
$$y = \tfrac{4}{3}x + 2$$

33.   Finding the slope: $m = \dfrac{-5 - (-1)}{3 - (-3)} = \dfrac{-5 + 1}{3 + 3} = \dfrac{-4}{6} = -\dfrac{2}{3}$

Using the point-slope formula:
$$y - (-5) = -\tfrac{2}{3}(x - 3)$$
$$y + 5 = -\tfrac{2}{3}x + 2$$
$$y = -\tfrac{2}{3}x - 3$$

35.   The y-intercept is 3, and the slope is: $m = \dfrac{3 - 0}{0 - (-1)} = \dfrac{3}{1} = 3$. The slope-intercept form is $y = 3x + 3$.

37.   The y-intercept is -1, and the slope is: $m = \dfrac{0 - (-1)}{4 - 0} = \dfrac{0 + 1}{4} = \dfrac{1}{4}$. The slope-intercept form is $y = \tfrac{1}{4}x - 1$.

39. **a.** Solving the equation:
$$-2x+1=6$$
$$-2x=5$$
$$x=-\tfrac{5}{2}$$

b. Writing in slope-intercept form:
$$-2x+y=6$$
$$y=2x+6$$

c. Substituting $x=0$:
$$-2(0)+y=6$$
$$0+y=6$$
$$y=6$$

d. Finding the slope:
$$-2x+y=6$$
$$y=2x+6$$

The slope is 2.

e. Graphing the line:

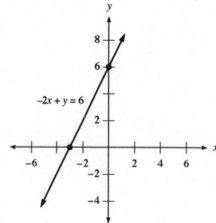

41. The slope is given by: $m=\dfrac{0-2}{3-0}=-\tfrac{2}{3}$. Since $b=2$, the equation is $y=-\tfrac{2}{3}x+2$.

43. The slope is given by: $m=\dfrac{0-(-5)}{-2-0}=-\tfrac{5}{2}$. Since $b=-5$, the equation is $y=-\tfrac{5}{2}x-5$.

45. Its equation is $x=3$.

47. **a.** After 5 years, the copier is worth $6,000.
    c. The slope of the line is –3000.
    e. The equation is $V=-3000t+21,000$.

    b. The copier is worth $12,000 after 3 years.
    d. The copier is decreasing in value by $3000 per year.

49. **a.** He will earn $3,000 for selling 1,000 shirts.
    c. The slope of the line is 2.
    e. The equation is $y=2x+1,000$.

    b. He must sell 500 shirts to earn $2,000 for a month.
    d. Kevin earns $2 for each shirt he sells.

51. Substituting $y=13$:
$$3x-5=13$$
$$3x=18$$
$$x=6$$

53. Substituting $y=-11$:
$$3x-5=-11$$
$$3x=-6$$
$$x=-2$$

55. Substituting $x=0$: $y=\tfrac{3}{7}(0)+4=0+4=4$

57. Substituting $x=-35$: $y=\tfrac{3}{7}(-35)+4=-15+4=-11$

59. Substituting $x=5$: $y=3(5)-5=15-5=10$

61. Substituting $x=-11$: $y=3(-11)-5=-33-5=-38$

63. Substituting $y=-9$:
$$\frac{x-6}{2}=-9$$
$$x-6=-18$$
$$x=-12$$

65. Substituting $y=-12$:
$$\frac{x-6}{2}=-12$$
$$x-6=-24$$
$$x=-18$$

67.   Graphing the line:

69.   Graphing the line:

71.   Graphing the line:

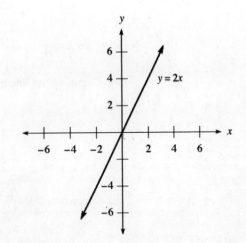

3.7 Linear Inequalities in Two Variables

1.   Checking the point $(0,0)$:

$$2(0) - 3(0) = 0 - 0 < 6 \quad \text{(true)}$$

Graphing the linear inequality:

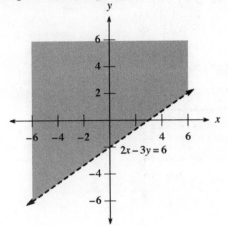

3.   Checking the point $(0,0)$:

$$0 - 2(0) = 0 - 0 \leq 4 \quad \text{(true)}$$

Graphing the linear inequality:

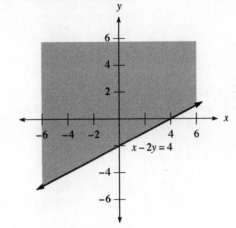

5. Checking the point $(0,0)$:
$$0-0 \le 2 \qquad \text{(true)}$$
Graphing the linear inequality:

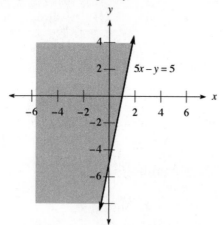

7. Checking the point $(0,0)$:
$$3(0)-4(0)=0-0 \ge 12 \quad \text{(false)}$$
Graphing the linear inequality:

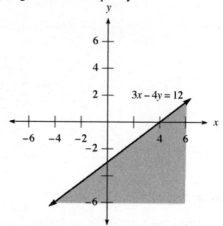

9. Checking the point $(0,0)$:
$$5(0)-0=0-0 \le 5 \qquad \text{(true)}$$
Graphing the linear inequality:

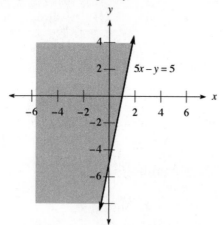

11. Checking the point $(0,0)$:
$$2(0)+6(0)=0+0 \le 12 \quad \text{(true)}$$
Graphing the linear inequality:

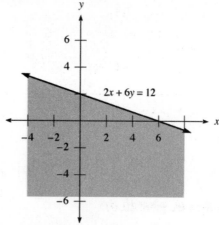

13. Graphing the linear inequality:

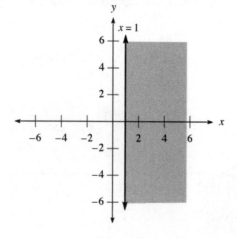

15. Graphing the linear inequality:

17. Graphing the linear inequality:

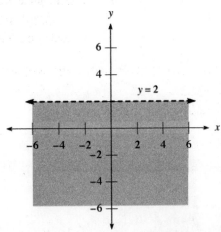

$y = 2$

19. Checking the point $(0,0)$:

$$2(0)+0 = 0+0 > 3 \qquad \text{(false)}$$

Graphing the linear inequality:

$2x+y=3$

21. Checking the point $(0,0)$:

$$0 \le 3(0)-1$$
$$0 \le -1 \qquad \text{(false)}$$

Graphing the linear inequality:

$y = 3x-1$

23. Checking the point $(0,0)$:

$$0 \le -\tfrac{1}{2}(0)+2$$
$$0 \le 2 \qquad \text{(true)}$$

Graphing the linear inequality:

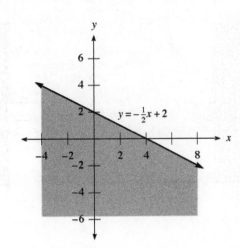

$y = -\tfrac{1}{2}x+2$

25.   **a.**   Solving the inequality:
$$4 + 3y < 12$$
$$3y < 8$$
$$y < \frac{8}{3}$$

b.   Solving the inequality:
$$4 - 3y < 12$$
$$-3y < 8$$
$$y \geq -\frac{8}{3}$$

c.   Solving for y:
$$4x + 3y = 12$$
$$3y = -4x + 12$$
$$y = -\frac{4}{3}x + 4$$

d.   Graphing the inequality:

The graph shows the line $y = -\frac{4}{3}x + 4$ as a dashed line with the region below and to the left shaded.

27.   **a.**   The slope is $\frac{2}{5}$ and the y-intercept is 2, so the equation is $y = \frac{2}{5}x + 2$.

b.   Since the shading is below the line, the inequality is $y < \frac{2}{5}x + 2$.

c.   Since the shading is above the line, the inequality is $y > \frac{2}{5}x + 2$.

29.   Simplifying the expression: $7 - 3(2x - 4) - 8 = 7 - 6x + 12 - 8 = -6x + 11$

31.   Solving the equation:
$$-\frac{3}{2}x = 12$$
$$-\frac{2}{3}\left(-\frac{3}{2}x\right) = -\frac{2}{3}(12)$$
$$x = -8$$

33.   Solving the equation:
$$8 - 2(x + 7) = 2$$
$$8 - 2x - 14 = 2$$
$$-2x - 6 = 2$$
$$-2x = 8$$
$$x = -4$$

35.   Solving for w:
$$2l + 2w = P$$
$$2w = P - 2l$$
$$w = \frac{P - 2l}{2}$$

37.   Solving the inequality:
$$3 - 2x > 5$$
$$3 - 3 - 2x > 5 - 3$$
$$-2x > 2$$
$$-\frac{1}{2}(-2x) < -\frac{1}{2}(2)$$
$$x < -1$$

39.   Solving for y:
$$3x - 2y \leq 12$$
$$3x - 3x - 2y \leq -3x + 12$$
$$-2y \leq -3x + 12$$
$$-\frac{1}{2}(-2y) \geq -\frac{1}{2}(-3x + 12)$$
$$y \geq \frac{3}{2}x - 6$$

Graphing the solution set:

41.   Let w represent the width and $3w+5$ represent the length. Using the perimeter formula:
$$2(w)+2(3w+5)=26$$
$$2w+6w+10=26$$
$$8w+10=26$$
$$8w=16$$
$$w=2$$
$$3w+5=3(2)+5=11$$
The width is 2 inches and the length is 11 inches.

Chapter 3 Review/Test

1.   Substituting $x=4$, $x=0$, $y=3$, and $y=0$:

$3(4)+y=6$	$3(0)+y=6$	$3x+3=6$	$3x+0=6$
$12+y=6$	$0+y=6$	$3x=3$	$3x=6$
$y=-6$	$y=6$	$x=1$	$x=2$

The ordered pairs are $(4,-6),(0,6),(1,3)$, and $(2,0)$.

2.   Substituting $x=5$, $x=0$, $y=2$, and $y=0$:

$2(5)-5y=20$	$2(0)-5y=20$	$2x-5(2)=20$	$2x-5(0)=20$
$10-5y=20$	$0-5y=20$	$2x-10=20$	$2x-0=20$
$-5y=10$	$-5y=20$	$2x=30$	$2x=20$
$y=-2$	$y=-4$	$x=15$	$x=10$

The ordered pairs are $(5,-2),(0,-4),(15,2)$, and $(10,0)$.

3.   Substituting $x=4$, $y=-2$, and $y=3$:

$y=2(4)-6$	$-2=2x-6$	$3=2x-6$
$y=8-6$	$4=2x$	$9=2x$
$y=2$	$x=2$	$x=\frac{9}{2}$

The ordered pairs are $(4,2),(2,-2)$, and $\left(\frac{9}{2},3\right)$.

4.   Substituting $x=2$, $y=0$, and $y=-3$:

$y=5(2)+3$	$0=5x+3$	$-3=5x+3$
$y=10+3$	$-3=5x$	$-6=5x$
$y=13$	$x=-\frac{3}{5}$	$x=-\frac{6}{5}$

The ordered pairs are $(2,13),\left(-\frac{3}{5},0\right)$, and $\left(-\frac{6}{5},-3\right)$.

5.   Substituting $x=2$, $x=-1$, and $x=-3$ results (in each case) in $y=-3$. The ordered pairs are $(2,-3),(-1,-3)$, and $(-3,-3)$.

6.   Substituting $y=5$, $y=0$, and $y=-1$ results (in each case) in $x=6$. The ordered pairs are $(6,5),(6,0)$, and $(6,-1)$.

7.   Substituting each ordered pair into the equation:
$$\left(-2,\tfrac{9}{2}\right):\quad 3(-2)-4\left(\tfrac{9}{2}\right)=-6-18=-24\neq12$$
$$(0,3):\quad 3(0)-4(3)=0-12=-12\neq12$$
$$\left(2,-\tfrac{3}{2}\right):\quad 3(2)-4\left(-\tfrac{3}{2}\right)=6+6=12$$
Only the ordered pair $\left(2,-\tfrac{3}{2}\right)$ is a solution.

8.   Substituting each ordered pair into the equation:
$$\left(-\tfrac{8}{3},-1\right):\ 3\left(-\tfrac{8}{3}\right)+7=-8+7=-1$$
$$\left(\tfrac{7}{3},0\right):\quad 3\left(\tfrac{7}{3}\right)+7=7+7=14\neq0$$
$$(-3,-2):\ 3(-3)+7=-9+7=-2$$
The ordered pairs $\left(-\tfrac{8}{3},-1\right)$ and $(-3,-2)$ are solutions.

9.   Graphing the ordered pair:

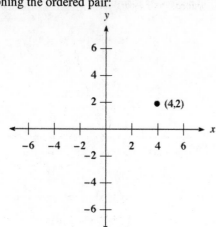

10.   Graphing the ordered pair:

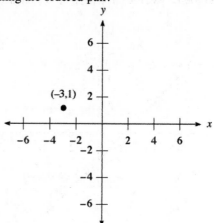

11.   Graphing the ordered pair:

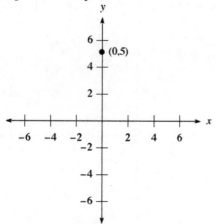

12.   Graphing the ordered pair:

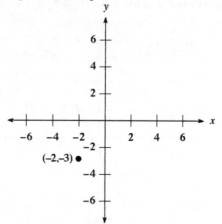

13.   Graphing the ordered pair:

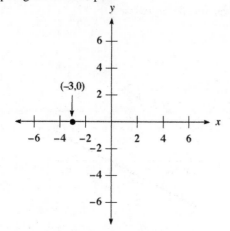

14.   Graphing the ordered pair:

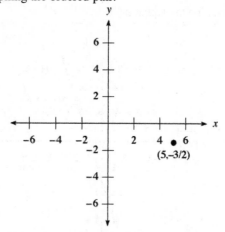

15. The ordered pairs are $(-2,0), (0,-2),$ and $(1,-3)$:

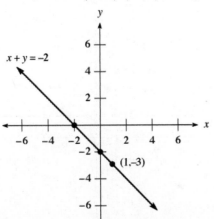

16. The ordered pairs are $(-1,-3), (1,3),$ and $(0,0)$:

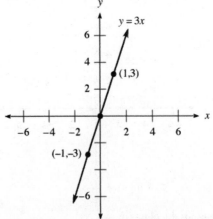

17. The ordered pairs are $(1,1), (0,-1),$ and $(-1,-3)$:

18. The ordered pairs are: $(-3,0), (-3,5),$ and $(-3,-5)$

19. Graphing the equation:

20. Graphing the equation:

21.   Graphing the equation:

22.   Graphing the equation:

23.   Graphing the equation:

24.   Graphing the equation:

25.   Graphing the equation:

26.   Graphing the equation:

27.   Graphing the equation:

28.   Graphing the equation:

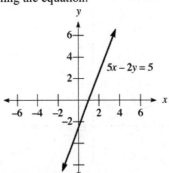

29.   To find the x-intercept, let $y = 0$:

$$3x - 0 = 6$$
$$3x = 6$$
$$x = 2$$

To find the y-intercept, let $x = 0$:

$$3(0) - y = 6$$
$$-y = 6$$
$$y = -6$$

30.   To find the x-intercept, let $y = 0$:

$$2x - 6(0) = 24$$
$$2x = 24$$
$$x = 12$$

To find the y-intercept, let $x = 0$:

$$2(0) - 6y = 24$$
$$-6y = 24$$
$$y = -4$$

31.   To find the x-intercept, let $y = 0$:

$$0 = x - 3$$
$$x = 3$$

To find the y-intercept, let $x = 0$:

$$y = 0 - 3$$
$$y = -3$$

32.   To find the x-intercept, let $y = 0$:

$$0 = 3x - 6$$
$$6 = 3x$$
$$x = 2$$

To find the y-intercept, let $x = 0$:

$$y = 3(0) - 6$$
$$y = -6$$

33.   Since the equation is $y = -5$, there is no x-intercept and the y-intercept is -5.

34.   Since the equation is $x = 4$, the x-intercept is 4 and there is no y-intercept.

35.   The slope is given by: $m = \dfrac{5 - 3}{3 - 2} = \dfrac{2}{1} = 2$

36.   The slope is given by: $m = \dfrac{-5 - 3}{6 - (-2)} = \dfrac{-8}{8} = -1$

37.   The slope is given by: $m = \dfrac{-8 - (-4)}{-3 - (-1)} = \dfrac{-8 + 4}{-3 + 1} = \dfrac{-4}{-2} = 2$

38.   The slope is given by: $m = \dfrac{2 - 4}{-\frac{1}{2} - \frac{1}{2}} = \dfrac{-2}{-1} = 2$

39.   Using the slope formula:

$$\frac{9 - 3}{x - 3} = 2$$
$$\frac{6}{x - 3} = 2$$
$$6 = 2(x - 3)$$
$$6 = 2x - 6$$
$$12 = 2x$$
$$x = 6$$

40.   Using the slope formula:

$$\frac{x + 5}{-5 - 5} = 2$$
$$\frac{x + 5}{-10} = 2$$
$$x + 5 = 2(-10)$$
$$x + 5 = -20$$
$$x = -25$$

41.   Using the point-slope formula:

$$y - 4 = -2\left(x - (-1)\right)$$
$$y - 4 = -2(x + 1)$$
$$y - 4 = -2x - 2$$
$$y = -2x + 2$$

42.   Using the point-slope formula:

$$y - 3 = \tfrac{1}{2}(x - 4)$$
$$y - 3 = \tfrac{1}{2}x - 2$$
$$y = \tfrac{1}{2}x + 1$$

43. Using the point-slope formula:

$$y - (-2) = -\tfrac{3}{4}(x - 3)$$
$$y + 2 = -\tfrac{3}{4}x + \tfrac{9}{4}$$
$$y = -\tfrac{3}{4}x + \tfrac{1}{4}$$

44. Using the point-slope formula:

$$y - 5 = 0(x - 3)$$
$$y - 5 = 0$$
$$y = 5$$

45. The slope-intercept form is $y = 3x + 2$.

46. The slope-intercept form is $y = -x + 6$.

47. The slope-intercept form is $y = -\tfrac{1}{3}x + \tfrac{3}{4}$.

48. The slope-intercept form is $y = 0$.

49. The equation is in slope-intercept form, so $m = 4$ and $b = -1$.

50. Solving for y:

$$2x + y = -5$$
$$y = -2x - 5$$

The equation is in slope-intercept form, so $m = -2$ and $b = -5$.

51. Solving for y:

$$6x + 3y = 9$$
$$3y = -6x + 9$$
$$y = -2x + 3$$

The equation is in slope-intercept form, so $m = -2$ and $b = 3$.

52. Solving for y:

$$5x + 2y = 8$$
$$2y = -5x + 8$$
$$y = -\tfrac{5}{2}x + 4$$

The equation is in slope-intercept form, so $m = -\tfrac{5}{2}$ and $b = 4$.

53. Checking the point $(0,0)$:

$$0 - 0 < 3 \qquad \text{(true)}$$

Graphing the linear inequality:

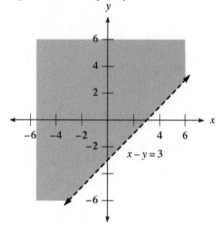

54. Graphing the linear inequality:

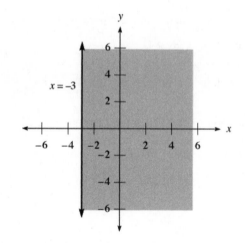

55.   Graphing the linear inequality:

$$y = -4$$

56.   Checking the point $(0,0)$:

$$0 \le -2(0) + 3$$
$$0 \le 3 \qquad \text{(true)}$$

Graphing the linear inequality:

$$y = -2x + 3$$

Chapter 4
Systems of Linear Equations

4.1 Solving Linear Systems by Graphing

1. Graphing both lines:

The intersection point is $(2,1)$.

3. Graphing both lines:

The intersection point is $(-1,2)$.

5. Graphing both lines:

The intersection point is $(3,5)$.

7. Graphing both lines:

The intersection point is $(4,3)$.

9.   Graphing both lines:

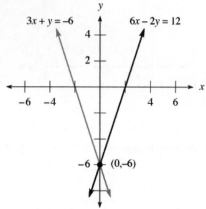

The intersection point is $(0,-6)$.

11.   Graphing both lines:

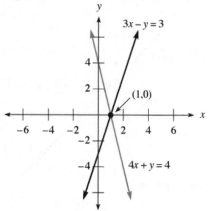

The intersection point is $(1,0)$.

13.   Graphing both lines:

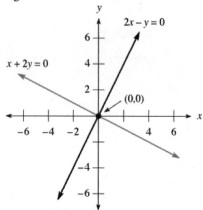

The intersection point is $(0,0)$.

15.   Graphing both lines:

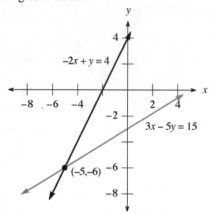

The intersection point is $(-5,-6)$.

17.   Graphing both lines:

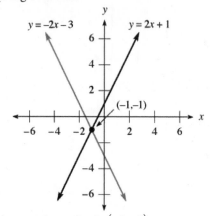

The intersection point is $(-1,-1)$.

19.   Graphing both lines:

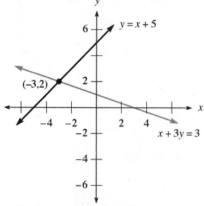

The intersection point is $(-3,2)$.

21. Graphing both lines:

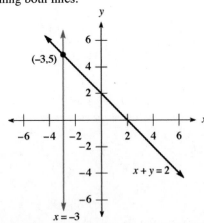

The intersection point is $(-3,5)$.

23. Graphing both lines:

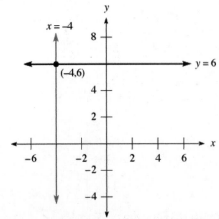

The intersection point is $(-4,6)$.

25. Graphing both lines:

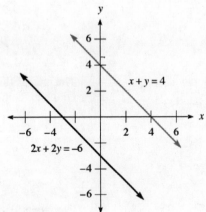

There is no intersection (the lines are parallel).

27. Graphing both lines:

The system is dependent (both lines are the same, they coincide).

29. **a.** Simplifying: $(3x - 4y) + (x - y) = 4x - 5y$

b. Substituting $x = 4$:

$$3(4) - 4y = 8$$
$$12 - 4y = 8$$
$$-4y = -4$$
$$y = 1$$

c. Substituting $x = 0$:

$$3(0) - 4y = 8$$
$$-4y = 8$$
$$y = -2$$

d. Graphing the line:

e. Graphing both lines:

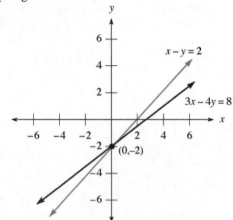

The intersection point is $(0, -2)$.

31. Graphing both lines:

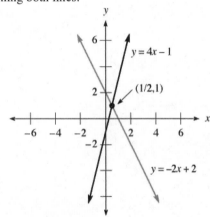

The intersection point is $\left(\frac{1}{2}, 1\right)$.

33. Graphing both lines:

The intersection point is $(2, 1)$.

35. **a.** If Jane worked 25 hours, she would earn the same amount at each position.

b. If Jane worked less than 20 hours, she should choose Gigi's since she earns more in that position.

c. If Jane worked more than 30 hours, she should choose Marcy's since she earns more in that position.

37. Simplifying: $6x + 100(0.04x + 0.75) = 6x + 4x + 75 = 10x + 75$

39. Simplifying: $13x - 1{,}000(0.002x + 0.035) = 13x - 2x - 35 = 11x - 35$

41. Simplifying: $16x - 10(1.7x - 5.8) = 16x - 17x + 58 = -x + 58$

43. Simplifying: $0.04x + 0.06(100 - x) = 0.04x + 6 - 0.06x = -0.02x + 6$

45. Simplifying: $0.025x - 0.028(1{,}000 + x) = 0.025x - 28 - 0.028x = -0.003x - 28$

47. Simplifying: $2.56x - 1.25(100 + x) = 2.56x - 125 - 1.25x = 1.31x - 125$

49. Simplifying: $(x + y) + (x - y) = x + y + x - y = 2x$

51. Simplifying: $(6x - 3y) + (x + 3y) = 6x - 3y + x + 3y = 7x$

53.   Simplifying: $(-12x - 20y) + (25x + 20y) = -12x - 20y + 25x + 20y = 13x$

55.   Simplifying: $-4(3x + 5y) = -12x - 20y$

57.   Simplifying: $12\left(\frac{1}{4}x + \frac{2}{3}y\right) = 12 \bullet \frac{1}{4}x + 12 \bullet \frac{2}{3}y = 3x + 8y$

59.   Simplifying: $-2(2x - y) = -4x + 2y$

61.   Substituting $x = 3$:

$$3 + y = 4$$
$$y = 1$$

63.   Substituting $x = 3$:

$$3 + 3y = 3$$
$$3y = 0$$
$$y = 0$$

65.   Substituting $x = 6$:

$$3(6) + 5y = -7$$
$$18 + 5y = -7$$
$$5y = -25$$
$$y = -5$$

4.2 The Elimination Method

1.   Adding the two equations:

$$2x = 4$$
$$x = 2$$

Substituting into the first equation:

$$2 + y = 3$$
$$y = 1$$

The solution is $(2, 1)$.

3.   Adding the two equations:

$$2y = 14$$
$$y = 7$$

Substituting into the first equation:

$$x + 7 = 10$$
$$x = 3$$

The solution is $(3, 7)$.

5.   Adding the two equations:

$$-2y = 10$$
$$y = -5$$

Substituting into the first equation:

$$x - (-5) = 7$$
$$x + 5 = 7$$
$$x = 2$$

The solution is $(2, -5)$.

7.   Adding the two equations:

$$4x = -4$$
$$x = -1$$

Substituting into the first equation:

$$-1 + y = -1$$
$$y = 0$$

The solution is $(-1, 0)$.

9.   Adding the two equations: $0 = 0$
The lines coincide (the system is dependent).

11.   Multiplying the first equation by 2:

$$6x - 2y = 8$$
$$2x + 2y = 24$$

Adding the two equations:

$$8x = 32$$
$$x = 4$$

Substituting into the first equation:

$$3(4) - y = 4$$
$$12 - y = 4$$
$$-y = -8$$
$$y = 8$$

The solution is $(4, 8)$.

13.   Multiplying the second equation by -3:

$$5x - 3y = -2$$
$$-30x + 3y = -3$$

Adding the two equations:

$$-25x = -5$$
$$x = \frac{1}{5}$$

Substituting into the first equation:

$$5\left(\frac{1}{5}\right) - 3y = -2$$
$$1 - 3y = -2$$
$$-3y = -3$$
$$y = 1$$

The solution is $\left(\frac{1}{5}, 1\right)$.

15. Multiplying the second equation by 4:
$$11x - 4y = 11$$
$$20x + 4y = 20$$
Adding the two equations:
$$31x = 31$$
$$x = 1$$
Substituting into the second equation:
$$5(1) + y = 5$$
$$5 + y = 5$$
$$y = 0$$
The solution is $(1, 0)$.

17. Multiplying the second equation by 3:
$$3x - 5y = 7$$
$$-3x + 3y = -3$$
Adding the two equations:
$$-2y = 4$$
$$y = -2$$
Substituting into the second equation:
$$-x - 2 = -1$$
$$-x = 1$$
$$x = -1$$
The solution is $(-1, -2)$.

19. Multiplying the first equation by –2:
$$2x + 16y = 2$$
$$-2x + 4y = 13$$
Adding the two equations:
$$20y = 15$$
$$y = \frac{3}{4}$$
Substituting into the first equation:
$$-x - 8\left(\frac{3}{4}\right) = -1$$
$$-x - 6 = -1$$
$$-x = 5$$
$$x = -5$$
The solution is $\left(-5, \frac{3}{4}\right)$.

21. Multiplying the first equation by 2:
$$-6x - 2y = 14$$
$$6x + 7y = 11$$
Adding the two equations:
$$5y = 25$$
$$y = 5$$
Substituting into the first equation:
$$-3x - 5 = 7$$
$$-3x = 12$$
$$x = -4$$
The solution is $(-4, 5)$.

23. Adding the two equations:
$$8x = -24$$
$$x = -3$$
Substituting into the second equation:
$$2(-3) + y = -16$$
$$-6 + y = -16$$
$$y = -10$$
The solution is $(-3, -10)$.

25. Multiplying the second equation by 3:
$$x + 3y = 9$$
$$6x - 3y = 12$$
Adding the two equations:
$$7x = 21$$
$$x = 3$$
Substituting into the first equation:
$$3 + 3y = 9$$
$$3y = 6$$
$$y = 2$$
The solution is $(3, 2)$.

27. Multiplying the second equation by 2:
$$x - 6y = 3$$
$$8x + 6y = 42$$
Adding the two equations:
$$9x = 45$$
$$x = 5$$
Substituting into the second equation:
$$4(5) + 3y = 21$$
$$20 + 3y = 21$$
$$3y = 1$$
$$y = \frac{1}{3}$$
The solution is $\left(5, \frac{1}{3}\right)$.

29. Multiplying the second equation by –3:
$$2x + 9y = 2$$
$$-15x - 9y = 24$$
Adding the two equations:
$$-13x = 26$$
$$x = -2$$
Substituting into the first equation:
$$2(-2) + 9y = 2$$
$$-4 + 9y = 2$$
$$9y = 6$$
$$y = \frac{2}{3}$$
The solution is $\left(-2, \frac{2}{3}\right)$.

31. To clear each equation of fractions, multiply the first equation by 12 and the second equation by 6:
$$12\left(\frac{1}{3}x + \frac{1}{4}y\right) = 12\left(\frac{7}{6}\right) \qquad\qquad 6\left(\frac{3}{2}x - \frac{1}{3}y\right) = 6\left(\frac{7}{3}\right)$$
$$4x + 3y = 14 \qquad\qquad\qquad\quad 9x - 2y = 14$$
The system of equations is:
$$4x + 3y = 14$$
$$9x - 2y = 14$$
Multiplying the first equation by 2 and the second equation by 3:
$$8x + 6y = 28$$
$$27x - 6y = 42$$
Adding the two equations:
$$35x = 70$$
$$x = 2$$
Substituting into $4x + 3y = 14$:
$$4(2) + 3y = 14$$
$$8 + 3y = 14$$
$$3y = 6$$
$$y = 2$$
The solution is $(2, 2)$.

33. Multiplying the first equation by –2:
$$-6x - 4y = 2$$
$$6x + 4y = 0$$
Adding the two equations:
$$0 = 2$$
Since this statement is false, the two lines are parallel, so the system has no solution.

35. Multiplying the first equation by 2 and the second equation by 3:
$$22x + 12y = 34$$
$$15x - 12y = 3$$
Adding the two equations:
$$37x = 37$$
$$x = 1$$
Substituting into the second equation:
$$5(1) - 4y = 1$$
$$5 - 4y = 1$$
$$-4y = -4$$
$$y = 1$$
The solution is $(1, 1)$.

37.   To clear each equation of fractions, multiply the first equation by 6 and the second equation by 6:

$$6\left(\tfrac{1}{2}x+\tfrac{1}{6}y\right)=6\left(\tfrac{1}{3}\right) \qquad\qquad 6\left(-x-\tfrac{1}{3}y\right)=6\left(-\tfrac{1}{6}\right)$$
$$3x+y=2 \qquad\qquad\qquad -6x-2y=-1$$

The system of equations is:
$$3x+y=2$$
$$-6x-2y=-1$$

Multiplying the first equation by 2:
$$6x+2y=4$$
$$-6x-2y=-1$$

Adding the two equations:
$$0=3$$

Since this statement is false, the two lines are parallel, so the system has no solution.

39.   Multiplying the first equation by –5:
$$-5x-5y=-110$$
$$5x+10y=170$$

Adding the two equations:
$$5y=60$$
$$y=12$$

Substituting into the first equation:
$$x+12=22$$
$$x=10$$

The solution is $(10,12)$.

41.   Multiplying the first equation by –5:
$$-5x-5y=-70$$
$$5x+25y=230$$

Adding the two equations:
$$20y=160$$
$$y=8$$

Substituting into the first equation:
$$x+8=14$$
$$x=6$$

The solution is $(6,8)$.

43.   Multiplying the first equation by –6:
$$-6x-6y=-90,000$$
$$6x+7y=98,000$$

Adding the two equations:
$$y=8,000$$

Substituting into the first equation:
$$x+8,000=15,000$$
$$x=7,000$$

The solution is $(7000,8000)$.

45.   Multiplying the first equation by –4:
$$-4x-4y=-44,000$$
$$4x+7y=68,000$$

Adding the two equations:
$$3y=24,000$$
$$y=8,000$$

Substituting into the first equation:
$$x+8,000=11,000$$
$$x=3,000$$

The solution is $(3000,8000)$.

47.   Multiplying the first equation by –5:
$$-5x-5y=-115$$
$$5x+10y=175$$

Adding the two equations:
$$5y=60$$
$$y=12$$

Substituting into the first equation:
$$x+12=23$$
$$x=11$$

The solution is $(11,12)$.

49.   Multiplying the second equation by 100 (to eliminate decimals):
$$x + y = 22$$
$$5x + 10y = 170$$
Multiplying the first equation by –5:
$$-5x - 5y = -110$$
$$5x + 10y = 170$$
Adding the two equations:
$$5y = 60$$
$$y = 12$$
Substituting into the first equation:
$$x + 12 = 22$$
$$x = 10$$
The solution is $(10, 12)$.

51.   Solving for y:
$$3x - y = 3$$
$$-y = -3x + 3$$
$$y = 3x - 3$$

The slope is 3 and the y-intercept is –3.

53.   Solving for y:
$$2x - 5y = 25$$
$$-5y = -2x + 25$$
$$y = \frac{2}{5}x - 5$$

The slope is $\frac{2}{5}$ and the y-intercept is –5.

55.   Finding the slope: $m = \dfrac{-5 - 3}{6 - (-2)} = \dfrac{-8}{8} = -1$

57.   Finding the slope: $m = \dfrac{-3 - 3}{2 - 5} = \dfrac{-6}{-3} = 2$

59.   Using the slope formula:
$$\frac{y - 5}{-4 - (-2)} = -3$$
$$\frac{y - 5}{-2} = -3$$
$$y - 5 = 6$$
$$y = 11$$

61.   Using the slope formula:
$$\frac{y - (-6)}{6 - 3} = 5$$
$$\frac{y + 6}{3} = 5$$
$$y + 6 = 15$$
$$y = 9$$

63.   Using the point-slope formula:
$$y - (-6) = 3(x - (-2))$$
$$y + 6 = 3(x + 2)$$
$$y + 6 = 3x + 6$$
$$y = 3x$$

65.   First find the slope: $m = \dfrac{1 - (-5)}{3 - (-3)} = \dfrac{1 + 5}{3 + 3} = \dfrac{6}{6} = 1$

Using the point-slope formula:
$$y - 1 = 1(x - 3)$$
$$y - 1 = x - 3$$
$$y = x - 2$$

67.   Solving the equation:
$$x + (2x - 1) = 2$$
$$x + 2x - 1 = 2$$
$$3x - 1 = 2$$
$$3x = 3$$
$$x = 1$$

69.   Solving the equation:
$$2(3y - 1) - 3y = 4$$
$$6y - 2 - 3y = 4$$
$$3y - 2 = 4$$
$$3y = 6$$
$$y = 2$$

71.   Solving the equation:
$$-2x + 3(5x - 1) = 10$$
$$-2x + 15x - 3 = 10$$
$$13x - 3 = 10$$
$$13x = 13$$
$$x = 1$$

73.   Solving for x:
$$x - 3y = -1$$
$$x = 3y - 1$$

75. Solving for y: $y = 2(1) - 1 = 2 - 1 = 1$

77. Solving for x: $x = 3(2) - 1 = 6 - 1 = 5$

79. Substituting $x = 13$: $y = 1.5(13) + 15 = 19.5 + 15 = 34.5$

81. Substituting $x = 12$: $y = 0.75(12) + 24.95 = 9 + 24.95 = 33.95$

4.3 The Substitution Method

1. Substituting into the first equation:
$$x + (2x - 1) = 11$$
$$3x - 1 = 11$$
$$3x = 12$$
$$x = 4$$
$$y = 2(4) - 1 = 7$$
The solution is $(4, 7)$.

3. Substituting into the first equation:
$$x + (5x + 2) = 20$$
$$6x + 2 = 20$$
$$6x = 18$$
$$x = 3$$
$$y = 5(3) + 2 = 17$$
The solution is $(3, 17)$.

5. Substituting into the first equation:
$$-2x + (-4x + 8) = -1$$
$$-6x + 8 = -1$$
$$-6x = -9$$
$$x = \frac{3}{2}$$
$$y = -4\left(\frac{3}{2}\right) + 8 = -6 + 8 = 2$$
The solution is $\left(\frac{3}{2}, 2\right)$.

7. Substituting into the first equation:
$$3(-y + 6) - 2y = -2$$
$$-3y + 18 - 2y = -2$$
$$-5y + 18 = -2$$
$$-5y = -20$$
$$y = 4$$
$$x = -4 + 6 = 2$$
The solution is $(2, 4)$.

9. Substituting into the first equation:
$$5x - 4(4) = -16$$
$$5x - 16 = -16$$
$$5x = 0$$
$$x = 0$$
The solution is $(0, 4)$.

11. Substituting into the first equation:
$$5x + 4(-3x) = 7$$
$$5x - 12x = 7$$
$$-7x = 7$$
$$x = -1$$
$$y = -3(-1) = 3$$
The solution is $(-1, 3)$.

13. Solving the second equation for x:
$$x - 2y = -1$$
$$x = 2y - 1$$
Substituting into the first equation:
$$(2y - 1) + 3y = 4$$
$$5y - 1 = 4$$
$$5y = 5$$
$$y = 1$$
$$x = 2(1) - 1 = 1$$
The solution is $(1, 1)$.

15. Solving the second equation for x:
$$x - 5y = 17$$
$$x = 5y + 17$$
Substituting into the first equation:
$$2(5y + 17) + y = 1$$
$$10y + 34 + y = 1$$
$$11y + 34 = 1$$
$$11y = -33$$
$$y = -3$$
$$x = 5(-3) + 17 = 2$$
The solution is $(2, -3)$.

17.   Solving the second equation for x:
$$x - 5y = -5$$
$$x = 5y - 5$$
Substituting into the first equation:
$$3(5y - 5) + 5y = -3$$
$$15y - 15 + 5y = -3$$
$$20y - 15 = -3$$
$$20y = 12$$
$$y = \tfrac{3}{5}$$
$$x = 5\left(\tfrac{3}{5}\right) - 5 = 3 - 5 = -2$$

The solution is $\left(-2, \tfrac{3}{5}\right)$.

19.   Solving the second equation for x:
$$x - 3y = -18$$
$$x = 3y - 18$$
Substituting into the first equation:
$$5(3y - 18) + 3y = 0$$
$$15y - 90 + 3y = 0$$
$$18y - 90 = 0$$
$$18y = 90$$
$$y = 5$$
$$x = 3(5) - 18 = -3$$

The solution is $(-3, 5)$.

21.   Solving the second equation for x:
$$x + 3y = 12$$
$$x = -3y + 12$$
Substituting into the first equation:
$$-3(-3y + 12) - 9y = 7$$
$$9y - 36 - 9y = 7$$
$$-36 = 7$$
Since this statement is false, there is no solution to the system. The two lines are parallel.

23.   Substituting into the first equation:
$$5x - 8(2x - 5) = 7$$
$$5x - 16x + 40 = 7$$
$$-11x + 40 = 7$$
$$-11x = -33$$
$$x = 3$$
$$y = 2(3) - 5 = 1$$

The solution is $(3, 1)$.

25.   Substituting into the first equation:
$$7(2y - 1) - 6y = -1$$
$$14y - 7 - 6y = -1$$
$$8y - 7 = -1$$
$$8y = 6$$
$$y = \tfrac{3}{4}$$
$$x = 2\left(\tfrac{3}{4}\right) - 1 = \tfrac{3}{2} - 1 = \tfrac{1}{2}$$

The solution is $\left(\tfrac{1}{2}, \tfrac{3}{4}\right)$.

27.   Substituting into the first equation:
$$-3x + 2(3x) = 6$$
$$-3x + 6x = 6$$
$$3x = 6$$
$$x = 2$$
$$y = 3(2) = 6$$

The solution is $(2, 6)$.

29.   Substituting into the first equation:
$$5(y) - 6y = -4$$
$$-y = -4$$
$$y = 4$$
$$x = 4$$

The solution is $(4, 4)$.

31.   Substituting into the first equation:
$$3x + 3(2x - 12) = 9$$
$$3x + 6x - 36 = 9$$
$$9x - 36 = 9$$
$$9x = 45$$
$$x = 5$$
$$y = 2(5) - 12 = -2$$

The solution is $(5, -2)$.

33.   Substituting into the first equation:
$$7x - 11(10) = 16$$
$$7x - 110 = 16$$
$$7x = 126$$
$$x = 18$$
$$y = 10$$

The solution is $(18, 10)$.

35.   Substituting into the first equation:
$$-4x + 4(x - 2) = -8$$
$$-4x + 4x - 8 = -8$$
$$-8 = -8$$
Since this statement is true, the system is dependent. The two lines coincide.

37. Substituting into the first equation:
$$2x + 5(12 - x) = 36$$
$$2x + 60 - 5x = 36$$
$$-3x + 60 = 36$$
$$-3x = -24$$
$$x = 8$$
$$y = 4$$
The solution is $(8, 4)$.

39. Substituting into the first equation:
$$5x + 2(18 - x) = 54$$
$$5x + 36 - 2x = 54$$
$$3x + 36 = 54$$
$$3x = 18$$
$$x = 6$$
$$y = 12$$
The solution is $(6, 12)$.

41. Substituting into the first equation:
$$2x + 2(2x) = 96$$
$$2x + 4x = 96$$
$$6x = 96$$
$$x = 16$$
$$y = 32$$
The solution is $(16, 32)$.

43. Substituting into the first equation:
$$0.05x + 0.10(22 - x) = 1.70$$
$$0.05x + 2.2 - 0.10x = 1.70$$
$$-0.05x + 2.2 = 1.7$$
$$-0.05x = -0.5$$
$$x = 10$$
$$y = 22 - 10 = 12$$
The solution is $(10, 12)$.

45. a. Solving the equation:
$$4x - 5 = 20$$
$$4x = 25$$
$$x = \frac{25}{4}$$

b. Solving for y:
$$4x - 5y = 20$$
$$-5y = -4x + 20$$
$$y = \frac{4}{5}x - 4$$

c. Solving for x:
$$x - y = 5$$
$$x = y + 5$$

d. Substituting into the first equation:
$$4(y + 5) - 5y = 20$$
$$4y + 20 - 5y = 20$$
$$-y + 20 = 20$$
$$-y = 0$$
$$y = 0$$
$$x = 0 + 5 = 5$$
The solution is $(5, 0)$.

47. a. At 1,000 miles the car and truck cost the same to operate.
b. If Daniel drives more than 1,200 miles, the car will be cheaper to operate.
c. If Daniel drives less than 800 miles, the truck will be cheaper to operate.
d. The graphs appear in the first quadrant only because all quantities are positive.

49. Simplifying: $6(3 + 4) + 5 = 6(7) + 5 = 42 + 5 = 47$

51. Simplifying: $1^2 + 2^2 + 3^2 = 1 + 4 + 9 = 14$

53. Simplifying: $5(6 + 3 \bullet 2) + 4 + 3 \bullet 2 = 5(6 + 6) + 4 + 3 \bullet 2 = 5(12) + 4 + 3 \bullet 2 = 60 + 4 + 6 = 70$

55. Simplifying: $(1^3 + 2^3) + [(2 \bullet 3) + (4 \bullet 5)] = (1 + 8) + [6 + 20] = 9 + 26 = 35$

57. Simplifying: $[2 \bullet 3 + 4 + 5] \div 3 = (6 + 4 + 5) \div 3 = 15 \div 3 = 5$

59. Simplifying: $6 \bullet 10^3 + 5 \bullet 10^2 + 4 \bullet 10^1 = 6,000 + 500 + 40 = 6,540$

61. Simplifying: $1 \bullet 10^3 + 7 \bullet 10^2 + 6 \bullet 10^1 + 0 = 1,000 + 700 + 60 + 0 = 1,760$

63. Simplifying: $4 \bullet 2 - 1 + 5 \bullet 3 - 2 = 8 - 1 + 15 - 2 = 20$

65. Simplifying: $(2^3 + 3^2) \bullet 4 - 5 = (8 + 9) \bullet 4 - 5 = 17 \bullet 4 - 5 = 68 - 5 = 63$

67. Simplifying: $2(2^2 + 3^2) + 3(3^2) = 2(4 + 9) + 3(9) = 2(13) + 3(9) = 26 + 27 = 53$

69. Let x and $5x + 8$ represent the two numbers. The equation is:

$$x + 5x + 8 = 26$$
$$6x + 8 = 26$$
$$6x = 18$$
$$x = 3$$
$$5x + 8 = 23$$

The two numbers are 3 and 23.

71. Let x represent the smaller number and $2x - 6$ represent the larger number. The equation is:

$$2x - 6 - x = 9$$
$$x - 6 = 9$$
$$x = 15$$
$$2x - 6 = 24$$

The two numbers are 15 and 24.

73. Let w represent the width and $3w + 5$ represent the length. Using the perimeter formula:

$$2(w) + 2(3w + 5) = 58$$
$$2w + 6w + 10 = 58$$
$$8w + 10 = 58$$
$$8w = 48$$
$$w = 6$$
$$3w + 5 = 23$$

The width is 6 inches and the length is 23 inches.

75. Completing the table:

	Nickels	Dimes
Number	$x + 4$	x
Value (cents)	$5(x + 4)$	$10x$

The equation is:

$$5(x + 4) + 10x = 170$$
$$5x + 20 + 10x = 170$$
$$15x + 20 = 170$$
$$15x = 150$$
$$x = 10$$
$$x + 4 = 14$$

John has 14 nickels and 10 dimes.

77. Substituting into the second equation:

$$x + (5x + 2) = 20$$
$$6x + 2 = 20$$
$$6x = 18$$
$$x = 3$$
$$y = 5(3) + 2 = 17$$

The solution is $(3, 17)$.

79. Multiplying the second equation by 100 (to eliminate decimals):

$$x + y = 15000$$
$$6x + 7y = 98000$$

Multiplying the first equation by –6:

$$-6x - 6y = -90000$$
$$6x + 7y = 98000$$

Adding the two equations:

$$y = 8000$$

Substituting into the first equation:

$$x + 8000 = 15000$$
$$x = 7000$$

The solution is $(7000, 8000)$.

4.4 Applications

1. Let x and y represent the two numbers. The system of equations is:

$$x + y = 25$$
$$y = 5 + x$$

Substituting into the first equation:

$$x + (5 + x) = 25$$
$$2x + 5 = 25$$
$$2x = 20$$
$$x = 10$$
$$y = 5 + 10 = 15$$

The two numbers are 10 and 15.

3. Let x and y represent the two numbers. The system of equations is:

$$x + y = 15$$
$$y = 4x$$

Substituting into the first equation:

$$x + 4x = 15$$
$$5x = 15$$
$$x = 3$$
$$y = 4(3) = 12$$

The two numbers are 3 and 12.

5. Let x represent the larger number and y represent the smaller number. The system of equations is:

$$x - y = 5$$
$$x = 2y + 1$$

Substituting into the first equation:

$$2y + 1 - y = 5$$
$$y + 1 = 5$$
$$y = 4$$
$$x = 2(4) + 1 = 9$$

The two numbers are 4 and 9.

7. Let x and y represent the two numbers. The system of equations is:

$$y = 4x + 5$$
$$x + y = 35$$

Substituting into the second equation:

$$x + 4x + 5 = 35$$
$$5x + 5 = 35$$
$$5x = 30$$
$$x = 6$$
$$y = 4(6) + 5 = 29$$

The two numbers are 6 and 29.

9. Let x represent the amount invested at 6% and y represent the amount invested at 8%. The system of equations is:
$$x + y = 20000$$
$$0.06x + 0.08y = 1380$$
Multiplying the first equation by –0.06:
$$-0.06x - 0.06y = -1200$$
$$0.06x + 0.08y = 1380$$
Adding the two equations:
$$0.02y = 180$$
$$y = 9000$$
Substituting into the first equation:
$$x + 9000 = 20000$$
$$x = 11000$$
Mr. Wilson invested $9,000 at 8% and $11,000 at 6%.

11. Let x represent the amount invested at 5% and y represent the amount invested at 6%. The system of equations is:
$$x = 4y$$
$$0.05x + 0.06y = 520$$
Substituting into the second equation:
$$0.05(4y) + 0.06y = 520$$
$$0.20y + 0.06y = 520$$
$$0.26y = 520$$
$$y = 2000$$
$$x = 4(2000) = 8000$$
She invested $8,000 at 5% and $2,000 at 6%.

13. Let x represent the number of nickels and y represent the number of quarters. The system of equations is:
$$x + y = 14$$
$$0.05x + 0.25y = 2.30$$
Multiplying the first equation by –0.05:
$$-0.05x - 0.05y = -0.7$$
$$0.05x + 0.25y = 2.30$$
Adding the two equations:
$$0.20y = 1.6$$
$$y = 8$$
Substituting into the first equation:
$$x + 8 = 14$$
$$x = 6$$
Ron has 6 nickels and 8 quarters.

15. Let x represent the number of dimes and y represent the number of quarters. The system of equations is:
$$x + y = 21$$
$$0.10x + 0.25y = 3.45$$
Multiplying the first equation by –0.10:
$$-0.10x - 0.10y = -2.10$$
$$0.10x + 0.25y = 3.45$$
Adding the two equations:
$$0.15y = 1.35$$
$$y = 9$$
Substituting into the first equation:
$$x + 9 = 21$$
$$x = 12$$
Tom has 12 dimes and 9 quarters.

17. Let x represent the liters of 50% alcohol solution and y represent the liters of 20% alcohol solution. The system of equations is:
$$x + y = 18$$
$$0.50x + 0.20y = 0.30(18)$$
Multiplying the first equation by –0.20:
$$-0.20x - 0.20y = -3.6$$
$$0.50x + 0.20y = 5.4$$
Adding the two equations:
$$0.30x = 1.8$$
$$x = 6$$
Substituting into the first equation:
$$6 + y = 18$$
$$y = 12$$
The mixture contains 6 liters of 50% alcohol solution and 12 liters of 20% alcohol solution.

19. Let x represent the gallons of 10% disinfectant and y represent the gallons of 7% disinfectant. The system of equations is:
$$x + y = 30$$
$$0.10x + 0.07y = 0.08(30)$$
Multiplying the first equation by –0.07:
$$-0.07x - 0.07y = -2.1$$
$$0.10x + 0.07y = 2.4$$
Adding the two equations:
$$0.03x = 0.3$$
$$x = 10$$
Substituting into the first equation:
$$10 + y = 30$$
$$y = 20$$
The mixture contains 10 gallons of 10% disinfectant and 20 gallons of 7% disinfectant.

21. Let x represent the number of adult tickets and y represent the number of kids tickets. The system of equations is:
$$x + y = 70$$
$$5.50x + 4.00y = 310$$
Multiplying the first equation by –4:
$$-4.00x - 4.00y = -280$$
$$5.50x + 4.00y = 310$$
Adding the two equations:
$$1.5x = 30$$
$$x = 20$$
Substituting into the first equation:
$$20 + y = 70$$
$$y = 50$$
The matinee had 20 adult tickets sold and 50 kids tickets sold.

23. Let x represent the width and y represent the length. The system of equations is:
$$2x + 2y = 96$$
$$y = 2x$$
Substituting into the first equation:
$$2x + 2(2x) = 96$$
$$2x + 4x = 96$$
$$6x = 96$$
$$x = 16$$
$$y = 2(16) = 32$$
The width is 16 feet and the length is 32 feet.

25.   Let x represent the number of $5 chips and y represent the number of $25 chips. The system of equations is:
$$x + y = 45$$
$$5x + 25y = 465$$
Multiplying the first equation by −5:
$$-5x - 5y = -225$$
$$5x + 25y = 465$$
Adding the two equations:
$$20y = 240$$
$$y = 12$$
Substituting into the first equation:
$$x + 12 = 45$$
$$x = 33$$
The gambler has 33 $5 chips and 12 $25 chips.

27.   Let x represent the number of shares of $11 stock and y represent the number of shares of $20 stock. The system of equations is:
$$x + y = 150$$
$$11x + 20y = 2550$$
Multiplying the first equation by −11:
$$-11x - 11y = -1650$$
$$11x + 20y = 2550$$
Adding the two equations:
$$9y = 900$$
$$y = 100$$
Substituting into the first equation:
$$x + 100 = 150$$
$$x = 50$$
She bought 50 shares at $11 and 100 shares at $20.

29.   Substituting $x = -2$, $x = 0$, and $x = 2$:
$$y = \tfrac{1}{2}(-2) + 3 = -1 + 3 = 2 \qquad y = \tfrac{1}{2}(0) + 3 = 0 + 3 = 3 \qquad y = \tfrac{1}{2}(2) + 3 = 1 + 3 = 4$$
The ordered pairs are $(-2, 2)$, $(0, 3)$, and $(2, 4)$.

31.   Graphing the line:

33.   Computing the slope: $m = \dfrac{1-5}{0-2} = \dfrac{-4}{-2} = 2$

35.   Using the point-slope formula:
$$y - 1 = \tfrac{1}{2}\left(x - (-2)\right)$$
$$y - 1 = \tfrac{1}{2}(x + 2)$$
$$y - 1 = \tfrac{1}{2}x + 1$$
$$y = \tfrac{1}{2}x + 2$$

37. Computing the slope: $m = \dfrac{1-5}{0-2} = \dfrac{-4}{-2} = 2$. Using the point-slope formula:

$$y - 1 = 2(x - 0)$$
$$y - 1 = 2x$$
$$y = 2x + 1$$

Chapter 4 Review/Test

1. Graphing both lines:

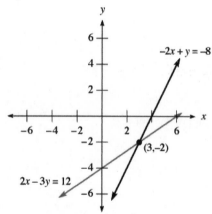

The intersection point is $(4, -2)$.

2. Graphing both lines:

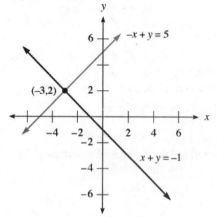

The intersection point is $(-3, 2)$.

3. Graphing both lines:

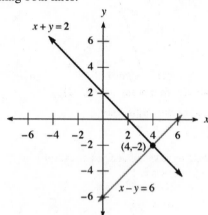

The intersection point is $(3, -2)$.

4. Graphing both lines:

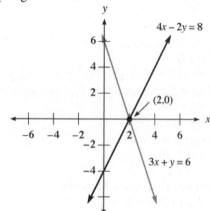

The intersection point is $(2, 0)$.

5. Graphing both lines:

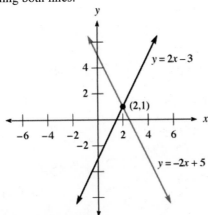

The intersection point is $(2,1)$.

7. Adding the two equations:
$$2x = 2$$
$$x = 1$$
Substituting into the second equation:
$$1 + y = -2$$
$$y = -3$$

The solution is $(1, -3)$.

9. Multiplying the first equation by 2:
$$10x - 6y = 4$$
$$-10x + 6y = -4$$
Adding the two equations:
$$0 = 0$$
Since this statement is true, the system is dependent. The two lines coincide.

10. Multiplying the first equation by 2 and the second equation by 3:
$$4x + 6y = -4$$
$$9x - 6y = 30$$
Adding the two equations:
$$13x = 26$$
$$x = 2$$
Substituting into the first equation:
$$2(2) + 3y = -2$$
$$4 + 3y = -2$$
$$3y = -6$$
$$y = -2$$
The solution is $(2, -2)$.

6. Graphing both lines:

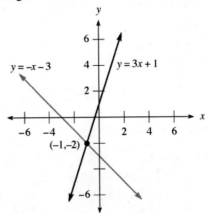

The intersection point is $(-1, -2)$.

8. Adding the two equations:
$$x = -2$$
Substituting into the second equation:
$$2(-2) + y = 1$$
$$-4 + y = 1$$
$$y = 5$$
The solution is $(-2, 5)$.

11.   Multiplying the second equation by –4:
$$-3x + 4y = 1$$
$$16x - 4y = 12$$
Adding the two equations:
$$13x = 13$$
$$x = 1$$
Substituting into the second equation:
$$-4(1) + y = -3$$
$$-4 + y = -3$$
$$y = 1$$
The solution is $(1,1)$.

12.   Multiplying the second equation by 2:
$$-4x - 2y = 3$$
$$4x + 2y = 2$$
Adding the two equations: $0 = 5$
Since this statement is false, there is no solution to the system. The two lines are parallel.

13.   Multiplying the first equation by 3 and the second equation by 5:
$$-6x + 15y = -33$$
$$35x - 15y = -25$$
Adding the two equations:
$$29x = -58$$
$$x = -2$$
Substituting into the first equation:
$$-2(-2) + 5y = -11$$
$$4 + 5y = -11$$
$$5y = -15$$
$$y = -3$$
The solution is $(-2,-3)$.

14.   Multiplying the first equation by 3 and the second equation by 2:
$$-6x + 15y = -45$$
$$6x - 8y = 38$$
Adding the two equations:
$$7y = -7$$
$$y = -1$$
Substituting into the second equation:
$$3x - 4(-1) = 19$$
$$3x + 4 = 19$$
$$3x = 15$$
$$x = 5$$
The solution is $(5,-1)$.

15.   Substituting into the first equation:
$$x + (-3x + 1) = 5$$
$$-2x + 1 = 5$$
$$-2x = 4$$
$$x = -2$$
Substituting into the second equation: $y = -3(-2) + 1 = 6 + 1 = 7$. The solution is $(-2, 7)$.

16. Substituting into the first equation:
$$x - (-2x - 10) = -2$$
$$x + 2x + 10 = -2$$
$$3x + 10 = -2$$
$$3x = -12$$
$$x = -4$$
Substituting into the second equation: $y = -2(-4) - 10 = 8 - 10 = -2$. The solution is $(-4, -2)$.

17. Substituting into the first equation:
$$4x - 3(3x + 7) = -16$$
$$4x - 9x - 21 = -16$$
$$-5x - 21 = -16$$
$$-5x = 5$$
$$x = -1$$
Substituting into the second equation: $y = 3(-1) + 7 = -3 + 7 = 4$. The solution is $(-1, 4)$.

18. Substituting into the first equation:
$$5x + 2(-8x + 10) = -2$$
$$5x - 16x + 20 = -2$$
$$-11x + 20 = -2$$
$$-11x = -22$$
$$x = 2$$
Substituting into the second equation: $y = -8(2) + 10 = -16 + 10 = -6$. The solution is $(2, -6)$.

19. Solving the first equation for x:
$$x - 4y = 2$$
$$x = 4y + 2$$
Substituting into the second equation:
$$-3(4y + 2) + 12y = -8$$
$$-12y - 6 + 12y = -8$$
$$-6 = -8$$
Since this statement is false, there is no solution to the system. The two lines are parallel.

20. Solving the second equation for y:
$$3x + y = -19$$
$$y = -3x - 19$$
Substituting into the first equation:
$$4x - 2(-3x - 19) = 8$$
$$4x + 6x + 38 = 8$$
$$10x + 38 = 8$$
$$10x = -30$$
$$x = -3$$
Substituting into $y = -3x - 19$: $y = -3(-3) - 19 = 9 - 19 = -10$. The solution is $(-3, -10)$.

21. Solving the second equation for x:
$$x + 6y = -11$$
$$x = -6y - 11$$
Substituting into the first equation:
$$10(-6y - 11) - 5y = 20$$
$$-60y - 110 - 5y = 20$$
$$-65y - 110 = 20$$
$$-65y = 130$$
$$y = -2$$
Substituting into $x = -6y - 11$: $x = -6(-2) - 11 = 12 - 11 = 1$. The solution is $(1, -2)$.

22. Solving the second equation for y:
$$-6x + 2y = -4$$
$$2y = 6x - 4$$
$$y = 3x - 2$$
Substituting into the first equation:
$$3x - (3x - 2) = 2$$
$$3x - 3x + 2 = 2$$
$$2 = 2$$
Since this statement is true, the system is dependent. The two lines coincide.

23. Let x represent the smaller number and y represent the larger number. The system of equations is:
$$x + y = 18$$
$$2x = 6 + y$$
Solving the first equation for y:
$$x + y = 18$$
$$y = -x + 18$$
Substituting into the second equation:
$$2x = 6 + (-x + 18)$$
$$2x = -x + 24$$
$$3x = 24$$
$$x = 8$$
Substituting into the first equation:
$$8 + y = 18$$
$$y = 10$$
The two numbers are 8 and 10.

24. Let x and y represent the two numbers. The system of equations is:
$$x - y = 16$$
$$x = 3y$$
Substituting into the first equation:
$$3y - y = 16$$
$$2y = 16$$
$$y = 8$$
$$x = 3(8) = 24$$
The two numbers are 24 and 8.

25. Let x represent the amount invested at 4% and y represent the amount invested at 5%. The system of equations is:
$$x + y = 12000$$
$$0.04x + 0.05y = 560$$
Multiplying the first equation by –0.04:
$$-0.04x - 0.04y = -480$$
$$0.04x + 0.05y = 560$$
Adding the two equations:
$$0.01y = 80$$
$$y = 8000$$
Substituting into the first equation:
$$x + 8000 = 12000$$
$$x = 4000$$
So $4,000 was invested at 4% and $8,000 was invested at 5%.

26. Let x represent the amount invested at 6% and y represent the amount invested at 8%. The system of equations is:
$$x + y = 14000$$
$$0.06x + 0.08y = 1060$$
Multiplying the first equation by –0.06:
$$-0.06x - 0.06y = -840$$
$$0.06x + 0.08y = 1060$$
Adding the two equations:
$$0.02y = 220$$
$$y = 11000$$
Substituting into the first equation:
$$x + 11000 = 14000$$
$$x = 3000$$
So $3,000 was invested at 6% and $11,000 was invested at 8%.

27. Let x represent the number of dimes and y represent the number of nickels. The system of equations is:
$$x + y = 17$$
$$0.10x + 0.05y = 1.35$$
Multiplying the first equation by –0.05:
$$-0.05x - 0.05y = -0.85$$
$$0.10x + 0.05y = 1.35$$
Adding the two equations:
$$0.05x = 0.50$$
$$x = 10$$
Substituting into the first equation:
$$10 + y = 17$$
$$y = 7$$
Barbara has 10 dimes and 7 nickels.

28. Let x represent the number of dimes and y represent the number of quarters. The system of equations is:
$$x + y = 15$$
$$0.10x + 0.25y = 2.40$$
Multiplying the first equation by –0.10:
$$-0.10x - 0.10y = -1.50$$
$$0.10x + 0.25y = 2.40$$
Adding the two equations:
$$0.15y = 0.9$$
$$y = 6$$
Substituting into the first equation:
$$x + 6 = 15$$
$$x = 9$$
Tom has 9 dimes and 6 quarters.

29.   Let x represent the liters of 20% alcohol solution and y represent the liters of 10% alcohol solution. The system of equations is:

$$x + y = 50$$
$$0.20x + 0.10y = 0.12(50)$$

Multiplying the first equation by –0.10:

$$-0.10x - 0.10y = -5$$
$$0.20x + 0.10y = 6$$

Adding the two equations:

$$0.10x = 1$$
$$x = 10$$

Substituting into the first equation:

$$10 + y = 50$$
$$y = 40$$

The solution contains 40 liters of 10% alcohol solution and 10 liters of 20% alcohol solution.

30.   Let x represent the liters of 25% alcohol solution and y represent the liters of 15% alcohol solution. The system of equations is:

$$x + y = 40$$
$$0.25x + 0.15y = 0.20(40)$$

Multiplying the first equation by –0.15:

$$-0.15x - 0.15y = -6$$
$$0.25x + 0.15y = 8$$

Adding the two equations:

$$0.10x = 2$$
$$x = 20$$

Substituting into the first equation:

$$20 + y = 40$$
$$y = 20$$

The solution contains 20 liters of 25% alcohol solution and 20 liters of 15% alcohol solution.

Chapter 5
Exponents and Polynomials

5.1 Multiplication with Exponents

1. The base is 4 and the exponent is 2. Evaluating the expression: $4^2 = 4 \cdot 4 = 16$

3. The base is 0.3 and the exponent is 2. Evaluating the expression: $0.3^2 = 0.3 \cdot 0.3 = 0.09$

5. The base is 4 and the exponent is 3. Evaluating the expression: $4^3 = 4 \cdot 4 \cdot 4 = 64$

7. The base is -5 and the exponent is 2. Evaluating the expression: $(-5)^2 = (-5) \cdot (-5) = 25$

9. The base is 2 and the exponent is 3. Evaluating the expression: $-2^3 = -2 \cdot 2 \cdot 2 = -8$

11. The base is 3 and the exponent is 4. Evaluating the expression: $3^4 = 3 \cdot 3 \cdot 3 \cdot 3 = 81$

13. The base is $\frac{2}{3}$ and the exponent is 2. Evaluating the expression: $\left(\frac{2}{3}\right)^2 = \left(\frac{2}{3}\right) \cdot \left(\frac{2}{3}\right) = \frac{4}{9}$

15. The base is $\frac{1}{2}$ and the exponent is 4. Evaluating the expression: $\left(\frac{1}{2}\right)^4 = \left(\frac{1}{2}\right) \cdot \left(\frac{1}{2}\right) \cdot \left(\frac{1}{2}\right) \cdot \left(\frac{1}{2}\right) = \frac{1}{16}$

17. a. Completing the table:

Number (x)	1	2	3	4	5	6	7
Square $\left(x^2\right)$	1	4	9	16	25	36	49

   b. For numbers larger than 1, the square of the number is larger than the number.

19. Simplifying the expression: $x^4 \cdot x^5 = x^{4+5} = x^9$ 21. Simplifying the expression: $y^{10} \cdot y^{20} = y^{10+20} = y^{30}$

23. Simplifying the expression: $2^5 \cdot 2^4 \cdot 2^3 = 2^{5+4+3} = 2^{12}$

25. Simplifying the expression: $x^4 \cdot x^6 \cdot x^8 \cdot x^{10} = x^{4+6+8+10} = x^{28}$

27. Simplifying the expression: $\left(x^2\right)^5 = x^{2 \cdot 5} = x^{10}$ 29. Simplifying the expression: $\left(5^4\right)^3 = 5^{4 \cdot 3} = 5^{12}$

31. Simplifying the expression: $\left(y^3\right)^3 = y^{3 \cdot 3} = y^9$ 33. Simplifying the expression: $\left(2^5\right)^{10} = 2^{5 \cdot 10} = 2^{50}$

35. Simplifying the expression: $\left(a^3\right)^x = a^{3x}$ 37. Simplifying the expression: $\left(b^x\right)^y = b^{xy}$

39. Simplifying the expression: $(4x)^2 = 4^2 \cdot x^2 = 16x^2$ 41. Simplifying the expression: $(2y)^5 = 2^5 \cdot y^5 = 32y^5$

43. Simplifying the expression: $(-3x)^4 = (-3)^4 \cdot x^4 = 81x^4$

45. Simplifying the expression: $(0.5ab)^2 = (0.5)^2 \cdot a^2 b^2 = 0.25a^2 b^2$

47. Simplifying the expression: $(4xyz)^3 = 4^3 \cdot x^3 y^3 z^3 = 64x^3 y^3 z^3$

49. Simplifying using properties of exponents: $\left(2x^4\right)^3 = 2^3\left(x^4\right)^3 = 8x^{12}$

51. Simplifying using properties of exponents: $\left(4a^3\right)^2 = 4^2\left(a^3\right)^2 = 16a^6$

53. Simplifying using properties of exponents: $\left(x^2\right)^3\left(x^4\right)^2 = x^6 \cdot x^8 = x^{14}$

55. Simplifying using properties of exponents: $\left(a^3\right)^1\left(a^2\right)^4 = a^3 \cdot a^8 = a^{11}$

57. Simplifying using properties of exponents: $(2x)^3(2x)^4 = (2x)^7 = 2^7 x^7 = 128x^7$

59. Simplifying using properties of exponents: $\left(3x^2\right)^3(2x)^4 = 3^3 x^6 \cdot 2^4 x^4 = 27x^6 \cdot 16x^4 = 432x^{10}$

61. Simplifying using properties of exponents: $\left(4x^2 y^3\right)^2 = 4^2 x^4 y^6 = 16x^4 y^6$

63. Simplifying using properties of exponents: $\left(\frac{2}{3}a^4 b^5\right)^3 = \left(\frac{2}{3}\right)^3 a^{12} b^{15} = \frac{8}{27}a^{12} b^{15}$

65. Writing as a perfect square: $x^4 = \left(x^2\right)^2$

67. Writing as a perfect square: $16x^2 = (4x)^2$

69. Writing as a perfect cube: $8 = (2)^3$

71. Writing as a perfect cube: $64x^3 = (4x)^3$

73. **a.** Substituting $x = 2$: $x^3 x^2 = (2)^3 (2)^2 = (8)(4) = 32$

    b. Substituting $x = 2$: $\left(x^3\right)^2 = \left(2^3\right)^2 = (8)^2 = 64$

    c. Substituting $x = 2$: $x^5 = (2)^5 = 32$

    d. Substituting $x = 2$: $x^6 = (2)^6 = 64$

75. **a.** Completing the table:

Number (x)	Square $\left(x^2\right)$
−3	9
−2	4
−1	1
0	0
1	1
2	4
3	9

77. Completing the table:

Number (x)	Square $\left(x^2\right)$
−2.5	6.25
−1.5	2.25
−0.5	0.25
0	0
0.5	0.25
1.5	2.25
2.5	6.25

b.   Constructing a line graph:

79. Writing in scientific notation: $43,200 = 4.32 \times 10^4$

81. Writing in scientific notation: $570 = 5.7 \times 10^2$

83. Writing in scientific notation: $238,000 = 2.38 \times 10^5$

85. Writing in expanded form: $2.49 \times 10^3 = 2,490$

87. Writing in expanded form: $3.52 \times 10^2 = 352$

89. Writing in expanded form: $2.8 \times 10^4 = 28,000$

91. The volume is given by: $V = (3 \text{ in.})^3 = 27 \text{ inches}^3$

93. The volume is given by: $V = (2.5 \text{ in.})^3 \approx 15.6 \text{ inches}^3$

95. The volume is given by: $V = (8 \text{ in.})(4.5 \text{ in.})(1 \text{ in.}) = 36 \text{ inches}^3$

97. Yes. If the box was 7 ft x 3 ft x 2 ft, you could fit inside of it.

99. Writing in scientific notation: $650,000,000 \text{ seconds} = 6.5 \times 10^8 \text{ seconds}$

101. Writing in expanded form: $7.4 \times 10^5 \text{ dollars} = \$740,000$

103. Writing in expanded form: $1.8 \times 10^5 \text{ dollars} = \$180,000$

105. Substitute $c = 8$, $b = 3.35$, and $s = 3.11$: $d = \pi \bullet 3.11 \bullet 8 \bullet \left(\frac{1}{2} \bullet 3.35\right)^2 \approx 219 \text{ inches}^3$

107. Substitute $c = 6$, $b = 3.59$, and $s = 2.99$: $d = \pi \bullet 2.99 \bullet 6 \bullet \left(\frac{1}{2} \bullet 3.59\right)^2 \approx 182 \text{ inches}^3$

109. Factoring into primes: $128 = 4 \bullet 32 = (2 \bullet 2) \bullet (4 \bullet 8) = (2 \bullet 2) \bullet (2 \bullet 2 \bullet 2 \bullet 2 \bullet 2) = 2^7$

111. Factoring into primes: $250 = 10 \bullet 25 = (2 \bullet 5) \bullet (5 \bullet 5) = 2 \bullet 5^3$

113. Factoring into primes: $720 = 10 \bullet 72 = (2 \bullet 5) \bullet (8 \bullet 9) = (2 \bullet 5) \bullet (2 \bullet 2 \bullet 2 \bullet 3 \bullet 3) = 2^4 \bullet 3^2 \bullet 5$

115. Factoring into primes: $820 = 10 \bullet 82 = (2 \bullet 5) \bullet (2 \bullet 41) = 2^2 \bullet 5 \bullet 41$

117. Factoring into primes: $6^3 = (2 \bullet 3)^3 = 2^3 \bullet 3^3$

119. Factoring into primes: $30^3 = (2 \bullet 3 \bullet 5)^3 = 2^3 \bullet 3^3 \bullet 5^3$

121. Factoring into primes: $25^3 = (5 \bullet 5)^3 = 5^3 \bullet 5^3 = 5^6$

123. Factoring into primes: $12^3 = (2 \bullet 2 \bullet 3)^3 = 2^3 \bullet 2^3 \bullet 3^3 = 2^6 \bullet 3^3$

125. Subtracting: $4 - 7 = 4 + (-7) = -3$

127. Subtracting: $4 - (-7) = 4 + 7 = 11$

129. Subtracting: $15 - 20 = 15 + (-20) = -5$

131. Subtracting: $-15 - (-20) = -15 + 20 = 5$

133. Simplifying: $2(3) - 4 = 6 - 4 = 2$

135. Simplifying: $4(3) - 3(2) = 12 - 6 = 6$

137. Simplifying: $2(5 - 3) = 2(2) = 4$

139. Simplifying: $5 + 4(-2) - 2(-3) = 5 - 8 + 6 = 3$

5.2 Division with Exponents

1. Writing with positive exponents: $3^{-2} = \dfrac{1}{3^2} = \dfrac{1}{9}$

3. Writing with positive exponents: $6^{-2} = \dfrac{1}{6^2} = \dfrac{1}{36}$

5. Writing with positive exponents: $8^{-2} = \dfrac{1}{8^2} = \dfrac{1}{64}$

7. Writing with positive exponents: $5^{-3} = \dfrac{1}{5^3} = \dfrac{1}{125}$

9. Writing with positive exponents: $2x^{-3} = 2 \cdot \dfrac{1}{x^3} = \dfrac{2}{x^3}$

11. Writing with positive exponents: $(2x)^{-3} = \dfrac{1}{(2x)^3} = \dfrac{1}{8x^3}$

13. Writing with positive exponents: $(5y)^{-2} = \dfrac{1}{(5y)^2} = \dfrac{1}{25y^2}$

15. Writing with positive exponents: $10^{-2} = \dfrac{1}{10^2} = \dfrac{1}{100}$

17. Completing the table:

Number (x)	Square $\left(x^2\right)$	Power of 2 $\left(2^x\right)$
-3	9	$\frac{1}{8}$
-2	4	$\frac{1}{4}$
-1	1	$\frac{1}{2}$
0	0	1
1	1	2
2	4	4
3	9	8

19. Simplifying: $\dfrac{5^1}{5^3} = 5^{1-3} = 5^{-2} = \dfrac{1}{5^2} = \dfrac{1}{25}$

21. Simplifying: $\dfrac{x^{10}}{x^4} = x^{10-4} = x^6$

23. Simplifying: $\dfrac{4^3}{4^0} = 4^{3-0} = 4^3 = 64$

25. Simplifying: $\dfrac{(2x)^7}{(2x)^4} = (2x)^{7-4} = (2x)^3 = 2^3 x^3 = 8x^3$

27. Simplifying: $\dfrac{6^{11}}{6} = \dfrac{6^{11}}{6^1} = 6^{11-1} = 6^{10} \quad (= 60,466,176)$

29. Simplifying: $\dfrac{6}{6^{11}} = \dfrac{6^1}{6^{11}} = 6^{1-11} = 6^{-10} = \dfrac{1}{6^{10}} \quad \left(= \dfrac{1}{60,466,176} \right)$

31. Simplifying: $\dfrac{2^{-5}}{2^3} = 2^{-5-3} = 2^{-8} = \dfrac{1}{2^8} = \dfrac{1}{256}$

33. Simplifying: $\dfrac{2^5}{2^{-3}} = 2^{5-(-3)} = 2^{5+3} = 2^8 = 256$

35. Simplifying: $\dfrac{(3x)^{-5}}{(3x)^{-8}} = (3x)^{-5-(-8)} = (3x)^{-5+8} = (3x)^3 = 3^3 x^3 = 27x^3$

37. Simplifying: $(3xy)^4 = 3^4 x^4 y^4 = 81x^4 y^4$

39. Simplifying: $10^0 = 1$

41. Simplifying: $\left(2a^2 b\right)^1 = 2a^2 b$

43. Simplifying: $\left(7y^3\right)^{-2} = \dfrac{1}{\left(7y^3\right)^2} = \dfrac{1}{49y^6}$

45. Simplifying: $x^{-3} \cdot x^{-5} = x^{-3-5} = x^{-8} = \dfrac{1}{x^8}$

47. Simplifying: $y^7 \cdot y^{-10} = y^{7-10} = y^{-3} = \dfrac{1}{y^3}$

49. Simplifying: $\dfrac{\left(x^2\right)^3}{x^4} = \dfrac{x^6}{x^4} = x^{6-4} = x^2$

51. Simplifying: $\dfrac{\left(a^4\right)^3}{\left(a^3\right)^2} = \dfrac{a^{12}}{a^6} = a^{12-6} = a^6$

53. Simplifying: $\dfrac{y^7}{\left(y^2\right)^8} = \dfrac{y^7}{y^{16}} = y^{7-16} = y^{-9} = \dfrac{1}{y^9}$

55. Simplifying: $\left(\dfrac{y^7}{y^2}\right)^8 = \left(y^{7-2}\right)^8 = \left(y^5\right)^8 = y^{40}$

57. Simplifying: $\dfrac{\left(x^{-2}\right)^3}{x^{-5}} = \dfrac{x^{-6}}{x^{-5}} = x^{-6-(-5)} = x^{-6+5} = x^{-1} = \dfrac{1}{x}$

59. Simplifying: $\left(\dfrac{x^{-2}}{x^{-5}}\right)^3 = \left(x^{-2+5}\right)^3 = \left(x^3\right)^3 = x^9$

61. Simplifying: $\dfrac{\left(a^3\right)^2\left(a^4\right)^5}{\left(a^5\right)^2} = \dfrac{a^6 \bullet a^{20}}{a^{10}} = \dfrac{a^{26}}{a^{10}} = a^{26-10} = a^{16}$

63. Simplifying: $\dfrac{\left(a^{-2}\right)^3\left(a^4\right)^2}{\left(a^{-3}\right)^{-2}} = \dfrac{a^{-6} \bullet a^8}{a^6} = \dfrac{a^2}{a^6} = a^{2-6} = a^{-4} = \dfrac{1}{a^4}$

65. **a.** Substituting $x = 2$: $\dfrac{x^7}{x^2} = \dfrac{2^7}{2^2} = \dfrac{128}{4} = 32$ **b.** Substituting $x = 2$: $x^5 = 2^5 = 32$

    c. Substituting $x = 2$: $\dfrac{x^2}{x^7} = \dfrac{2^2}{2^7} = \dfrac{4}{128} = \dfrac{1}{32}$ **d.** Substituting $x = 2$: $x^{-5} = 2^{-5} = \dfrac{1}{2^5} = \dfrac{1}{32}$

67. **a.** Writing as a perfect square: $\dfrac{1}{25} = \left(\dfrac{1}{5}\right)^2$ **b.** Writing as a perfect square: $\dfrac{1}{64} = \left(\dfrac{1}{8}\right)^2$

    c. Writing as a perfect square: $\dfrac{1}{x^2} = \left(\dfrac{1}{x}\right)^2$ **d.** Writing as a perfect square: $\dfrac{1}{x^4} = \left(\dfrac{1}{x^2}\right)^2$

Number (x)	Power of 2 $\left(2^x\right)$
−3	$\frac{1}{8}$
−2	$\frac{1}{4}$
−1	$\frac{1}{2}$
0	1
1	2
2	4
3	8

69.   Completing the table:

Constructing the line graph:

71.   Writing in scientific notation: $0.0048 = 4.8 \times 10^{-3}$ **73.**   Writing in scientific notation: $25 = 2.5 \times 10^1$

75.   Writing in scientific notation: $0.000009 = 9 \times 10^{-6}$

Expanded Form	Scientific Notation $\left(n \times 10^r\right)$
0.000357	3.57×10^{-4}
0.00357	3.57×10^{-3}
0.0357	3.57×10^{-2}
0.357	3.57×10^{-1}
3.57	3.57×10^{0}
35.7	3.57×10^{1}
357	3.57×10^{2}
3,570	3.57×10^{3}
35,700	3.57×10^{4}

77.   Completing the table:

79.   Writing in expanded form: $4.23 \times 10^{-3} = 0.00423$ **81.**   Writing in expanded form: $8 \times 10^{-5} = 0.00008$

83.   Writing in expanded form: $4.2 \times 10^0 = 4.2$

85.   Writing in expanded form: 2×10^{-3} seconds $= 0.002$ seconds

87.   Writing each number in scientific notation:

$0.0024 = 2.4 \times 10^{-3}$, $0.0051 = 5.1 \times 10^{-3}$, $0.0096 = 9.6 \times 10^{-3}$, $0.0099 = 9.9 \times 10^{-3}$, $0.0111 = 1.11 \times 10^{-2}$

89.   Writing in scientific notation: $25 \times 10^3 = 2.5 \times 10^4$ **91.**   Writing in scientific notation: $23.5 \times 10^4 = 2.35 \times 10^5$

93.   Writing in scientific notation: $0.82 \times 10^{-3} = 8.2 \times 10^{-4}$

95. The area of the smaller square is $(10 \text{ in.})^2 = 100 \text{ inches}^2$, while the area of the larger square is

$(20 \text{ in.})^2 = 400 \text{ inches}^2$. It would take 4 smaller squares to cover the larger square.

97. The area of the smaller square is x^2, while the area of the larger square is $(2x)^2 = 4x^2$. It would take 4 smaller squares to cover the larger square.

99. The volume of the smaller box is $(6 \text{ in.})^3 = 216 \text{ inches}^3$, while the volume of the larger box is

$(12 \text{ in.})^3 = 1,728 \text{ inches}^3$. Thus 8 smaller boxes will fit inside the larger box ($8 \cdot 216 = 1,728$).

101. The volume of the smaller box is x^3, while the volume of the larger box is $(2x)^3 = 8x^3$. Thus 8 smaller boxes will fit inside the larger box.

103. Simplifying: $4x + 3x = (4+3)x = 7x$

105. Simplifying: $5a - 3a = (5-3)a = 2a$

107. Simplifying: $4y + 5y + y = (4+5+1)y = 10y$

109. Simplifying: $3(4.5) = 13.5$

111. Simplifying: $\frac{4}{5}(10) = \frac{40}{5} = 8$

113. Simplifying: $6.8(3.9) = 26.52$

115. Simplifying: $-3 + 15 = 12$

117. Simplifying: $x^5 \cdot x^3 = x^{5+3} = x^8$

119. Simplifying: $\frac{x^3}{x^2} = x^{3-2} = x$

121. Simplifying: $\frac{y^3}{y^5} = y^{3-5} = y^{-2} = \frac{1}{y^2}$

123. Writing in expanded form: $3.4 \times 10^2 = 340$

5.3 Operations with Monomials

1. Multiplying the monomials: $(3x^4)(4x^3) = 12x^{4+3} = 12x^7$

3. Multiplying the monomials: $(-2y^4)(8y^7) = -16y^{4+7} = -16y^{11}$

5. Multiplying the monomials: $(8x)(4x) = 32x^{1+1} = 32x^2$

7. Multiplying the monomials: $(10a^3)(10a)(2a^2) = 200a^{3+1+2} = 200a^6$

9. Multiplying the monomials: $(6ab^2)(-4a^2b) = -24a^{1+2}b^{2+1} = -24a^3b^3$

11. Multiplying the monomials: $(4x^2y)(3x^3y^3)(2xy^4) = 24x^{2+3+1}y^{1+3+4} = 24x^6y^8$

13. Dividing the monomials: $\frac{15x^3}{5x^2} = \frac{15}{5} \cdot \frac{x^3}{x^2} = 3x$

15. Dividing the monomials: $\frac{18y^9}{3y^{12}} = \frac{18}{3} \cdot \frac{y^9}{y^{12}} = 6 \cdot \frac{1}{y^3} = \frac{6}{y^3}$

17. Dividing the monomials: $\frac{32a^3}{64a^4} = \frac{32}{64} \cdot \frac{a^3}{a^4} = \frac{1}{2} \cdot \frac{1}{a} = \frac{1}{2a}$

19. Dividing the monomials: $\frac{21a^2b^3}{-7ab^5} = \frac{21}{-7} \cdot \frac{a^2}{a} \cdot \frac{b^3}{b^5} = -3 \cdot a \cdot \frac{1}{b^2} = -\frac{3a}{b^2}$

21. Dividing the monomials: $\frac{3x^3y^2z}{27xy^2z^3} = \frac{3}{27} \cdot \frac{x^3}{x} \cdot \frac{y^2}{y^2} \cdot \frac{z}{z^3} = \frac{1}{9} \cdot x^2 \cdot \frac{1}{z^2} = \frac{x^2}{9z^2}$

23. Completing the table:

a	b	ab	$\dfrac{a}{b}$	$\dfrac{b}{a}$
10	$5x$	$50x$	$\dfrac{2}{x}$	$\dfrac{x}{2}$
$20x^3$	$6x^2$	$120x^5$	$\dfrac{10x}{3}$	$\dfrac{3}{10x}$
$25x^5$	$5x^4$	$125x^9$	$5x$	$\dfrac{1}{5x}$
$3x^{-2}$	$3x^2$	9	$\dfrac{1}{x^4}$	x^4
$-2y^4$	$8y^7$	$-16y^{11}$	$-\dfrac{1}{4y^3}$	$-4y^3$

25. Finding the product: $\left(3\times10^3\right)\left(2\times10^5\right)=6\times10^8$

27. Finding the product: $\left(3.5\times10^4\right)\left(5\times10^{-6}\right)=17.5\times10^{-2}=1.75\times10^{-1}$

29. Finding the product: $\left(5.5\times10^{-3}\right)\left(2.2\times10^{-4}\right)=12.1\times10^{-7}=1.21\times10^{-6}$

31. Finding the quotient: $\dfrac{8.4\times10^5}{2\times10^2}=4.2\times10^3$ **33.** Finding the quotient: $\dfrac{6\times10^8}{2\times10^{-2}}=3\times10^{10}$

35. Finding the quotient: $\dfrac{2.5\times10^{-6}}{5\times10^{-4}}=0.5\times10^{-2}=5.0\times10^{-3}$

37. Combining the monomials: $3x^2+5x^2=(3+5)x^2=8x^2$

39. Combining the monomials: $8x^5-19x^5=(8-19)x^5=-11x^5$

41. Combining the monomials: $2a+a-3a=(2+1-3)a=0a=0$

43. Combining the monomials: $10x^3-8x^3+2x^3=(10-8+2)x^3=4x^3$

45. Combining the monomials: $20ab^2-19ab^2+30ab^2=(20-19+30)ab^2=31ab^2$

47. Completing the table:

a	b	ab	$a+b$
$5x$	$3x$	$15x^2$	$8x$
$4x^2$	$2x^2$	$8x^4$	$6x^2$
$3x^3$	$6x^3$	$18x^6$	$9x^3$
$2x^4$	$-3x^4$	$-6x^8$	$-x^4$
x^5	$7x^5$	$7x^{10}$	$8x^5$

49. Simplifying the expression: $\dfrac{\left(3x^2\right)\left(8x^5\right)}{6x^4}=\dfrac{24x^7}{6x^4}=\dfrac{24}{6}\cdot\dfrac{x^7}{x^4}=4x^3$

51. Simplifying the expression: $\dfrac{\left(9a^2b\right)\left(2a^3b^4\right)}{18a^5b^7}=\dfrac{18a^5b^5}{18a^5b^7}=\dfrac{18}{18}\cdot\dfrac{a^5}{a^5}\cdot\dfrac{b^5}{b^7}=1\cdot\dfrac{1}{b^2}=\dfrac{1}{b^2}$

53. Simplifying the expression: $\dfrac{\left(4x^3y^2\right)\left(9x^4y^{10}\right)}{\left(3x^5y\right)\left(2x^6y\right)}=\dfrac{36x^7y^{12}}{6x^{11}y^2}=\dfrac{36}{6}\cdot\dfrac{x^7}{x^{11}}\cdot\dfrac{y^{12}}{y^2}=6\cdot\dfrac{1}{x^4}\cdot y^{10}=\dfrac{6y^{10}}{x^4}$

55. Simplifying the expression: $\dfrac{\left(6\times10^8\right)\left(3\times10^5\right)}{9\times10^7}=\dfrac{18\times10^{13}}{9\times10^7}=2\times10^6$

57.   Simplifying the expression: $\dfrac{\left(5\times10^3\right)\left(4\times10^{-5}\right)}{2\times10^{-2}} = \dfrac{20\times10^{-2}}{2\times10^{-2}} = 10 = 1\times10^1$

59.   Simplifying the expression: $\dfrac{\left(2.8\times10^{-7}\right)\left(3.6\times10^4\right)}{2.4\times10^3} = \dfrac{10.08\times10^{-3}}{2.4\times10^3} = 4.2\times10^{-6}$

61.   Simplifying the expression: $\dfrac{18x^4}{3x} + \dfrac{21x^7}{7x^4} = 6x^3 + 3x^3 = 9x^3$

63.   Simplifying the expression: $\dfrac{45a^6}{9a^4} - \dfrac{50a^8}{2a^6} = 5a^2 - 25a^2 = -20a^2$

65.   Simplifying the expression: $\dfrac{6x^7y^4}{3x^2y^2} + \dfrac{8x^5y^8}{2y^6} = 2x^5y^2 + 4x^5y^2 = 6x^5y^2$

67.   Applying the distributive property: $xy\left(x+\dfrac{1}{y}\right) = xy\bullet x + xy\bullet\dfrac{1}{y} = x^2y + x$

69.   Applying the distributive property: $xy\left(\dfrac{1}{y}+\dfrac{1}{x}\right) = xy\bullet\dfrac{1}{y} + xy\bullet\dfrac{1}{x} = x + y$

71.   Applying the distributive property: $x^2\left(1-\dfrac{4}{x^2}\right) = x^2\bullet1 - x^2\bullet\dfrac{4}{x^2} = x^2 - 4$

73.   Applying the distributive property: $x^2\left(1-\dfrac{1}{x}-\dfrac{6}{x^2}\right) = x^2\bullet1 - x^2\bullet\dfrac{1}{x} - x^2\bullet\dfrac{6}{x^2} = x^2 - x - 6$

75.   Applying the distributive property: $x^2\left(1-\dfrac{5}{x}\right) = x^2\bullet1 - x^2\bullet\dfrac{5}{x} = x^2 - 5x$

77.   Applying the distributive property: $x^2\left(1-\dfrac{8}{x}\right) = x^2\bullet1 - x^2\bullet\dfrac{8}{x} = x^2 - 8x$

79.   **a.**   Dividing by $5a^2$: $\dfrac{10a^2}{5a^2} = 2$ **81.**   **a.**   Dividing by $8x^2y$: $\dfrac{24x^3y^2}{8x^2y} = 3xy$

      b.   Dividing by $5a^2$: $\dfrac{-15a^2b}{5a^2} = -3b$ **b.**   Dividing by $8x^2y$: $\dfrac{16x^2y^2}{8x^2y} = 2y$

      c.   Dividing by $5a^2$: $\dfrac{25a^2b^2}{5a^2} = 5b^2$ **c.**   Dividing by $8x^2y$: $\dfrac{-4x^2y^3}{8x^2y} = -\dfrac{y^2}{2}$

83.   Evaluating when $x=-2$: $4x = 4(-2) = -8$
85.   Evaluating when $x=-2$: $-2x+5 = -2(-2)+5 = 4+5 = 9$
87.   Evaluating when $x=-2$: $x^2+5x+6 = (-2)^2+5(-2)+6 = 4-10+6 = 0$

89.   The ordered pairs are $(-2,-2)$, $(0,2)$, and $(2,6)$:

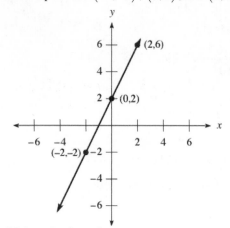

91.   The ordered pairs are $(-3,0)$, $(0,1)$, and $(3,2)$:

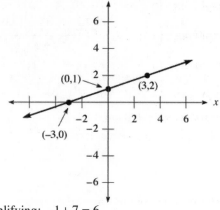

93.   Simplifying: $3-8=-5$

95.   Simplifying: $-1+7=6$

97.   Simplifying: $3(5)^2+1=3(25)+1=75+1=76$

99.   Simplifying: $2x^2+4x^2=6x^2$

101.   Simplifying: $-5x+7x=2x$

103.   Simplifying: $-(2x+9)=-2x-9$

105.   Substituting $x=4$: $2x+3=2(4)+3=8+3=11$

5.4 Addition and Subtraction of Polynomials

1.   This is a trinomial of degree 3.

3.   This is a trinomial of degree 3.

5.   This is a binomial of degree 1.

7.   This is a binomial of degree 2.

9.   This is a monomial of degree 2.

11.   This is a monomial of degree 0.

13.   Combining the polynomials: $(2x^2+3x+4)+(3x^2+2x+5)=(2x^2+3x^2)+(3x+2x)+(4+5)=5x^2+5x+9$

15.   Combining the polynomials: $(3a^2-4a+1)+(2a^2-5a+6)=(3a^2+2a^2)+(-4a-5a)+(1+6)=5a^2-9a+7$

17.   Combining the polynomials: $x^2+4x+2x+8=x^2+(4x+2x)+8=x^2+6x+8$

19.   Combining the polynomials: $6x^2-3x-10x+5=6x^2+(-3x-10x)+5=6x^2-13x+5$

21.   Combining the polynomials: $x^2-3x+3x-9=x^2+(-3x+3x)-9=x^2-9$

23.   Combining the polynomials: $3y^2-5y-6y+10=3y^2+(-5y-6y)+10=3y^2-11y+10$

25.   Combining the polynomials:
$$(6x^3-4x^2+2x)+(9x^2-6x+3)=6x^3+(-4x^2+9x^2)+(2x-6x)+3=6x^3+5x^2-4x+3$$

27.   Combining the polynomials:
$$\left(\tfrac{2}{3}x^2-\tfrac{1}{5}x-\tfrac{3}{4}\right)+\left(\tfrac{4}{3}x^2-\tfrac{4}{5}x+\tfrac{7}{4}\right)=\left(\tfrac{2}{3}x^2+\tfrac{4}{3}x^2\right)+\left(-\tfrac{1}{5}x-\tfrac{4}{5}x\right)+\left(-\tfrac{3}{4}+\tfrac{7}{4}\right)=2x^2-x+1$$

29.   Combining the polynomials: $(a^2-a-1)-(-a^2+a+1)=a^2-a-1+a^2-a-1=2a^2-2a-2$

31.   Combining the polynomials:
$$\left(\tfrac{5}{9}x^3+\tfrac{1}{3}x^2-2x+1\right)-\left(\tfrac{2}{3}x^3+x^2+\tfrac{1}{2}x-\tfrac{3}{4}\right)=\tfrac{5}{9}x^3+\tfrac{1}{3}x^2-2x+1-\tfrac{2}{3}x^3-x^2-\tfrac{1}{2}x+\tfrac{3}{4}$$
$$=-\tfrac{1}{9}x^3-\tfrac{2}{3}x^2-\tfrac{5}{2}x+\tfrac{7}{4}$$

33.   Combining the polynomials:
$$(4y^2-3y+2)+(5y^2+12y-4)-(13y^2-6y+20)=4y^2-3y+2+5y^2+12y-4-13y^2+6y-20$$
$$=(4y^2+5y^2-13y^2)+(-3y+12y+6y)+(2-4-20)$$
$$=-4y^2+15y-22$$

35.   Simplifying: $\left(x^2 - 5x\right) - \left(x^2 - 3x\right) = x^2 - 5x - x^2 + 3x = -2x$

37.   Simplifying: $\left(6x^2 - 11x\right) - \left(6x^2 - 15x\right) = 6x^2 - 11x - 6x^2 + 15x = 4x$

39.   Simplifying: $\left(x^3 + 3x^2 + 9x\right) - \left(3x^2 + 9x + 27\right) = x^3 + 3x^2 + 9x - 3x^2 - 9x - 27 = x^3 - 27$

41.   Simplifying: $\left(x^3 + 4x^2 + 4x\right) + \left(2x^2 + 8x + 8\right) = x^3 + 4x^2 + 4x + 2x^2 + 8x + 8 = x^3 + 6x^2 + 12x + 8$

43.   Simplifying: $\left(x^2 - 4\right) - \left(x^2 - 4x + 4\right) = x^2 - 4 - x^2 + 4x - 4 = 4x - 8$

45.   Performing the subtraction:
$$\left(11x^2 - 10x + 13\right) - \left(10x^2 + 23x - 50\right) = 11x^2 - 10x + 13 - 10x^2 - 23x + 50$$
$$= \left(11x^2 - 10x^2\right) + \left(-10x - 23x\right) + \left(13 + 50\right)$$
$$= x^2 - 33x + 63$$

47.   Performing the subtraction:
$$\left(11y^2 + 11y + 11\right) - \left(3y^2 + 7y - 15\right) = 11y^2 + 11y + 11 - 3y^2 - 7y + 15$$
$$= \left(11y^2 - 3y^2\right) + \left(11y - 7y\right) + \left(11 + 15\right)$$
$$= 8y^2 + 4y + 26$$

49.   Performing the addition:
$$\left(25x^2 - 50x + 75\right) + \left(50x^2 - 100x - 150\right) = \left(25x^2 + 50x^2\right) + \left(-50x - 100x\right) + \left(75 - 150\right) = 75x^2 - 150x - 75$$

51.   Performing the operations:
$$\left(3x - 2\right) + \left(11x + 5\right) - \left(2x + 1\right) = 3x - 2 + 11x + 5 - 2x - 1 = \left(3x + 11x - 2x\right) + \left(-2 + 5 - 1\right) = 12x + 2$$

53.   Evaluating when $x = 3$: $x^2 - 2x + 1 = (3)^2 - 2(3) + 1 = 9 - 6 + 1 = 4$

55.   **a.**   Substituting $p = 5$: $100p^2 - 1,300p + 4,000 = 100(5)^2 - 1,300(5) + 4,000 = 2,500 - 6,500 + 4,000 = 0$

   b.   Substituting $p = 8$: $100p^2 - 1,300p + 4,000 = 100(8)^2 - 1,300(8) + 4,000 = 6,400 - 10,400 + 4,000 = 0$

57.   **a.**   Substituting $x = 8$: $600 + 1,000x - 100x^2 = 600 + 1,000(8) - 100(8)^2 = 600 + 8,000 - 6,400 = 2,200$

   b.   Substituting $x = -2$: $600 + 1,000x - 100x^2 = 600 + 1,000(-2) - 100(-2)^2 = 600 - 2,000 - 400 = -1,800$

59.   Finding the volume of the cylinder and sphere:
$$V_{\text{cylinder}} = \pi\left(3^2\right)(6) = 54\pi \qquad\qquad V_{\text{sphere}} = \tfrac{4}{3}\pi\left(3^3\right) = 36\pi$$

   Subtracting to find the amount of space to pack: $V = 54\pi - 36\pi = 18\pi$ inches^3

61.   Multiplying: $3x(-5x) = -15x^2$

63.   Multiplying: $2x\left(3x^2\right) = 6x^{1+2} = 6x^3$

65.   Multiplying: $3x^2\left(2x^2\right) = 6x^{2+2} = 6x^4$

67.   Simplifying: $(-5)(-1) = 5$

69.   Simplifying: $(-1)(6) = -6$

71.   Simplifying: $(5x)(-4x) = -20x^2$

73.   Simplifying: $3x(-7) = -21x$

75.   Simplifying: $5x + (-3x) = 2x$

77.   Multiplying: $3(2x - 6) = 6x - 18$

5.5 Multiplication with Polynomials

1. Using the distributive property: $2x(3x+1) = 2x(3x) + 2x(1) = 6x^2 + 2x$

3. Using the distributive property: $2x^2\left(3x^2 - 2x + 1\right) = 2x^2\left(3x^2\right) - 2x^2\left(2x\right) + 2x^2\left(1\right) = 6x^4 - 4x^3 + 2x^2$

5. Using the distributive property: $2ab\left(a^2 - ab + 1\right) = 2ab\left(a^2\right) - 2ab\left(ab\right) + 2ab\left(1\right) = 2a^3b - 2a^2b^2 + 2ab$

7. Using the distributive property: $y^2\left(3y^2 + 9y + 12\right) = y^2\left(3y^2\right) + y^2\left(9y\right) + y^2\left(12\right) = 3y^4 + 9y^3 + 12y^2$

9. Using the distributive property:
$$4x^2y\left(2x^3y + 3x^2y^2 + 8y^3\right) = 4x^2y\left(2x^3y\right) + 4x^2y\left(3x^2y^2\right) + 4x^2y\left(8y^3\right) = 8x^5y^2 + 12x^4y^3 + 32x^2y^4$$

11. Multiplying using the FOIL method: $(x+3)(x+4) = x^2 + 3x + 4x + 12 = x^2 + 7x + 12$

13. Multiplying using the FOIL method: $(x+6)(x+1) = x^2 + 6x + 1x + 6 = x^2 + 7x + 6$

15. Multiplying using the FOIL method: $\left(x+\frac{1}{2}\right)\left(x+\frac{3}{2}\right) = x^2 + \frac{1}{2}x + \frac{3}{2}x + \frac{3}{4} = x^2 + 2x + \frac{3}{4}$

17. Multiplying using the FOIL method: $(a+5)(a-3) = a^2 + 5a - 3a - 15 = a^2 + 2a - 15$

19. Multiplying using the FOIL method: $(x-a)(y+b) = xy + bx - ay - ab$

21. Multiplying using the FOIL method: $(x+6)(x-6) = x^2 + 6x - 6x - 36 = x^2 - 36$

23. Multiplying using the FOIL method: $\left(y+\frac{5}{6}\right)\left(y-\frac{5}{6}\right) = y^2 + \frac{5}{6}y - \frac{5}{6}y - \frac{25}{36} = y^2 - \frac{25}{36}$

25. Multiplying using the FOIL method: $(2x-3)(x-4) = 2x^2 - 3x - 8x + 12 = 2x^2 - 11x + 12$

27. Multiplying using the FOIL method: $(a+2)(2a-1) = 2a^2 + 4a - a - 2 = 2a^2 + 3a - 2$

29. Multiplying using the FOIL method: $(2x-5)(3x-2) = 6x^2 - 15x - 4x + 10 = 6x^2 - 19x + 10$

31. Multiplying using the FOIL method: $(2x+3)(a+4) = 2ax + 3a + 8x + 12$

33. Multiplying using the FOIL method: $(5x-4)(5x+4) = 25x^2 - 20x + 20x - 16 = 25x^2 - 16$

35. Multiplying using the FOIL method: $\left(2x-\frac{1}{2}\right)\left(x+\frac{3}{2}\right) = 2x^2 - \frac{1}{2}x + 3x - \frac{3}{4} = 2x^2 + \frac{5}{2}x - \frac{3}{4}$

37. Multiplying using the FOIL method: $(1-2a)(3-4a) = 3 - 6a - 4a + 8a^2 = 3 - 10a + 8a^2$

39. The product is $(x+2)(x+3) = x^2 + 5x + 6$:

	x	3
x	x^2	$3x$
2	$2x$	6

41. The product is $(x+1)(2x+2) = 2x^2 + 4x + 2$:

	x	x	2
x	x^2	x^2	$2x$
1	x	x	2

43. Multiplying using the column method:

$$
\begin{array}{r}
a^2 - 3a + 2 \\
a - 3 \\
\hline
a^3 - 3a^2 + 2a \\
-3a^2 + 9a - 6 \\
\hline
a^3 - 6a^2 + 11a - 6
\end{array}
$$

45. Multiplying using the column method:

$$
\begin{array}{r}
x^2 - 2x + 4 \\
x + 2 \\
\hline
x^3 - 2x^2 + 4x \\
2x^2 - 4x + 8 \\
\hline
x^3 + 8
\end{array}
$$

47.   Multiplying using the column method:

$$\begin{array}{r} x^2 \ +8x\ +9 \\ 2x\ +1 \\ \hline 2x^3+16x^2+18x \\ x^2+8x+9 \\ \hline 2x^3+17x^2+26x+9 \end{array}$$

49.   Multiplying using the column method:

$$\begin{array}{r} 5x^2+2x+1 \\ x^2-3x+5 \\ \hline 5x^4+2x^3+\ x^2 \\ -15x^3-6x^2-3x \\ 25x^2+10x+5 \\ \hline 5x^4-13x^3+20x^2+7x+5 \end{array}$$

51.   Multiplying using the FOIL method: $\left(x^2+3\right)\left(2x^2-5\right)=2x^4-5x^2+6x^2-15=2x^4+x^2-15$

53.   Multiplying using the FOIL method: $\left(3a^4+2\right)\left(2a^2+5\right)=6a^6+15a^4+4a^2+10$

55.   First multiply two polynomials using the FOIL method: $(x+3)(x+4)=x^2+3x+4x+12=x^2+7x+12$

   Now using the column method:

$$\begin{array}{r} x^2+7x\ +12 \\ x\ +5 \\ \hline x^3+7x^2+12x \\ 5x^2+35x+60 \\ \hline x^3+12x^2+47x+60 \end{array}$$

57.   Simplifying: $(x-3)(x-2)+2=x^2-3x-2x+6+2=x^2-5x+8$

59.   Simplifying: $(2x-3)(4x+3)+4=8x^2+6x-12x-9+4=8x^2-6x-5$

61.   Simplifying: $(x+4)(x-5)+(-5)(2)=x^2-5x+4x-20-10=x^2-x-30$

63.   Simplifying: $2(x-3)+x(x+2)=2x-6+x^2+2x=x^2+4x-6$

65.   Simplifying: $3x(x+1)-2x(x-5)=3x^2+3x-2x^2+10x=x^2+13x$

67.   Simplifying: $x(x+2)-3=x^2+2x-3$ **69.**   Simplifying: $a(a-3)+6=a^2-3a+6$

71.   **a.**   Finding the product: $(x+1)(x-1)=x^2-x+x-1=x^2-1$

   b.   Finding the product: $(x+1)(x+1)=x^2+x+x+1=x^2+2x+1$

   c.   Finding the product: $(x+1)\left(x^2+2x+1\right)=x^3+2x^2+x+x^2+2x+1=x^3+3x^2+3x+1$

   d.   Finding the product:

$$(x+1)\left(x^3+3x^2+3x+1\right)=x^4+3x^3+3x^2+x+x^3+3x^2+3x+1=x^4+4x^3+6x^2+4x+1$$

73.   **a.**   Finding the product: $(x+1)(x-1)=x^2-x+x-1=x^2-1$

   b.   Finding the product: $(x+1)(x-2)=x^2-2x+x-2=x^2-x-2$

   c.   Finding the product: $(x+1)(x-3)=x^2-3x+x-3=x^2-2x-3$

   d.   Finding the product: $(x+1)(x-4)=x^2-4x+x-4=x^2-3x-4$

75.   The expression is equal to 0 if $x=0$.   **77.**   The expression is equal to 0 if $x=-5$.

79.   The expression is equal to 0 if $x=3$ or $x=-2$.   **81.**   The expression is never equal to 0.

83.   The expression is equal to 0 if $x=a$ or $x=b$.

85.   Let x represent the width and $2x+5$ represent the length. The area is given by: $A=x(2x+5)=2x^2+5x$

87.   Let x and $x+1$ represent the width and length, respectively. The area is given by: $A=x(x+1)=x^2+x$

89.   The revenue is: $R=xp=(1,200-100p)p=1,200p-100p^2$

91. Graphing each line:

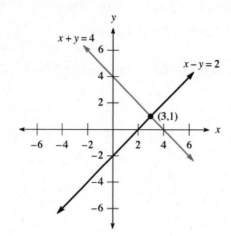

The intersection point is $(3,1)$.

93. Multiplying the second equation by –2:
$$3x + 2y = 1$$
$$-4x - 2y = -6$$
Adding the two equations:
$$-x = -5$$
$$x = 5$$
Substituting into the second equation:
$$2(5) + y = 3$$
$$10 + y = 3$$
$$y = -7$$
The solution is $(5, -7)$.

95. Substituting into the first equation:
$$x + 5x + 2 = 20$$
$$6x + 2 = 20$$
$$6x = 18$$
$$x = 3$$
Substituting into the second equation: $y = 5(3) + 2 = 15 + 2 = 17$. The solution is $(3, 17)$.

97. Let x represent the amount invested at 8% and y represent the amount invested at 10%. The system of equations is:
$$x + y = 1200$$
$$0.08x + 0.10y = 104$$
Multiplying the first equation by –0.08:
$$-0.08x - 0.08y = -96$$
$$0.08x + 0.10y = 104$$
Adding the two equations:
$$0.02y = 8$$
$$y = 400$$
Substituting into the first equation:
$$x + 400 = 1200$$
$$x = 800$$
So $800 is invested at 8% and $400 is invested at 10%.

99. Simplifying: $13 \bullet 13 = 169$

101. Simplifying: $2(x)(-5) = -10x$

103. Simplifying: $6x + (-6x) = 0$

105. Simplifying: $(2x)(-3) + (2x)(3) = -6x + 6x = 0$

107. Multiplying: $-4(3x - 4) = -12x + 16$

109. Multiplying: $(x-1)(x+2) = x^2 + 2x - x - 2 = x^2 + x - 2$

111. Multiplying: $(x+3)(x+3) = x^2 + 3x + 3x + 9 = x^2 + 6x + 9$

5.6 Binomial Squares and Other Special Products

1. Multiplying using the FOIL method: $(x-2)^2 = (x-2)(x-2) = x^2 - 2x - 2x + 4 = x^2 - 4x + 4$

3. Multiplying using the FOIL method: $(a+3)^2 = (a+3)(a+3) = a^2 + 3a + 3a + 9 = a^2 + 6a + 9$

5. Multiplying using the FOIL method: $(x-5)^2 = (x-5)(x-5) = x^2 - 5x - 5x + 25 = x^2 - 10x + 25$

7. Multiplying using the FOIL method: $\left(a - \frac{1}{2}\right)^2 = \left(a - \frac{1}{2}\right)\left(a - \frac{1}{2}\right) = a^2 - \frac{1}{2}a - \frac{1}{2}a + \frac{1}{4} = a^2 - a + \frac{1}{4}$

9. Multiplying using the FOIL method: $(x+10)^2 = (x+10)(x+10) = x^2 + 10x + 10x + 100 = x^2 + 20x + 100$

11. Multiplying using the square of binomial formula: $(a+0.8)^2 = a^2 + 2(a)(0.8) + (0.8)^2 = a^2 + 1.6a + 0.64$

13. Multiplying using the square of binomial formula: $(2x-1)^2 = (2x)^2 - 2(2x)(1) + (1)^2 = 4x^2 - 4x + 1$

15. Multiplying using the square of binomial formula: $(4a+5)^2 = (4a)^2 + 2(4a)(5) + (5)^2 = 16a^2 + 40a + 25$

17. Multiplying using the square of binomial formula: $(3x-2)^2 = (3x)^2 - 2(3x)(2) + (2)^2 = 9x^2 - 12x + 4$

19. Multiplying using the square of binomial formula: $(3a+5b)^2 = (3a)^2 + 2(3a)(5b) + (5b)^2 = 9a^2 + 30ab + 25b^2$

21. Multiplying using the square of binomial formula: $(4x-5y)^2 = (4x)^2 - 2(4x)(5y) + (5y)^2 = 16x^2 - 40xy + 25y^2$

23. Multiplying using the square of binomial formula: $(7m+2n)^2 = (7m)^2 + 2(7m)(2n) + (2n)^2 = 49m^2 + 28mn + 4n^2$

25. Multiplying using the square of binomial formula:

$$(6x-10y)^2 = (6x)^2 - 2(6x)(10y) + (10y)^2 = 36x^2 - 120xy + 100y^2$$

27. Multiplying using the square of binomial formula: $\left(x^2+5\right)^2 = \left(x^2\right)^2 + 2\left(x^2\right)(5) + (5)^2 = x^4 + 10x^2 + 25$

29. Multiplying using the square of binomial formula: $\left(a^2+1\right)^2 = \left(a^2\right)^2 + 2\left(a^2\right)(1) + (1)^2 = a^4 + 2a^2 + 1$

31. Multiplying using the square of binomial formula: $\left(y+\frac{3}{2}\right)^2 = (y)^2 + 2(y)\left(\frac{3}{2}\right) + \left(\frac{3}{2}\right)^2 = y^2 + 3y + \frac{9}{4}$

33. Multiplying using the square of binomial formula: $\left(a+\frac{1}{2}\right)^2 = (a)^2 + 2(a)\left(\frac{1}{2}\right) + \left(\frac{1}{2}\right)^2 = a^2 + a + \frac{1}{4}$

35. Multiplying using the square of binomial formula: $\left(x+\frac{3}{4}\right)^2 = (x)^2 + 2(x)\left(\frac{3}{4}\right) + \left(\frac{3}{4}\right)^2 = x^2 + \frac{3}{2}x + \frac{9}{16}$

37. Multiplying using the square of binomial formula: $\left(t+\frac{1}{5}\right)^2 = (t)^2 + 2(t)\left(\frac{1}{5}\right) + \left(\frac{1}{5}\right)^2 = t^2 + \frac{2}{5}t + \frac{1}{25}$

39. Completing the table:

x	$(x+3)^2$	x^2+9	x^2+6x+9
1	16	10	16
2	25	13	25
3	36	18	36
4	49	25	49

41. Completing the table:

a	b	$(a+b)^2$	a^2+b^2	a^2+ab+b^2	$a^2+2ab+b^2$
1	1	4	2	3	4
3	5	64	34	49	64
3	4	49	25	37	49
4	5	81	41	61	81

43. Multiplying using the FOIL method: $(a+5)(a-5) = a^2 + 5a - 5a - 25 = a^2 - 25$

45. Multiplying using the FOIL method: $(y-1)(y+1) = y^2 - y + y - 1 = y^2 - 1$

47. Multiplying using the difference of squares formula: $(9+x)(9-x) = (9)^2 - (x)^2 = 81 - x^2$

49. Multiplying using the difference of squares formula: $(2x+5)(2x-5) = (2x)^2 - (5)^2 = 4x^2 - 25$

51. Multiplying using the difference of squares formula: $\left(4x+\frac{1}{3}\right)\left(4x-\frac{1}{3}\right) = (4x)^2 - \left(\frac{1}{3}\right)^2 = 16x^2 - \frac{1}{9}$

53. Multiplying using the difference of squares formula: $(2a+7)(2a-7) = (2a)^2 - (7)^2 = 4a^2 - 49$

55. Multiplying using the difference of squares formula: $(6-7x)(6+7x) = (6)^2 - (7x)^2 = 36 - 49x^2$

57. Multiplying using the difference of squares formula: $(x^2+3)(x^2-3) = (x^2)^2 - (3)^2 = x^4 - 9$

59. Multiplying using the difference of squares formula: $(a^2+4)(a^2-4) = (a^2)^2 - (4)^2 = a^4 - 16$

61. Multiplying using the difference of squares formula: $(5y^4-8)(5y^4+8) = (5y^4)^2 - (8)^2 = 25y^8 - 64$

63. Multiplying and simplifying: $(x+3)(x-3) + (x+5)(x-5) = (x^2-9) + (x^2-25) = 2x^2 - 34$

65. Multiplying and simplifying:
$$(2x+3)^2 - (4x-1)^2 = (4x^2+12x+9) - (16x^2-8x+1) = 4x^2 + 12x + 9 - 16x^2 + 8x - 1 = -12x^2 + 20x + 8$$

67. Multiplying and simplifying:
$$(a+1)^2 - (a+2)^2 + (a+3)^2 = (a^2+2a+1) - (a^2+4a+4) + (a^2+6a+9)$$
$$= a^2 + 2a + 1 - a^2 - 4a - 4 + a^2 + 6a + 9$$
$$= a^2 + 4a + 6$$

69. Multiplying and simplifying:
$$(2x+3)^3 = (2x+3)(2x+3)^2$$
$$= (2x+3)(4x^2+12x+9)$$
$$= 8x^3 + 24x^2 + 18x + 12x^2 + 36x + 27$$
$$= 8x^3 + 36x^2 + 54x + 27$$

71. a.   Evaluating when $x = 6$: $x^2 - 25 = (6)^2 - 25 = 36 - 25 = 11$

   b.   Evaluating when $x = 6$: $(x-5)^2 = (6-5)^2 = (1)^2 = 1$

   c.   Evaluating when $x = 6$: $(x+5)(x-5) = (6+5)(6-5) = (11)(1) = 11$

73. a.   Evaluating when $x = -2$: $(x+3)^2 = (-2+3)^2 = (1)^2 = 1$

   b.   Evaluating when $x = -2$: $x^2 + 9 = (-2)^2 + 9 = 4 + 9 = 13$

   c.   Evaluating when $x = -2$: $x^2 + 6x + 9 = (-2)^2 + 6(-2) + 9 = 4 - 12 + 9 = 1$

75. Let x and $x + 1$ represent the two integers. The expression can be written as:
$$(x)^2 + (x+1)^2 = x^2 + (x^2+2x+1) = 2x^2 + 2x + 1$$

77. Let x, $x + 1$, and $x + 2$ represent the three integers. The expression can be written as:
$$(x)^2 + (x+1)^2 + (x+2)^2 = x^2 + (x^2 + 2x + 1) + (x^2 + 4x + 4) = 3x^2 + 6x + 5$$

79. Verifying the areas: $(a+b)^2 = a^2 + ab + ab + b^2 = a^2 + 2ab + b^2$

81. Graphing both lines:

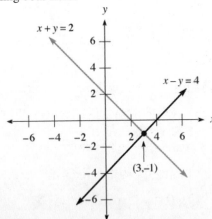

The intersection point is $(3, -1)$.

83. Graphing both lines:

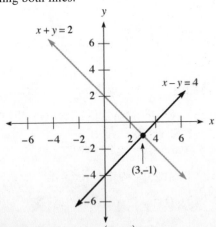

The intersection point is $(-1, 1)$.

85. Simplifying: $\dfrac{10x^3}{5x} = 2x^{3-1} = 2x^2$

87. Simplifying: $\dfrac{3x^2}{3} = x^2$

89. Simplifying: $\dfrac{9x^2}{3x} = 3x^{2-1} = 3x$

91. Simplifying: $\dfrac{24x^3y^2}{8x^2y} = 3x^{3-2}y^{2-1} = 3xy$

93. Simplifying: $\dfrac{15x^2y}{3xy} = \dfrac{15}{3} \cdot \dfrac{x^2}{x} \cdot \dfrac{y}{y} = 5x$

95. Simplifying: $\dfrac{35a^6b^8}{70a^2b^{10}} = \dfrac{35}{70} \cdot \dfrac{a^6}{a^2} \cdot \dfrac{b^8}{b^{10}} = \dfrac{1}{2} \cdot a^4 \cdot \dfrac{1}{b^2} = \dfrac{a^4}{2b^2}$

5.7 Dividing a Polynomial by a Monomial

1. Performing the division: $\dfrac{5x^2 - 10x}{5x} = \dfrac{5x^2}{5x} - \dfrac{10x}{5x} = x - 2$

3. Performing the division: $\dfrac{15x - 10x^3}{5x} = \dfrac{15x}{5x} - \dfrac{10x^3}{5x} = 3 - 2x^2$

5. Performing the division: $\dfrac{25x^2y - 10xy}{5x} = \dfrac{25x^2y}{5x} - \dfrac{10xy}{5x} = 5xy - 2y$

7. Performing the division: $\dfrac{35x^5 - 30x^4 + 25x^3}{5x} = \dfrac{35x^5}{5x} - \dfrac{30x^4}{5x} + \dfrac{25x^3}{5x} = 7x^4 - 6x^3 + 5x^2$

9. Performing the division: $\dfrac{50x^5 - 25x^3 + 5x}{5x} = \dfrac{50x^5}{5x} - \dfrac{25x^3}{5x} + \dfrac{5x}{5x} = 10x^4 - 5x^2 + 1$

11. Performing the division: $\dfrac{8a^2 - 4a}{-2a} = \dfrac{8a^2}{-2a} + \dfrac{-4a}{-2a} = -4a + 2$

13. Performing the division: $\dfrac{16a^5 + 24a^4}{-2a} = \dfrac{16a^5}{-2a} + \dfrac{24a^4}{-2a} = -8a^4 - 12a^3$

15. Performing the division: $\dfrac{8ab + 10a^2}{-2a} = \dfrac{8ab}{-2a} + \dfrac{10a^2}{-2a} = -4b - 5a$

17. Performing the division: $\dfrac{12a^3b - 6a^2b^2 + 14ab^3}{-2a} = \dfrac{12a^3b}{-2a} + \dfrac{-6a^2b^2}{-2a} + \dfrac{14ab^3}{-2a} = -6a^2b + 3ab^2 - 7b^3$

19. Performing the division: $\dfrac{a^2 + 2ab + b^2}{-2a} = \dfrac{a^2}{-2a} + \dfrac{2ab}{-2a} + \dfrac{b^2}{-2a} = -\dfrac{a}{2} - b - \dfrac{b^2}{2a}$

21. Performing the division: $\dfrac{6x + 8y}{2} = \dfrac{6x}{2} + \dfrac{8y}{2} = 3x + 4y$

23. Performing the division: $\dfrac{7y - 21}{-7} = \dfrac{7y}{-7} + \dfrac{-21}{-7} = -y + 3$

25. Performing the division: $\dfrac{2x^2 + 16x - 18}{2} = \dfrac{2x^2}{2} + \dfrac{16x}{2} - \dfrac{18}{2} = x^2 + 8x - 9$

27. Performing the division: $\dfrac{3y^2 - 9y + 3}{3} = \dfrac{3y^2}{3} - \dfrac{9y}{3} + \dfrac{3}{3} = y^2 - 3y + 1$

29. Performing the division: $\dfrac{x^2y - x^3y^2}{-x^2y} = \dfrac{x^2y}{-x^2y} + \dfrac{-x^3y^2}{-x^2y} = -1 + xy$

31. Performing the division: $\dfrac{a^2b^2 - ab^2}{-ab^2} = \dfrac{a^2b^2}{-ab^2} + \dfrac{-ab^2}{-ab^2} = -a + 1$

33. Performing the division: $\dfrac{x^3 - 3x^2y + xy^2}{x} = \dfrac{x^3}{x} - \dfrac{3x^2y}{x} + \dfrac{xy^2}{x} = x^2 - 3xy + y^2$

35. Performing the division: $\dfrac{10a^2 - 15a^2b + 25a^2b^2}{5a^2} = \dfrac{10a^2}{5a^2} - \dfrac{15a^2b}{5a^2} + \dfrac{25a^2b^2}{5a^2} = 2 - 3b + 5b^2$

37. Performing the division: $\dfrac{26x^2y^2 - 13xy}{-13xy} = \dfrac{26x^2y^2}{-13xy} + \dfrac{-13xy}{-13xy} = -2xy + 1$

39. Performing the division: $\dfrac{4x^2y^2 - 2xy}{4xy} = \dfrac{4x^2y^2}{4xy} - \dfrac{2xy}{4xy} = xy - \dfrac{1}{2}$

41. Performing the division: $\dfrac{5a^2x - 10ax^2 + 15a^2x^2}{20a^2x^2} = \dfrac{5a^2x}{20a^2x^2} - \dfrac{10ax^2}{20a^2x^2} + \dfrac{15a^2x^2}{20a^2x^2} = \dfrac{1}{4x} - \dfrac{1}{2a} + \dfrac{3}{4}$

43. Performing the division: $\dfrac{16x^5 + 8x^2 + 12x}{12x^3} = \dfrac{16x^5}{12x^3} + \dfrac{8x^2}{12x^3} + \dfrac{12x}{12x^3} = \dfrac{4x^2}{3} + \dfrac{2}{3x} + \dfrac{1}{x^2}$

45. Performing the division: $\dfrac{9a^{5m} - 27a^{3m}}{3a^{2m}} = \dfrac{9a^{5m}}{3a^{2m}} - \dfrac{27a^{3m}}{3a^{2m}} = 3a^{5m-2m} - 9a^{3m-2m} = 3a^{3m} - 9a^m$

47. Performing the division:
$$\dfrac{10x^{5m} - 25x^{3m} + 35x^m}{5x^m} = \dfrac{10x^{5m}}{5x^m} - \dfrac{25x^{3m}}{5x^m} + \dfrac{35x^m}{5x^m} = 2x^{5m-m} - 5x^{3m-m} + 7x^{m-m} = 2x^{4m} - 5x^{2m} + 7$$

49. Simplifying and then dividing:
$$\dfrac{2x^3(3x+2) - 3x^2(2x-4)}{2x^2} = \dfrac{6x^4 + 4x^3 - 6x^3 + 12x^2}{2x^2}$$
$$= \dfrac{6x^4 - 2x^3 + 12x^2}{2x^2}$$
$$= \dfrac{6x^4}{2x^2} - \dfrac{2x^3}{2x^2} + \dfrac{12x^2}{2x^2}$$
$$= 3x^2 - x + 6$$

51. Simplifying and then dividing:
$$\dfrac{(x+2)^2 - (x-2)^2}{2x} = \dfrac{\left(x^2 + 4x + 4\right) - \left(x^2 - 4x + 4\right)}{2x} = \dfrac{x^2 + 4x + 4 - x^2 + 4x - 4}{2x} = \dfrac{8x}{2x} = 4$$

53. Simplifying and then dividing:
$$\frac{(x+5)^2+(x+5)(x-5)}{2x}=\frac{\left(x^2+10x+25\right)+\left(x^2-25\right)}{2x}=\frac{2x^2+10x}{2x}=\frac{2x^2}{2x}+\frac{10x}{2x}=x+5$$

55. **a.** Evaluating when $x=2$: $2x+3=2(2)+3=4+3=7$

   b. Evaluating when $x=2$: $\dfrac{10x+15}{5}=\dfrac{10(2)+15}{5}=\dfrac{20+15}{5}=\dfrac{35}{5}=7$

   c. Evaluating when $x=2$: $10x+3=10(2)+3=20+3=23$

57. **a.** Evaluating when $x=10$: $\dfrac{3x+8}{2}=\dfrac{3(10)+8}{2}=\dfrac{30+8}{2}=\dfrac{38}{2}=19$

   b. Evaluating when $x=10$: $3x+4=3(10)+4=30+4=34$

   c. Evaluating when $x=10$: $\frac{3}{2}x+4=\frac{3}{2}(10)+4=15+4=19$

59. **a.** Evaluating when $x=2$: $2x^2-3x=2(2)^2-3(2)=2(4)-3(2)=8-6=2$

   b. Evaluating when $x=2$: $\dfrac{10x^3-15x^2}{5x}=\dfrac{10(2)^3-15(2)^2}{5(2)}=\dfrac{10(8)-15(4)}{5(2)}=\dfrac{80-60}{10}=\dfrac{20}{10}=2$

61. Adding the two equations:
$$2x=14$$
$$x=7$$
Substituting into the first equation:
$$7+y=6$$
$$y=-1$$
The solution is $(7,-1)$.

63. Multiplying the second equation by 3:
$$2x-3y=-5$$
$$3x+3y=15$$
Adding the two equations:
$$5x=10$$
$$x=2$$
Substituting into the second equation:
$$2+y=5$$
$$y=3$$
The solution is $(2,3)$.

65. Substituting into the first equation:
$$x+2x-1=2$$
$$3x-1=2$$
$$3x=3$$
$$x=1$$
Substituting into the second equation: $y=2(1)-1=2-1=1$. The solution is $(1,1)$.

67. Substituting into the first equation:
$$4x+2(-2x+4)=8$$
$$4x-4x+8=8$$
$$8=8$$
Since this statement is true, the system is dependent. The two lines coincide.

69. Dividing:

$$
\begin{array}{r}
146 \\
27\overline{)3962} \\
\underline{27} \\
126 \\
\underline{108} \\
182 \\
\underline{162} \\
20
\end{array}
$$

The quotient is $146\frac{20}{27}$.

71. Dividing: $\dfrac{2x^2+5x}{x}=\dfrac{2x^2}{x}+\dfrac{5x}{x}=2x+5$

73. Multiplying: $(x-3)x=x^2-3x$

75. Multiplying: $2x^2(x-5)=2x^3-10x^2$

77. Subtracting: $(x^2-5x)-(x^2-3x)=x^2-5x-x^2+3x=-2x$

79. Subtracting: $(-2x+8)-(-2x+6)=-2x+8+2x-6=2$

5.8 Dividing a Polynomial by a Polynomial

1. Using long division:

$$
\begin{array}{r}
x-2 \\
x-3\overline{)x^2-5x+6} \\
\underline{x^2-3x} \\
-2x+6 \\
\underline{-2x+6} \\
0
\end{array}
$$

The quotient is $x-2$.

3. Using long division:

$$
\begin{array}{r}
a+4 \\
a+5\overline{)a^2+9a+20} \\
\underline{a^2+5a} \\
4a+20 \\
\underline{4a+20} \\
0
\end{array}
$$

The quotient is $a+4$.

5. Using long division:

$$
\begin{array}{r}
x-3 \\
x-3\overline{)x^2-6x+9} \\
\underline{x^2-3x} \\
-3x+9 \\
\underline{-3x+9} \\
0
\end{array}
$$

The quotient is $x-3$.

7. Using long division:

$$
\begin{array}{r}
x+3 \\
2x-1\overline{)2x^2+5x-3} \\
\underline{2x^2-\ x} \\
6x-3 \\
\underline{6x-3} \\
0
\end{array}
$$

The quotient is $x+3$.

9. Using long division:

$$
\begin{array}{r}
a-5 \\
2a+1\overline{)2a^2-9a-5} \\
\underline{2a^2+\ a} \\
-10a-5 \\
\underline{-10a-5} \\
0
\end{array}
$$

The quotient is $a-5$.

11. Using long division:

$$
\begin{array}{r}
x+2 \\
x+3\overline{)x^2+5x+8} \\
\underline{x^2+3x} \\
2x+8 \\
\underline{2x+6} \\
2
\end{array}
$$

The quotient is $x+2+\dfrac{2}{x+3}$.

13. Using long division:

$$a + 5 \overline{\smash{\big)}\, a^2 + 3a + 2} \quad \begin{array}{r} a - 2 \end{array}$$

$$\begin{array}{r}
\underline{a^2 + 5a} \\
-2a + 2 \\
\underline{-2a - 10} \\
12
\end{array}$$

The quotient is $a - 2 + \dfrac{12}{a + 5}$.

15. Using long division:

$$x - 2 \overline{\smash{\big)}\, x^2 + 2x + 1} \quad \begin{array}{r} x + 4 \end{array}$$

$$\begin{array}{r}
\underline{x^2 - 2x} \\
4x + 1 \\
\underline{4x - 8} \\
9
\end{array}$$

The quotient is $x + 4 + \dfrac{9}{x - 2}$.

17. Using long division:

$$x + 1 \overline{\smash{\big)}\, x^2 + 5x - 6} \quad \begin{array}{r} x + 4 \end{array}$$

$$\begin{array}{r}
\underline{x^2 + x} \\
4x - 6 \\
\underline{4x + 4} \\
-10
\end{array}$$

The quotient is $x + 4 + \dfrac{-10}{x + 1}$.

19. Using long division:

$$a + 2 \overline{\smash{\big)}\, a^2 + 3a + 1} \quad \begin{array}{r} a + 1 \end{array}$$

$$\begin{array}{r}
\underline{a^2 + 2a} \\
a + 1 \\
\underline{a + 2} \\
-1
\end{array}$$

The quotient is $a + 1 + \dfrac{-1}{a + 2}$.

21. Using long division:

$$2x + 4 \overline{\smash{\big)}\, 2x^2 - 2x + 5} \quad \begin{array}{r} x - 3 \end{array}$$

$$\begin{array}{r}
\underline{2x^2 + 4x} \\
-6x + 5 \\
\underline{-6x - 12} \\
17
\end{array}$$

The quotient is $x - 3 + \dfrac{17}{2x + 4}$.

23. Using long division:

$$2a + 3 \overline{\smash{\big)}\, 6a^2 + 5a + 1} \quad \begin{array}{r} 3a - 2 \end{array}$$

$$\begin{array}{r}
\underline{6a^2 + 9a} \\
-4a + 1 \\
\underline{-4a - 6} \\
7
\end{array}$$

The quotient is $3a - 2 + \dfrac{7}{2a + 3}$.

25. Using long division:

$$3a - 5 \overline{\smash{\big)}\, 6a^3 - 13a^2 - 4a + 15} \quad \begin{array}{r} 2a^2 - a - 3 \end{array}$$

$$\begin{array}{r}
\underline{6a^3 - 10a^2} \\
-3a^2 - 4a \\
\underline{-3a^2 + 5a} \\
-9a + 15 \\
\underline{-9a + 15} \\
0
\end{array}$$

The quotient is $2a^2 - a - 3$.

27. Using long division:

$$x + 1 \overline{\smash{\big)}\, x^3 + 0x^2 + 4x + 5} \quad \begin{array}{r} x^2 - x + 5 \end{array}$$

$$\begin{array}{r}
\underline{x^3 + x^2} \\
-x^2 + 4x \\
\underline{-x^2 - x} \\
5x + 5 \\
\underline{5x + 5} \\
0
\end{array}$$

The quotient is $x^2 - x + 5$.

29.   Using long division:

$$\begin{array}{r} x^2 + x + 1 \\ x-1\overline{\smash{)}\,x^3 + 0x^2 + 0x - 1} \\ \underline{x^3 -\ x^2} \\ x^2 + 0x \\ \underline{x^2 -\ x} \\ x - 1 \\ \underline{x - 1} \\ 0 \end{array}$$

The quotient is $x^2 + x + 1$.

31.   Using long division:

$$\begin{array}{r} x^2 + 2x + 4 \\ x-2\overline{\smash{)}\,x^3 + 0x^2 + 0x - 8} \\ \underline{x^3 - 2x^2} \\ 2x^2 + 0x \\ \underline{2x^2 - 4x} \\ 4x - 8 \\ \underline{4x - 8} \\ 0 \end{array}$$

The quotient is $x^2 + 2x + 4$.

33.
- **a.** Evaluating when $x = 3$: $x^2 + 2x + 4 = (3)^2 + 2(3) + 4 = 9 + 6 + 4 = 19$
- **b.** Evaluating when $x = 3$: $\dfrac{x^3 - 8}{x - 2} = \dfrac{(3)^3 - 8}{3 - 2} = \dfrac{27 - 8}{3 - 2} = \dfrac{19}{1} = 19$
- **c.** Evaluating when $x = 3$: $x^2 - 4 = (3)^2 - 4 = 9 - 4 = 5$

35.
- **a.** Evaluating when $x = 4$: $x + 3 = 4 + 3 = 7$
- **b.** Evaluating when $x = 4$: $\dfrac{x^2 + 9}{x + 3} = \dfrac{(4)^2 + 9}{4 + 3} = \dfrac{16 + 9}{4 + 3} = \dfrac{25}{7}$
- **c.** Evaluating when $x = 4$: $x - 3 + \dfrac{18}{x + 3} = 4 - 3 + \dfrac{18}{4 + 3} = 1 + \dfrac{18}{7} = \dfrac{7}{7} + \dfrac{18}{7} = \dfrac{25}{7}$

37.   The monthly payment is: $\dfrac{\$5,162}{12} \approx \$430.17$

39.   The monthly payment is: $\dfrac{\$3,482}{12} \approx \$290.17$

41.   Let x and y represent the two numbers. The system of equations is:

$$x + y = 25$$
$$y = 4x$$

Substituting into the first equation:

$$x + 4x = 25$$
$$5x = 25$$
$$x = 5$$
$$y = 4(5) = 20$$

The two numbers are 5 and 20.

43.   Let x represent the amount invested at 8% and y represent the amount invested at 9%. The system of equations is:

$$x + y = 1200$$
$$0.08x + 0.09y = 100$$

Multiplying the first equation by –0.08:

$$-0.08x - 0.08y = -96$$
$$0.08x + 0.09y = 100$$

Adding the two equations:

$$0.01y = 4$$
$$y = 400$$

Substituting into the first equation:

$$x + 400 = 1200$$
$$x = 800$$

You have $800 invested at 8% and $400 invested at 9%.

45. Let x represent the number of \$5 bills and $x + 4$ represent the number of \$10 bills. The equation is:

$$5(x) + 10(x + 4) = 160$$
$$5x + 10x + 40 = 160$$
$$15x + 40 = 160$$
$$15x = 120$$
$$x = 8$$
$$x + 4 = 12$$

You have 8 \$5 bills and 12 \$10 bills.

47. Let x represent the gallons of 20% antifreeze and y represent the gallons of 60% antifreeze. The system of equations is:

$$x + y = 16$$
$$0.20x + 0.60y = 0.35(16)$$

Multiplying the first equation by –0.20:

$$-0.20x - 0.20y = -3.2$$
$$0.20x + 0.60y = 5.6$$

Adding the two equations:

$$0.40y = 2.4$$
$$y = 6$$

Substituting into the first equation:

$$x + 6 = 16$$
$$x = 10$$

The mixture contains 10 gallons of 20% antifreeze and 6 gallons of 60% antifreeze.

Chapter 5 Review/Test

1. Simplifying the expression: $(-1)^3 = (-1)(-1)(-1) = -1$ **2.** Simplifying the expression: $-8^2 = -8 \cdot 8 = -64$

3. Simplifying the expression: $\left(\frac{3}{7}\right)^2 = \left(\frac{3}{7}\right) \cdot \left(\frac{3}{7}\right) = \frac{9}{49}$ **4.** Simplifying the expression: $y^3 \cdot y^9 = y^{3+9} = y^{12}$

5. Simplifying the expression: $x^{15} \cdot x^7 \cdot x^5 \cdot x^3 = x^{15+7+5+3} = x^{30}$

6. Simplifying the expression: $\left(x^7\right)^5 = x^{7 \cdot 5} = x^{35}$

7. Simplifying the expression: $\left(2^6\right)^4 = 2^{6 \cdot 4} = 2^{24}$ **8.** Simplifying the expression: $(3y)^3 = 3^3 \cdot y^3 = 27y^3$

9. Simplifying the expression: $(-2xyz)^3 = (-2)^3 \cdot x^3 y^3 z^3 = -8x^3 y^3 z^3$

10. Writing with positive exponents: $7^{-2} = \frac{1}{7^2} = \frac{1}{49}$ **11.** Writing with positive exponents: $4x^{-5} = 4 \cdot \frac{1}{x^5} = \frac{4}{x^5}$

12. Writing with positive exponents: $(3y)^{-3} = \frac{1}{(3y)^3} = \frac{1}{3^3 y^3} = \frac{1}{27y^3}$

13. Simplifying the expression: $\frac{a^9}{a^3} = a^{9-3} = a^6$

14. Simplifying the expression: $\left(\frac{x^3}{x^5}\right)^2 = \left(x^{3-5}\right)^2 = \left(x^{-2}\right)^2 = x^{-4} = \frac{1}{x^4}$

15. Simplifying the expression: $\frac{x^9}{x^{-6}} = x^{9-(-6)} = x^{9+6} = x^{15}$

16. Simplifying the expression: $\frac{x^{-7}}{x^{-2}} = x^{-7-(-2)} = x^{-7+2} = x^{-5} = \frac{1}{x^5}$

17. Simplifying the expression: $(-3xy)^0 = 1$

18. Simplifying the expression: $3^0 - 5^1 + 5^0 = 1 - 5 + 1 = -3$

19.   Simplifying the expression: $\left(3x^3y^2\right)^2 = 3^2\,x^6y^4 = 9x^6y^4$

20.   Simplifying the expression: $\left(2a^3b^2\right)^4\left(2a^5b^6\right)^2 = 2^4\,a^{12}b^8 \bullet 2^2\,a^{10}b^{12} = 2^6\,a^{22}b^{20} = 64a^{22}b^{20}$

21.   Simplifying the expression: $\left(-3xy^2\right)^{-3} = \dfrac{1}{\left(-3xy^2\right)^3} = \dfrac{1}{(-3)^3\,x^3y^6} = -\dfrac{1}{27x^3y^6}$

22.   Simplifying the expression: $\dfrac{\left(b^3\right)^4\left(b^2\right)^5}{\left(b^7\right)^3} = \dfrac{b^{12}\bullet b^{10}}{b^{21}} = \dfrac{b^{22}}{b^{21}} = b$

23.   Simplifying the expression: $\dfrac{\left(x^{-3}\right)^3\left(x^6\right)^{-1}}{\left(x^{-5}\right)^{-4}} = \dfrac{x^{-9}\bullet x^{-6}}{x^{20}} = \dfrac{x^{-15}}{x^{20}} = x^{-15-20} = x^{-35} = \dfrac{1}{x^{35}}$

24.   Simplifying the expression: $\dfrac{\left(2x^4\right)\left(15x^9\right)}{6x^6} = \dfrac{30x^{13}}{6x^6} = \dfrac{30}{6}\bullet\dfrac{x^{13}}{x^6} = 5x^7$

25.   Simplifying the expression: $\dfrac{\left(10x^3y^5\right)\left(21x^2y^6\right)}{\left(7xy^3\right)\left(5x^9y\right)} = \dfrac{210x^5y^{11}}{35x^{10}y^4} = \dfrac{210}{35}\bullet\dfrac{x^5}{x^{10}}\bullet\dfrac{y^{11}}{y^4} = 6\bullet\dfrac{1}{x^5}\bullet y^7 = \dfrac{6y^7}{x^5}$

26.   Simplifying the expression: $\dfrac{21a^{10}}{3a^4} - \dfrac{18a^{17}}{6a^{11}} = 7a^6 - 3a^6 = 4a^6$

27.   Simplifying the expression: $\dfrac{8x^8y^3}{2x^3y} - \dfrac{10x^6y^9}{5xy^7} = 4x^5y^2 - 2x^5y^2 = 2x^5y^2$

28.   Finding the product: $\left(3.2\times10^3\right)\left(2\times10^4\right) = 6.4\times10^7$ **29.**   Finding the quotient: $\dfrac{4.6\times10^5}{2\times10^{-3}} = 2.3\times10^8$

30.   Finding the value: $\dfrac{\left(4\times10^6\right)\left(6\times10^5\right)}{3\times10^8} = \dfrac{24\times10^{11}}{3\times10^8} = 8\times10^3$

31.   Performing the operations: $\left(3a^2 - 5a + 5\right) + \left(5a^2 - 7a - 8\right) = \left(3a^2 + 5a^2\right) + \left(-5a - 7a\right) + \left(5 - 8\right) = 8a^2 - 12a - 3$

32.   Performing the operations:
$$\left(-7x^2 + 3x - 6\right) - \left(8x^2 - 4x + 7\right) + \left(3x^2 - 2x - 1\right) = -7x^2 + 3x - 6 - 8x^2 + 4x - 7 + 3x^2 - 2x - 1$$
$$= \left(-7x^2 - 8x^2 + 3x^2\right) + \left(3x + 4x - 2x\right) + \left(-6 - 7 - 1\right)$$
$$= -12x^2 + 5x - 14$$

33.   Performing the operations:
$$\left(4x^2 - 3x - 2\right) - \left(8x^2 + 3x - 2\right) = 4x^2 - 3x - 2 - 8x^2 - 3x + 2$$
$$= \left(4x^2 - 8x^2\right) + \left(-3x - 3x\right) + \left(-2 + 2\right)$$
$$= -4x^2 - 6x$$

34.   Evaluating when $x = 3$: $2x^2 - 3x + 5 = 2(3)^2 - 3(3) + 5 = 2(9) - 9 + 5 = 18 - 9 + 5 = 14$

35.   Multiplying: $3x(4x - 7) = 3x(4x) - 3x(7) = 12x^2 - 21x$

36.   Multiplying: $8x^3y\left(3x^2y - 5xy^2 + 4y^3\right) = 8x^3y\left(3x^2y\right) - 8x^3y\left(5xy^2\right) + 8x^3y\left(4y^3\right) = 24x^5y^2 - 40x^4y^3 + 32x^3y^4$

37. Multiplying using the column method:

$$a^2 + 5a - 4$$
$$\underline{\qquad a + 1}$$
$$a^3 + 5a^2 - 4a$$
$$\underline{\quad a^2 + 5a - 4}$$
$$a^3 + 6a^2 + \ a - 4$$

38. Multiplying using the column method:

$$x^2 - 5x + 25$$
$$\underline{\qquad x + 5}$$
$$x^3 - 5x^2 + 25x$$
$$\underline{\quad 5x^2 - 25x + 125}$$
$$x^3 + 125$$

39. Multiplying using the FOIL method: $(3x - 7)(2x - 5) = 6x^2 - 14x - 15x + 35 = 6x^2 - 29x + 35$

40. Multiplying using the difference of squares formula: $\left(5y + \frac{1}{5}\right)\left(5y - \frac{1}{5}\right) = (5y)^2 - \left(\frac{1}{5}\right)^2 = 25y^2 - \frac{1}{25}$

41. Multiplying using the difference of squares formula: $\left(a^2 - 3\right)\left(a^2 + 3\right) = \left(a^2\right)^2 - (3)^2 = a^4 - 9$

42. Multiplying using the square of binomial formula: $(a - 5)^2 = a^2 - 2(a)(5) + (5)^2 = a^2 - 10a + 25$

43. Multiplying using the square of binomial formula: $(3x + 4)^2 = (3x)^2 + 2(3x)(4) + (4)^2 = 9x^2 + 24x + 16$

44. Multiplying using the square of binomial formula: $\left(y^2 + 3\right)^2 = \left(y^2\right)^2 + 2\left(y^2\right)(3) + (3)^2 = y^4 + 6y^2 + 9$

45. Performing the division: $\dfrac{10ab + 20a^2}{-5a} = \dfrac{10ab}{-5a} + \dfrac{20a^2}{-5a} = -2b - 4a$

46. Performing the division: $\dfrac{40x^5y^4 - 32x^3y^3 - 16x^2y}{-8xy} = \dfrac{40x^5y^4}{-8xy} + \dfrac{-32x^3y^3}{-8xy} + \dfrac{-16x^2y}{-8xy} = -5x^4y^3 + 4x^2y^2 + 2x$

47. Using long division:

$$
\begin{array}{r}
x + 9 \\
x + 6 \overline{\smash{\big)}\, x^2 + 15x + 54} \\
\underline{x^2 + \ 6x} \\
9x + 54 \\
\underline{9x + 54} \\
0
\end{array}
$$

The quotient is $x + 9$.

48. Using long division:

$$
\begin{array}{r}
2x + 5 \\
3x - 1 \overline{\smash{\big)}\, 6x^2 + 13x - 5} \\
\underline{6x^2 - \ 2x} \\
15x - 5 \\
\underline{15x - 5} \\
0
\end{array}
$$

The quotient is $2x + 5$.

49. Using long division:

$$
\begin{array}{r}
x^2 - 4x + 16 \\
x + 4 \overline{\smash{\big)}\, x^3 + 0x^2 + 0x + 64} \\
\underline{x^3 + 4x^2} \\
-4x^2 + \ 0x \\
\underline{-4x^2 - 16x} \\
16x + 64 \\
\underline{16x + 64} \\
0
\end{array}
$$

The quotient is $x^2 - 4x + 16$.

50. Using long division:

$$
\begin{array}{r}
x - 3 \\
3x + 2 \overline{\smash{\big)}\, 3x^2 - 7x + 10} \\
\underline{3x^2 + 2x} \\
-9x + 10 \\
\underline{-9x - 6} \\
16
\end{array}
$$

The quotient is $x - 3 + \dfrac{16}{3x + 2}$.

51. Using long division:

$$\begin{array}{r} x^2 - 4x + 5 \\ 2x+1{\overline{\smash{\big)}\,2x^3 - 7x^2 + 6x + 10}} \\ \underline{2x^3 + x^2} \\ -8x^2 + 6x \\ \underline{-8x^2 - 4x} \\ 10x + 10 \\ \underline{10x + 5} \\ 5 \end{array}$$

The quotient is $x^2 - 4x + 5 + \dfrac{5}{2x+1}$.

Chapter 6
Factoring

6.1 The Greatest Common Factor and Factoring by Grouping

1. Factoring out the greatest common factor: $15x + 25 = 5(3x + 5)$

3. Factoring out the greatest common factor: $6a + 9 = 3(2a + 3)$

5. Factoring out the greatest common factor: $4x - 8y = 4(x - 2y)$

7. Factoring out the greatest common factor: $3x^2 - 6x - 9 = 3(x^2 - 2x - 3)$

9. Factoring out the greatest common factor: $3a^2 - 3a - 60 = 3(a^2 - a - 20)$

11. Factoring out the greatest common factor: $24y^2 - 52y + 24 = 4(6y^2 - 13y + 6)$

13. Factoring out the greatest common factor: $9x^2 - 8x^3 = x^2(9 - 8x)$

15. Factoring out the greatest common factor: $13a^2 - 26a^3 = 13a^2(1 - 2a)$

17. Factoring out the greatest common factor: $21x^2 y - 28xy^2 = 7xy(3x - 4y)$

19. Factoring out the greatest common factor: $22a^2 b^2 - 11ab^2 = 11ab^2(2a - 1)$

21. Factoring out the greatest common factor: $7x^3 + 21x^2 - 28x = 7x(x^2 + 3x - 4)$

23. Factoring out the greatest common factor: $121y^4 - 11x^4 = 11(11y^4 - x^4)$

25. Factoring out the greatest common factor: $100x^4 - 50x^3 + 25x^2 = 25x^2(4x^2 - 2x + 1)$

27. Factoring out the greatest common factor: $8a^2 + 16b^2 + 32c^2 = 8(a^2 + 2b^2 + 4c^2)$

29. Factoring out the greatest common factor: $4a^2 b - 16ab^2 + 32a^2 b^2 = 4ab(a - 4b + 8ab)$

31. Factoring out the greatest common factor: $121a^3 b^2 - 22a^2 b^3 + 33a^3 b^3 = 11a^2 b^2(11a - 2b + 3ab)$

33. Factoring out the greatest common factor: $12x^2 y^3 - 72x^5 y^3 - 36x^4 y^4 = 12x^2 y^3(1 - 6x^3 - 3x^2 y)$

35. Factoring by grouping: $xy + 5x + 3y + 15 = x(y + 5) + 3(y + 5) = (y + 5)(x + 3)$

37. Factoring by grouping: $xy + 6x + 2y + 12 = x(y + 6) + 2(y + 6) = (y + 6)(x + 2)$

39. Factoring by grouping: $ab + 7a - 3b - 21 = a(b + 7) - 3(b + 7) = (b + 7)(a - 3)$

41. Factoring by grouping: $ax - bx + ay - by = x(a - b) + y(a - b) = (a - b)(x + y)$

43.   Factoring by grouping: $2ax + 6x - 5a - 15 = 2x(a+3) - 5(a+3) = (a+3)(2x-5)$

45.   Factoring by grouping: $3xb - 4b - 6x + 8 = b(3x-4) - 2(3x-4) = (3x-4)(b-2)$

47.   Factoring by grouping: $x^2 + ax + 2x + 2a = x(x+a) + 2(x+a) = (x+a)(x+2)$

49.   Factoring by grouping: $x^2 - ax - bx + ab = x(x-a) - b(x-a) = (x-a)(x-b)$

51.   Factoring by grouping: $ax + ay + bx + by + cx + cy = a(x+y) + b(x+y) + c(x+y) = (x+y)(a+b+c)$

53.   Factoring by grouping: $6x^2 + 9x + 4x + 6 = 3x(2x+3) + 2(2x+3) = (2x+3)(3x+2)$

55.   Factoring by grouping: $20x^2 - 2x + 50x - 5 = 2x(10x-1) + 5(10x-1) = (10x-1)(2x+5)$

57.   Factoring by grouping: $20x^2 + 4x + 25x + 5 = 4x(5x+1) + 5(5x+1) = (5x+1)(4x+5)$

59.   Factoring by grouping: $x^3 + 2x^2 + 3x + 6 = x^2(x+2) + 3(x+2) = (x+2)(x^2+3)$

61.   Factoring by grouping: $6x^3 - 4x^2 + 15x - 10 = 2x^2(3x-2) + 5(3x-2) = (3x-2)(2x^2+5)$

63.   Its greatest common factor is $3 \cdot 2 = 6$.

65.   The correct factoring is: $12x^2 + 6x + 3 = 3(4x^2 + 2x + 1)$

67.   Dividing: $\dfrac{y^3 - 16y^2 + 64y}{y} = \dfrac{y^3}{y} - \dfrac{16y^2}{y} + \dfrac{64y}{y} = y^2 - 16y + 64$

69.   Dividing: $\dfrac{-12x^4 + 48x^3 + 144x^2}{-12x^2} = \dfrac{-12x^4}{-12x^2} + \dfrac{48x^3}{-12x^2} + \dfrac{144x^2}{-12x^2} = x^2 - 4x - 12$

71.   Dividing: $\dfrac{-18y^5 + 63y^4 - 108y^3}{-9y^3} = \dfrac{-18y^5}{-9y^3} + \dfrac{63y^4}{-9y^3} - \dfrac{108y^3}{-9y^3} = 2y^2 - 7y + 12$

73.   Subtracting the polynomials: $(5x^2 + 5x - 4) - (3x^2 - 2x + 7) = 5x^2 + 5x - 4 - 3x^2 + 2x - 7 = 2x^2 + 7x - 11$

75.   Subtracting the polynomials: $(7x+3) - (4x-5) = 7x + 3 - 4x + 5 = 3x + 8$

77.   Subtracting the polynomials: $(5x^2 - 5) - (2x^2 - 4x) = 5x^2 - 5 - 2x^2 + 4x = 3x^2 + 4x - 5$

79.   Multiplying using the FOIL method: $(x-7)(x+2) = x^2 - 7x + 2x - 14 = x^2 - 5x - 14$

81.   Multiplying using the FOIL method: $(x-3)(x+2) = x^2 - 3x + 2x - 6 = x^2 - x - 6$

83.   Multiplying using the column method:

$$
\begin{array}{r}
x^2 - 3x + 9 \\
x + 3 \\
\hline
x^3 - 3x^2 + 9x \\
3x^2 - 9x + 27 \\
\hline
x^3 + 27
\end{array}
$$

85.   Multiplying using the column method:

$$
\begin{array}{r}
x^2 + 4x - 3 \\
2x + 1 \\
\hline
2x^3 + 8x^2 - 6x \\
x^2 + 4x - 3 \\
\hline
2x^3 + 9x^2 - 2x - 3
\end{array}
$$

87.   Multiplying: $3x^4(6x^3 - 4x^2 + 2x) = 3x^4 \cdot 6x^3 - 3x^4 \cdot 4x^2 + 3x^4 \cdot 2x = 18x^7 - 12x^6 + 6x^5$

89.   Multiplying: $\left(x + \frac{1}{3}\right)\left(x + \frac{2}{3}\right) = x^2 + \frac{2}{3}x + \frac{1}{3}x + \frac{2}{9} = x^2 + x + \frac{2}{9}$

91.   Multiplying: $(6x + 4y)(2x - 3y) = 12x^2 - 18xy + 8xy - 12y^2 = 12x^2 - 10xy - 12y^2$

93.   Multiplying: $(9a + 1)(9a - 1) = 81a^2 - 9a + 9a - 1 = 81a^2 - 1$

95.   Multiplying: $(x - 9)(x - 9) = x^2 - 9x - 9x + 81 = x^2 - 18x + 81$

97.   Multiplying: $(x + 2)(x^2 - 2x + 4) = x(x^2 - 2x + 4) + 2(x^2 - 2x + 4) = x^3 - 2x^2 + 4x + 2x^2 - 4x + 8 = x^3 + 8$

6.2 Factoring Trinomials

1.   Factoring the trinomial: $x^2 + 7x + 12 = (x+3)(x+4)$ **3.**   Factoring the trinomial: $x^2 + 3x + 2 = (x+2)(x+1)$

5.   Factoring the trinomial: $a^2 + 10a + 21 = (a+7)(a+3)$ **7.**   Factoring the trinomial: $x^2 - 7x + 10 = (x-5)(x-2)$

9.   Factoring the trinomial: $y^2 - 10y + 21 = (y-7)(y-3)$ **11.**   Factoring the trinomial: $x^2 - x - 12 = (x-4)(x+3)$

13.   Factoring the trinomial: $y^2 + y - 12 = (y+4)(y-3)$ **15.**   Factoring the trinomial: $x^2 + 5x - 14 = (x+7)(x-2)$

17.   Factoring the trinomial: $r^2 - 8r - 9 = (r-9)(r+1)$ **19.**   Factoring the trinomial: $x^2 - x - 30 = (x-6)(x+5)$

21.   Factoring the trinomial: $a^2 + 15a + 56 = (a+7)(a+8)$ **23.**   Factoring the trinomial: $y^2 - y - 42 = (y-7)(y+6)$

25.   Factoring the trinomial: $x^2 + 13x + 42 = (x+7)(x+6)$

27.   Factoring the trinomial: $2x^2 + 6x + 4 = 2\left(x^2 + 3x + 2\right) = 2(x+2)(x+1)$

29.   Factoring the trinomial: $3a^2 - 3a - 60 = 3\left(a^2 - a - 20\right) = 3(a-5)(a+4)$

31.   Factoring the trinomial: $100x^2 - 500x + 600 = 100\left(x^2 - 5x + 6\right) = 100(x-3)(x-2)$

33.   Factoring the trinomial: $100p^2 - 1300p + 4000 = 100\left(p^2 - 13p + 40\right) = 100(p-8)(p-5)$

35.   Factoring the trinomial: $x^4 - x^3 - 12x^2 = x^2\left(x^2 - x - 12\right) = x^2(x-4)(x+3)$

37.   Factoring the trinomial: $2r^3 + 4r^2 - 30r = 2r\left(r^2 + 2r - 15\right) = 2r(r+5)(r-3)$

39.   Factoring the trinomial: $2y^4 - 6y^3 - 8y^2 = 2y^2\left(y^2 - 3y - 4\right) = 2y^2(y-4)(y+1)$

41.   Factoring the trinomial: $x^5 + 4x^4 + 4x^3 = x^3\left(x^2 + 4x + 4\right) = x^3(x+2)(x+2) = x^3(x+2)^2$

43.   Factoring the trinomial: $3y^4 - 12y^3 - 15y^2 = 3y^2\left(y^2 - 4y - 5\right) = 3y^2(y-5)(y+1)$

45.   Factoring the trinomial: $4x^4 - 52x^3 + 144x^2 = 4x^2\left(x^2 - 13x + 36\right) = 4x^2(x-9)(x-4)$

47.   Factoring the trinomial: $x^2 + 5xy + 6y^2 = (x+2y)(x+3y)$

49.   Factoring the trinomial: $x^2 - 9xy + 20y^2 = (x-4y)(x-5y)$

51.   Factoring the trinomial: $a^2 + 2ab - 8b^2 = (a+4b)(a-2b)$

53.   Factoring the trinomial: $a^2 - 10ab + 25b^2 = (a-5b)(a-5b) = (a-5b)^2$

55.   Factoring the trinomial: $a^2 + 10ab + 25b^2 = (a+5b)(a+5b) = (a+5b)^2$

57.   Factoring the trinomial: $x^2 + 2xa - 48a^2 = (x+8a)(x-6a)$

59.   Factoring the trinomial: $x^2 - 5xb - 36b^2 = (x-9b)(x+4b)$

61.   Factoring the trinomial: $x^4 - 5x^2 + 6 = \left(x^2 - 2\right)\left(x^2 - 3\right)$

63.   Factoring the trinomial: $x^2 - 80x - 2000 = (x-100)(x+20)$

65.   Factoring the trinomial: $x^2 - x + \frac{1}{4} = \left(x - \frac{1}{2}\right)\left(x - \frac{1}{2}\right) = \left(x - \frac{1}{2}\right)^2$

67.   Factoring the trinomial: $x^2 + 0.6x + 0.08 = (x+0.4)(x+0.2)$

69.   We can use long division to find the other factor:

$$
\begin{array}{r}
x+16 \\
x+8\overline{\smash{\big)}\ x^2+24x+128} \\
\underline{x^2+\ 8x} \\
16x+128 \\
\underline{16x+128} \\
0
\end{array}
$$

The other factor is $x+16$.

71.   Using FOIL to multiply out the factors: $(4x+3)(x-1)=4x^2+3x-4x-3=4x^2-x-3$

73.   Simplifying; $\left(-\frac{2}{5}\right)^2=\left(-\frac{2}{5}\right)\left(-\frac{2}{5}\right)=\frac{4}{25}$ 75.   Simplifying; $\left(3a^3\right)^2\left(2a^2\right)^3=\left(9a^6\right)\left(8a^6\right)=72a^{12}$

77.   Simplifying; $\dfrac{(4x)^{-7}}{(4x)^{-5}}=(4x)^{-2}=\dfrac{1}{(4x)^2}=\dfrac{1}{16x^2}$ 79.   Simplifying; $\dfrac{12a^5b^3}{72a^2b^5}=\dfrac{1}{6}a^{5-2}b^{3-5}=\dfrac{1}{6}a^3b^{-2}=\dfrac{a^3}{6b^2}$

81.   Simplifying; $\dfrac{15x^{-5}y^3}{45x^2y^5}=\dfrac{1}{3}x^{-5-2}y^{3-5}=\dfrac{1}{3}x^{-7}y^{-2}=\dfrac{1}{3x^7y^2}$

83.   Simplifying; $\left(-7x^3y\right)\left(3xy^4\right)=-21x^{3+1}y^{1+4}=-21x^4y^5$

85.   Simplifying; $\left(-5a^3b^{-1}\right)\left(4a^{-2}b^4\right)=-20a^{3-2}b^{-1+4}=-20ab^3$

87.   Simplifying; $\left(9a^2b^3\right)\left(-3a^3b^5\right)=-27a^{2+3}b^{3+5}=-27a^5b^8$

89.   Multiplying using the FOIL method: $(6a+1)(a+2)=6a^2+a+12a+2=6a^2+13a+2$

91.   Multiplying using the FOIL method: $(3a+2)(2a+1)=6a^2+4a+3a+2=6a^2+7a+2$

93.   Multiplying using the FOIL method: $(6a+2)(a+1)=6a^2+2a+6a+2=6a^2+8a+2$

6.3 More Trinomials to Factor

1.   Factoring the trinomial: $2x^2+7x+3=(2x+1)(x+3)$

3.   Factoring the trinomial: $2a^2-a-3=(2a-3)(a+1)$

5.   Factoring the trinomial: $3x^2+2x-5=(3x+5)(x-1)$

7.   Factoring the trinomial: $3y^2-14y-5=(3y+1)(y-5)$

9.   Factoring the trinomial: $6x^2+13x+6=(3x+2)(2x+3)$

11.   Factoring the trinomial: $4x^2-12xy+9y^2=(2x-3y)(2x-3y)=(2x-3y)^2$

13.   Factoring the trinomial: $4y^2-11y-3=(4y+1)(y-3)$

15.   Factoring the trinomial: $20x^2-41x+20=(4x-5)(5x-4)$

17.   Factoring the trinomial: $20a^2+48ab-5b^2=(10a-b)(2a+5b)$

19.   Factoring the trinomial: $20x^2-21x-5=(4x-5)(5x+1)$

21.   Factoring the trinomial: $12m^2+16m-3=(6m-1)(2m+3)$

23.   Factoring the trinomial: $20x^2+37x+15=(4x+5)(5x+3)$

25.   Factoring the trinomial: $12a^2-25ab+12b^2=(3a-4b)(4a-3b)$

27.   Factoring the trinomial: $3x^2-xy-14y^2=(3x-7y)(x+2y)$

29. Factoring the trinomial: $14x^2 + 29x - 15 = (2x+5)(7x-3)$

31. Factoring the trinomial: $6x^2 - 43x + 55 = (3x-5)(2x-11)$

33. Factoring the trinomial: $15t^2 - 67t + 38 = (5t-19)(3t-2)$

35. Factoring the trinomial: $4x^2 + 2x - 6 = 2(2x^2 + x - 3) = 2(2x+3)(x-1)$

37. Factoring the trinomial: $24a^2 - 50a + 24 = 2(12a^2 - 25a + 12) = 2(4a-3)(3a-4)$

39. Factoring the trinomial: $10x^3 - 23x^2 + 12x = x(10x^2 - 23x + 12) = x(5x-4)(2x-3)$

41. Factoring the trinomial: $6x^4 - 11x^3 - 10x^2 = x^2(6x^2 - 11x - 10) = x^2(3x+2)(2x-5)$

43. Factoring the trinomial: $10a^3 - 6a^2 - 4a = 2a(5a^2 - 3a - 2) = 2a(5a+2)(a-1)$

45. Factoring the trinomial: $15x^3 - 102x^2 - 21x = 3x(5x^2 - 34x - 7) = 3x(5x+1)(x-7)$

47. Factoring the trinomial: $35y^3 - 60y^2 - 20y = 5y(7y^2 - 12y - 4) = 5y(7y+2)(y-2)$

49. Factoring the trinomial: $15a^4 - 2a^3 - a^2 = a^2(15a^2 - 2a - 1) = a^2(5a+1)(3a-1)$

51. Factoring the trinomial: $24x^2y - 6xy - 45y = 3y(8x^2 - 2x - 15) = 3y(4x+5)(2x-3)$

53. Factoring the trinomial: $12x^2y - 34xy^2 + 14y^3 = 2y(6x^2 - 17xy + 7y^2) = 2y(2x-y)(3x-7y)$

55. Evaluating each expression when $x = 2$:
$$2x^2 + 7x + 3 = 2(2)^2 + 7(2) + 3 = 8 + 14 + 3 = 25$$
$$(2x+1)(x+3) = (2 \cdot 2 + 1)(2+3) = (5)(5) = 25$$

57. Multiplying using the difference of squares formula: $(2x+3)(2x-3) = (2x)^2 - (3)^2 = 4x^2 - 9$

59. Multiplying using the difference of squares formula:
$$(x+3)(x-3)(x^2+9) = (x^2-9)(x^2+9) = (x^2)^2 - (9)^2 = x^4 - 81$$

61. Factoring: $12x^2 - 71x + 105 = (x-3)(12x-35)$. The other factor is $12x - 35$.

63. Factoring: $54x^2 + 111x + 56 = (6x+7)(9x+8)$. The other factor is $9x + 8$.

65. Factoring: $16t^2 - 64t + 48 = 16(t^2 - 4t + 3) = 16(t-1)(t-3)$

67. Factoring: $h = 8 + 62t - 16t^2 = 2(4 + 31t - 8t^2) = 2(4-t)(1+8t)$. Now completing the table:

Time t (seconds)	Height h (feet)
0	8
1	54
2	68
3	50
4	0

69. Simplifying: $(6x^3 - 4x^2 + 2x) + (9x^2 - 6x + 3) = 6x^3 - 4x^2 + 2x + 9x^2 - 6x + 3 = 6x^3 + 5x^2 - 4x + 3$

71. Simplifying: $(-7x^4 + 4x^3 - 6x) + (8x^4 + 7x^3 - 9) = -7x^4 + 4x^3 - 6x + 8x^4 + 7x^3 - 9 = x^4 + 11x^3 - 6x - 9$

73. Simplifying: $(2x^5 + 3x^3 + 4x) + (5x^3 - 6x - 7) = 2x^5 + 3x^3 + 4x + 5x^3 - 6x - 7 = 2x^5 + 8x^3 - 2x - 7$

75. Simplifying: $(-8x^5 - 5x^4 + 7) + (7x^4 + 2x^2 + 5) = -8x^5 - 5x^4 + 7 + 7x^4 + 2x^2 + 5 = -8x^5 + 2x^4 + 2x^2 + 12$

77. Simplifying: $\dfrac{24x^3y^7}{6x^{-2}y^4} + \dfrac{27x^{-2}y^{10}}{9x^{-7}y^7} = 4x^5y^3 + 3x^5y^3 = 7x^5y^3$

79. Simplifying: $\dfrac{18a^5b^9}{3a^3b^6} - \dfrac{48a^{-3}b^{-1}}{16a^{-5}b^{-4}} = 6a^2b^3 - 3a^2b^3 = 3a^2b^3$

81. Multiplying: $(x+3)(x-3) = x^2 - (3)^2 = x^2 - 9$ 83. Multiplying: $(x+5)(x-5) = x^2 - (5)^2 = x^2 - 25$

85. Multiplying: $(x+7)(x-7) = x^2 - (7)^2 = x^2 - 49$ 87. Multiplying: $(x+9)(x-9) = x^2 - (9)^2 = x^2 - 81$

89. Multiplying: $(2x-3y)(2x+3y) = (2x)^2 - (3y)^2 = 4x^2 - 9y^2$

91. Multiplying: $\left(x^2+4\right)(x+2)(x-2) = \left(x^2+4\right)\left(x^2-4\right) = \left(x^2\right)^2 - (4)^2 = x^4 - 16$

93. Multiplying: $(x+3)^2 = x^2 + 2(x)(3) + (3)^2 = x^2 + 6x + 9$

95. Multiplying: $(x+5)^2 = x^2 + 2(x)(5) + (5)^2 = x^2 + 10x + 25$

97. Multiplying: $(x+7)^2 = x^2 + 2(x)(7) + (7)^2 = x^2 + 14x + 49$

99. Multiplying: $(x+9)^2 = x^2 + 2(x)(9) + (9)^2 = x^2 + 18x + 81$

101. Multiplying: $(2x+3)^2 = (2x)^2 + 2(2x)(3) + (3)^2 = 4x^2 + 12x + 9$

103. Multiplying: $(4x-2y)^2 = (4x)^2 - 2(4x)(2y) + (2y)^2 = 16x^2 - 16xy + 4y^2$

6.4 The Difference of Two Squares

1. Factoring the binomial: $x^2 - 9 = (x+3)(x-3)$ 3. Factoring the binomial: $a^2 - 36 = (a+6)(a-6)$

5. Factoring the binomial: $x^2 - 49 = (x+7)(x-7)$

7. Factoring the binomial: $4a^2 - 16 = 4\left(a^2 - 4\right) = 4(a+2)(a-2)$

9. The expression $9x^2 + 25$ cannot be factored.

11. Factoring the binomial: $25x^2 - 169 = (5x+13)(5x-13)$

13. Factoring the binomial: $9a^2 - 16b^2 = (3a+4b)(3a-4b)$

15. Factoring the binomial: $9 - m^2 = (3+m)(3-m)$ 17. Factoring the binomial: $25 - 4x^2 = (5+2x)(5-2x)$

19. Factoring the binomial: $2x^2 - 18 = 2\left(x^2 - 9\right) = 2(x+3)(x-3)$

21. Factoring the binomial: $32a^2 - 128 = 32\left(a^2 - 4\right) = 32(a+2)(a-2)$

23. Factoring the binomial: $8x^2y - 18y = 2y\left(4x^2 - 9\right) = 2y(2x+3)(2x-3)$

25. Factoring the binomial: $a^4 - b^4 = \left(a^2+b^2\right)\left(a^2-b^2\right) = \left(a^2+b^2\right)(a+b)(a-b)$

27. Factoring the binomial: $16m^4 - 81 = \left(4m^2+9\right)\left(4m^2-9\right) = \left(4m^2+9\right)(2m+3)(2m-3)$

29. Factoring the binomial: $3x^3y - 75xy^3 = 3xy\left(x^2 - 25y^2\right) = 3xy(x+5y)(x-5y)$

31. Factoring the trinomial: $x^2 - 2x + 1 = (x-1)(x-1) = (x-1)^2$

33. Factoring the trinomial: $x^2 + 2x + 1 = (x+1)(x+1) = (x+1)^2$

35. Factoring the trinomial: $a^2 - 10a + 25 = (a-5)(a-5) = (a-5)^2$

37. Factoring the trinomial: $y^2 + 4y + 4 = (y+2)(y+2) = (y+2)^2$

39.   Factoring the trinomial: $x^2 - 4x + 4 = (x-2)(x-2) = (x-2)^2$

41.   Factoring the trinomial: $m^2 - 12m + 36 = (m-6)(m-6) = (m-6)^2$

43.   Factoring the trinomial: $4a^2 + 12a + 9 = (2a+3)(2a+3) = (2a+3)^2$

45.   Factoring the trinomial: $49x^2 - 14x + 1 = (7x-1)(7x-1) = (7x-1)^2$

47.   Factoring the trinomial: $9y^2 - 30y + 25 = (3y-5)(3y-5) = (3y-5)^2$

49.   Factoring the trinomial: $x^2 + 10xy + 25y^2 = (x+5y)(x+5y) = (x+5y)^2$

51.   Factoring the trinomial: $9a^2 + 6ab + b^2 = (3a+b)(3a+b) = (3a+b)^2$

53.   Factoring the trinomial: $y^2 - 3y + \frac{9}{4} = \left(y - \frac{3}{2}\right)\left(y - \frac{3}{2}\right) = \left(y - \frac{3}{2}\right)^2$

55.   Factoring the trinomial: $a^2 + a + \frac{1}{4} = \left(a + \frac{1}{2}\right)\left(a + \frac{1}{2}\right) = \left(a + \frac{1}{2}\right)^2$

57.   Factoring the trinomial: $x^2 - 7x + \frac{49}{4} = \left(x - \frac{7}{2}\right)\left(x - \frac{7}{2}\right) = \left(x - \frac{7}{2}\right)^2$

59.   Factoring the trinomial: $x^2 - \frac{3}{4}x + \frac{9}{64} = \left(x - \frac{3}{8}\right)\left(x - \frac{3}{8}\right) = \left(x - \frac{3}{8}\right)^2$

61.   Factoring the trinomial: $t^2 - \frac{2}{5}t + \frac{1}{25} = \left(t - \frac{1}{5}\right)\left(t - \frac{1}{5}\right) = \left(t - \frac{1}{5}\right)^2$

63.   Factoring the trinomial: $3a^2 + 18a + 27 = 3\left(a^2 + 6a + 9\right) = 3(a+3)(a+3) = 3(a+3)^2$

65.   Factoring the trinomial: $2x^2 + 20xy + 50y^2 = 2\left(x^2 + 10xy + 25y^2\right) = 2(x+5y)(x+5y) = 2(x+5y)^2$

67.   Factoring the trinomial: $5x^3 + 30x^2y + 45xy^2 = 5x\left(x^2 + 6xy + 9y^2\right) = 5x(x+3y)(x+3y) = 5x(x+3y)^2$

69.   Factoring by grouping: $x^2 + 6x + 9 - y^2 = (x+3)^2 - y^2 = (x+3+y)(x+3-y)$

71.   Factoring by grouping: $x^2 + 2xy + y^2 - 9 = (x+y)^2 - 9 = (x+y+3)(x+y-3)$

73.   Since $(x+7)^2 = x^2 + 14x + 49$, the value is $b = 14$.   **75.**   Since $(x+5)^2 = x^2 + 10x + 25$, the value is $c = 25$.

77.   Dividing: $\dfrac{24y^3 - 36y^2 - 18y}{6y} = \dfrac{24y^3}{6y} - \dfrac{36y^2}{6y} - \dfrac{18y}{6y} = 4y^2 - 6y - 3$

79.   Dividing: $\dfrac{48x^7 - 36x^5 + 12x^2}{4x^2} = \dfrac{48x^7}{4x^2} - \dfrac{36x^5}{4x^2} + \dfrac{12x^2}{4x^2} = 12x^5 - 9x^3 + 3$

81.   Dividing: $\dfrac{18x^7 + 12x^6 - 6x^5}{-3x^4} = \dfrac{18x^7}{-3x^4} + \dfrac{12x^6}{-3x^4} - \dfrac{6x^5}{-3x^4} = -6x^3 - 4x^2 + 2x$

83.   Dividing: $\dfrac{-42x^5 + 24x^4 - 66x^2}{6x^2} = \dfrac{-42x^5}{6x^2} + \dfrac{24x^4}{6x^2} - \dfrac{66x^2}{6x^2} = -7x^3 + 4x^2 - 11$

85.   Using long division:

$$
\begin{array}{r}
x - 2 \\
x-3\overline{)x^2 - 5x + 8} \\
\underline{x^2 - 3x} \\
-2x + 8 \\
\underline{-2x + 6} \\
2
\end{array}
$$

The quotient is $x - 2 + \dfrac{2}{x-3}$.

87.   Using long division:

$$
\begin{array}{r}
3x - 2 \\
2x+3\overline{)6x^2 + 5x + 3} \\
\underline{6x^2 + 9x} \\
-4x + 3 \\
\underline{-4x - 6} \\
9
\end{array}
$$

The quotient is $3x - 2 + \dfrac{9}{2x+3}$.

89. **a.** Multiplying: $1^3 = 1$ **b.** Multiplying: $2^3 = 8$

c. Multiplying: $3^3 = 27$ **d.** Multiplying: $4^3 = 64$

e. Multiplying: $5^3 = 125$

91. **a.** Multiplying: $x(x^2 - x + 1) = x^3 - x^2 + x$ **b.** Multiplying: $1(x^2 - x + 1) = x^2 - x + 1$

c. Multiplying: $(x+1)(x^2 - x + 1) = x(x^2 - x + 1) + 1(x^2 - x + 1) = x^3 - x^2 + x + x^2 - x + 1 = x^3 + 1$

93. **a.** Multiplying: $x(x^2 - 2x + 4) = x^3 - 2x^2 + 4x$ **b.** Multiplying: $2(x^2 - 2x + 4) = 2x^2 - 4x + 8$

c. Multiplying: $(x+2)(x^2 - 2x + 4) = x(x^2 - 2x + 4) + 2(x^2 - 2x + 4) = x^3 - 2x^2 + 4x + 2x^2 - 4x + 8 = x^3 + 8$

95. **a.** Multiplying: $x(x^2 - 3x + 9) = x^3 - 3x^2 + 9x$ **b.** Multiplying: $3(x^2 - 3x + 9) = 3x^2 - 9x + 27$

c. Multiplying:
$$(x+3)(x^2 - 3x + 9) = x(x^2 - 3x + 9) + 3(x^2 - 3x + 9) = x^3 - 3x^2 + 9x + 3x^2 - 9x + 27 = x^3 + 27$$

97. **a.** Multiplying: $x(x^2 - 4x + 16) = x^3 - 4x^2 + 16x$ **b.** Multiplying: $4(x^2 - 4x + 16) = 4x^2 - 16x + 64$

c. Multiplying:
$$(x+4)(x^2 - 4x + 16) = x(x^2 - 4x + 16) + 4(x^2 - 4x + 16) = x^3 - 4x^2 + 16x + 4x^2 - 16x + 64 = x^3 + 64$$

99. **a.** Multiplying: $x(x^2 - 5x + 25) = x^3 - 5x^2 + 25x$ **b.** Multiplying: $5(x^2 - 5x + 25) = 5x^2 - 25x + 125$

c. Multiplying:
$$(x+5)(x^2 - 5x + 25) = x(x^2 - 5x + 25) + 5(x^2 - 5x + 25) = x^3 - 5x^2 + 25x + 5x^2 - 25x + 125 = x^3 + 125$$

6.5 The Sum and Difference of Two Cubes

1. Factoring: $x^3 - y^3 = (x - y)(x^2 + xy + y^2)$ **3.** Factoring: $a^3 + 8 = (a + 2)(a^2 - 2a + 4)$

5. Factoring: $27 + x^3 = (3 + x)(9 - 3x + x^2)$ **7.** Factoring: $y^3 - 1 = (y - 1)(y^2 + y + 1)$

9. Factoring: $y^3 - 64 = (y - 4)(y^2 + 4y + 16)$ **11.** Factoring: $125h^3 - t^3 = (5h - t)(25h^2 + 5ht + t^2)$

13. Factoring: $x^3 - 216 = (x - 6)(x^2 + 6x + 36)$

15. Factoring: $2y^3 - 54 = 2(y^3 - 27) = 2(y - 3)(y^2 + 3y + 9)$

17. Factoring: $2a^3 - 128b^3 = 2(a^3 - 64b^3) = 2(a - 4b)(a^2 + 4ab + 16b^2)$

19. Factoring: $2x^3 + 432y^3 = 2(x^3 + 216y^3) = 2(x + 6y)(x^2 - 6xy + 36y^2)$

21. Factoring: $10a^3 - 640b^3 = 10(a^3 - 64b^3) = 10(a - 4b)(a^2 + 4ab + 16b^2)$

23. Factoring: $10r^3 - 1250 = 10(r^3 - 125) = 10(r - 5)(r^2 + 5r + 25)$

25. Factoring: $64 + 27a^3 = (4 + 3a)(16 - 12a + 9a^2)$ **27.** Factoring: $8x^3 - 27y^3 = (2x - 3y)(4x^2 + 6xy + 9y^2)$

29. Factoring: $t^3 + \frac{1}{27} = \left(t + \frac{1}{3}\right)\left(t^2 - \frac{1}{3}t + \frac{1}{9}\right)$ **31.** Factoring: $27x^3 - \frac{1}{27} = \left(3x - \frac{1}{3}\right)\left(9x^2 + x + \frac{1}{9}\right)$

33. Factoring: $64a^3 + 125b^3 = (4a + 5b)(16a^2 - 20ab + 25b^2)$

35. Factoring: $\frac{1}{8}x^3 - \frac{1}{27}y^3 = \left(\frac{1}{2}x - \frac{1}{3}y\right)\left(\frac{1}{4}x^2 + \frac{1}{6}xy + \frac{1}{9}y^2\right)$

37. Factoring: $a^6 - b^6 = (a^3 + b^3)(a^3 - b^3) = (a + b)(a^2 - ab + b^2)(a - b)(a^2 + ab + b^2)$

39.   Factoring: $64x^6 - y^6 = \left(8x^3 + y^3\right)\left(8x^3 - y^3\right) = (2x+y)\left(4x^2 - 2xy + y^2\right)(2x-y)\left(4x^2 + 2xy + y^2\right)$

41.   Factoring: $x^6 - (5y)^6 = \left(x^3 + (5y)^3\right)\left(x^3 - (5y)^3\right) = (x+5y)\left(x^2 - 5xy + 25y^2\right)(x-5y)\left(x^2 + 5xy + 25y^2\right)$

43.   Solving for x:
$$2x - 6y = 8$$
$$2x = 6y + 8$$
$$x = 3y + 4$$

45.   Solving for x:
$$4x - 6y = 8$$
$$4x = 6y + 8$$
$$x = \tfrac{3}{2}y + 2$$

47.   Solving for y:
$$3x - 6y = -18$$
$$-6y = -3x - 18$$
$$y = \tfrac{1}{2}x + 3$$

49.   Solving for y:
$$4x - 6y = 24$$
$$-6y = -4x + 24$$
$$y = \tfrac{2}{3}x - 4$$

51.   Multiplying: $2x^3(x+2)(x-2) = 2x^3\left(x^2 - 4\right) = 2x^5 - 8x^3$

53.   Multiplying: $3x^2(x-3)^2 = 3x^2\left(x^2 - 6x + 9\right) = 3x^4 - 18x^3 + 27x^2$

55.   Multiplying: $y\left(y^2 + 25\right) = y^3 + 25y$

57.   Multiplying: $(5a-2)(3a+1) = 15a^2 + 5a - 6a - 2 = 15a^2 - a - 2$

59.   Multiplying: $4x^2(x-5)(x+2) = 4x^2\left(x^2 - 3x - 10\right) = 4x^4 - 12x^3 - 40x^2$

61.   Multiplying: $2ab^3\left(b^2 - 4b + 1\right) = 2ab^5 - 8ab^4 + 2ab^3$

6.6 Factoring: A General Review

1.   Factoring the polynomial: $x^2 - 81 = (x+9)(x-9)$

3.   Factoring the polynomial: $x^2 + 2x - 15 = (x+5)(x-3)$

5.   Factoring the polynomial: $x^2 + 6x + 9 = (x+3)(x+3) = (x+3)^2$

7.   Factoring the polynomial: $y^2 - 10y + 25 = (y-5)(y-5) = (y-5)^2$

9.   Factoring the polynomial: $2a^3b + 6a^2b + 2ab = 2ab\left(a^2 + 3a + 1\right)$

11.   The polynomial $x^2 + x + 1$ cannot be factored.

13.   Factoring the polynomial: $12a^2 - 75 = 3\left(4a^2 - 25\right) = 3(2a+5)(2a-5)$

15.   Factoring the polynomial: $9x^2 - 12xy + 4y^2 = (3x-2y)(3x-2y) = (3x-2y)^2$

17.   Factoring the polynomial: $4x^3 + 16xy^2 = 4x\left(x^2 + 4y^2\right)$

19.   Factoring the polynomial: $2y^3 + 20y^2 + 50y = 2y\left(y^2 + 10y + 25\right) = 2y(y+5)(y+5) = 2y(y+5)^2$

21.   Factoring the polynomial: $a^6 + 4a^4b^2 = a^4\left(a^2 + 4b^2\right)$

23.   Factoring the polynomial: $xy + 3x + 4y + 12 = x(y+3) + 4(y+3) = (y+3)(x+4)$

25.   Factoring the polynomial: $x^3 - 27 = (x-3)\left(x^2 + 3x + 9\right)$

27.   Factoring the polynomial: $xy - 5x + 2y - 10 = x(y-5) + 2(y-5) = (y-5)(x+2)$

29.   Factoring the polynomial: $5a^2 + 10ab + 5b^2 = 5\left(a^2 + 2ab + b^2\right) = 5(a+b)(a+b) = 5(a+b)^2$

31.   The polynomial $x^2 + 49$ cannot be factored.

33. Factoring the polynomial: $3x^2 + 15xy + 18y^2 = 3\left(x^2 + 5xy + 6y^2\right) = 3(x+2y)(x+3y)$

35. Factoring the polynomial: $2x^2 + 15x - 38 = (2x+19)(x-2)$

37. Factoring the polynomial: $100x^2 - 300x + 200 = 100\left(x^2 - 3x + 2\right) = 100(x-2)(x-1)$

39. Factoring the polynomial: $x^2 - 64 = (x+8)(x-8)$

41. Factoring the polynomial: $x^2 + 3x + ax + 3a = x(x+3) + a(x+3) = (x+3)(x+a)$

43. Factoring the polynomial: $49a^7 - 9a^5 = a^5\left(49a^2 - 9\right) = a^5(7a+3)(7a-3)$

45. The polynomial $49x^2 + 9y^2$ cannot be factored.

47. Factoring the polynomial: $25a^3 + 20a^2 + 3a = a\left(25a^2 + 20a + 3\right) = a(5a+3)(5a+1)$

49. Factoring the polynomial: $xa - xb + ay - by = x(a-b) + y(a-b) = (a-b)(x+y)$

51. Factoring the polynomial: $48a^4b - 3a^2b = 3a^2b\left(16a^2 - 1\right) = 3a^2b(4a+1)(4a-1)$

53. Factoring the polynomial: $20x^4 - 45x^2 = 5x^2\left(4x^2 - 9\right) = 5x^2(2x+3)(2x-3)$

55. Factoring the polynomial: $3x^2 + 35xy - 82y^2 = (3x+41y)(x-2y)$

57. Factoring the polynomial: $16x^5 - 44x^4 + 30x^3 = 2x^3\left(8x^2 - 22x + 15\right) = 2x^3(2x-3)(4x-5)$

59. Factoring the polynomial: $2x^2 + 2ax + 3x + 3a = 2x(x+a) + 3(x+a) = (x+a)(2x+3)$

61. Factoring the polynomial: $y^4 - 1 = \left(y^2 + 1\right)\left(y^2 - 1\right) = \left(y^2 + 1\right)(y+1)(y-1)$

63. Factoring the polynomial:

$$12x^4y^2 + 36x^3y^3 + 27x^2y^4 = 3x^2y^2\left(4x^2 + 12xy + 9y^2\right) = 3x^2y^2(2x+3y)(2x+3y) = 3x^2y^2(2x+3y)^2$$

65. Factoring the polynomial: $16t^2 - 64t + 48 = 16\left(t^2 - 4t + 3\right) = 16(t-1)(t-3)$

67. Factoring the polynomial: $54x^2 + 111x + 56 = (6x+7)(9x+8)$

69. Solving the equation:
$$-2(x+4) = -10$$
$$-2x - 8 = -10$$
$$-2x = -2$$
$$x = 1$$

71. Solving the equation:
$$\tfrac{3}{5}x + 4 = 22$$
$$\tfrac{3}{5}x = 18$$
$$x = \tfrac{5}{3}(18) = 30$$

73. Solving the equation:
$$6x - 4(9 - x) = -96$$
$$6x - 36 + 4x = -96$$
$$10x - 36 = -96$$
$$10x = -60$$
$$x = -6$$

75. Solving the equation:
$$2x - 3(4x - 7) = -3x$$
$$2x - 12x + 21 = -3x$$
$$-10x + 21 = -3x$$
$$21 = 7x$$
$$x = 3$$

77. Solving the equation:
$$\tfrac{1}{2}x - \tfrac{5}{12} = \tfrac{1}{12}x + \tfrac{5}{12}$$
$$12\left(\tfrac{1}{2}x - \tfrac{5}{12}\right) = 12\left(\tfrac{1}{12}x + \tfrac{5}{12}\right)$$
$$6x - 5 = x + 5$$
$$5x - 5 = 5$$
$$5x = 10$$
$$x = 2$$

79. Solving the equation:
$$3x - 6 = 9$$
$$3x = 15$$
$$x = 5$$

81. Solving the equation:
$$2x + 3 = 0$$
$$2x = -3$$
$$x = -\frac{3}{2}$$

83. Solving the equation:
$$4x + 3 = 0$$
$$4x = -3$$
$$x = -\frac{3}{4}$$

6.7 Solving Equations by Factoring

1. Setting each factor equal to 0:
$$x + 2 = 0 \qquad\qquad x - 1 = 0$$
$$x = -2 \qquad\qquad\quad x = 1$$
The solutions are -2 and 1.

3. Setting each factor equal to 0:
$$a - 4 = 0 \qquad\qquad a - 5 = 0$$
$$a = 4 \qquad\qquad\quad a = 5$$
The solutions are 4 and 5.

5. Setting each factor equal to 0:
$$x = 0 \qquad\quad x + 1 = 0 \qquad\qquad x - 3 = 0$$
$$x = -1 \qquad\qquad\quad x = 3$$
The solutions are $0, -1$ and 3.

7. Setting each factor equal to 0:
$$3x + 2 = 0 \qquad\qquad 2x + 3 = 0$$
$$3x = -2 \qquad\qquad\quad 2x = -3$$
$$x = -\frac{2}{3} \qquad\qquad\quad x = -\frac{3}{2}$$
The solutions are $-\frac{2}{3}$ and $-\frac{3}{2}$.

9. Setting each factor equal to 0:
$$m = 0 \qquad\quad 3m + 4 = 0 \qquad\qquad 3m - 4 = 0$$
$$3m = -4 \qquad\qquad\quad 3m = 4$$
$$m = -\frac{4}{3} \qquad\qquad\quad m = \frac{4}{3}$$
The solutions are $0, -\frac{4}{3}$ and $\frac{4}{3}$.

11. Setting each factor equal to 0:
$$2y = 0 \qquad\quad 3y + 1 = 0 \qquad\qquad 5y + 3 = 0$$
$$y = 0 \qquad\quad 3y = -1 \qquad\qquad\quad 5y = -3$$
$$y = -\frac{1}{3} \qquad\qquad\quad y = -\frac{3}{5}$$
The solutions are $0, -\frac{1}{3}$ and $-\frac{3}{5}$.

13. Solving by factoring:
$$x^2 + 3x + 2 = 0$$
$$(x + 2)(x + 1) = 0$$
$$x = -2, -1$$

15. Solving by factoring:
$$x^2 - 9x + 20 = 0$$
$$(x - 4)(x - 5) = 0$$
$$x = 4, 5$$

17. Solving by factoring:
$$a^2 - 2a - 24 = 0$$
$$(a - 6)(a + 4) = 0$$
$$a = 6, -4$$

19. Solving by factoring:
$$100x^2 - 500x + 600 = 0$$
$$100\left(x^2 - 5x + 6\right) = 0$$
$$100(x - 2)(x - 3) = 0$$
$$x = 2, 3$$

21.   Solving by factoring:
$$x^2 = -6x - 9$$
$$x^2 + 6x + 9 = 0$$
$$(x+3)^2 = 0$$
$$x + 3 = 0$$
$$x = -3$$

23.   Solving by factoring:
$$a^2 - 16 = 0$$
$$(a+4)(a-4) = 0$$
$$a = -4, 4$$

25.   Solving by factoring:
$$2x^2 + 5x - 12 = 0$$
$$(2x-3)(x+4) = 0$$
$$x = \frac{3}{2}, -4$$

27.   Solving by factoring:
$$9x^2 + 12x + 4 = 0$$
$$(3x+2)^2 = 0$$
$$3x + 2 = 0$$
$$x = -\frac{2}{3}$$

29.   Solving by factoring:
$$a^2 + 25 = 10a$$
$$a^2 - 10a + 25 = 0$$
$$(a-5)^2 = 0$$
$$a - 5 = 0$$
$$a = 5$$

31.   Solving by factoring:
$$2x^2 = 3x + 20$$
$$2x^2 - 3x - 20 = 0$$
$$(2x+5)(x-4) = 0$$
$$x = -\frac{5}{2}, 4$$

33.   Solving by factoring:
$$3m^2 = 20 - 7m$$
$$3m^2 + 7m - 20 = 0$$
$$(3m-5)(m+4) = 0$$
$$m = \frac{5}{3}, -4$$

35.   Solving by factoring:
$$4x^2 - 49 = 0$$
$$(2x+7)(2x-7) = 0$$
$$x = -\frac{7}{2}, \frac{7}{2}$$

37.   Solving by factoring:
$$x^2 + 6x = 0$$
$$x(x+6) = 0$$
$$x = 0, -6$$

39.   Solving by factoring:
$$x^2 - 3x = 0$$
$$x(x-3) = 0$$
$$x = 0, 3$$

41.   Solving by factoring:
$$2x^2 = 8x$$
$$2x^2 - 8x = 0$$
$$2x(x-4) = 0$$
$$x = 0, 4$$

43.   Solving by factoring:
$$3x^2 = 15x$$
$$3x^2 - 15x = 0$$
$$3x(x-5) = 0$$
$$x = 0, 5$$

45.   Solving by factoring:
$$1,400 = 400 + 700x - 100x^2$$
$$100x^2 - 700x + 1,000 = 0$$
$$100(x^2 - 7x + 10) = 0$$
$$100(x-5)(x-2) = 0$$
$$x = 2, 5$$

47.   Solving by factoring:
$$6x^2 = -5x + 4$$
$$6x^2 + 5x - 4 = 0$$
$$(3x+4)(2x-1) = 0$$
$$x = -\frac{4}{3}, \frac{1}{2}$$

49.   Solving by factoring:
$$x(2x-3) = 20$$
$$2x^2 - 3x = 20$$
$$2x^2 - 3x - 20 = 0$$
$$(2x+5)(x-4) = 0$$
$$x = -\frac{5}{2}, 4$$

51.   Solving by factoring:
$$t(t+2) = 80$$
$$t^2 + 2t = 80$$
$$t^2 + 2t - 80 = 0$$
$$(t+10)(t-8) = 0$$
$$t = -10, 8$$

53. Solving by factoring:
$$4000 = (1300 - 100p)p$$
$$4000 = 1300p - 100p^2$$
$$100p^2 - 1300p + 4000 = 0$$
$$100(p^2 - 13p + 40) = 0$$
$$100(p - 8)(p - 5) = 0$$
$$p = 5, 8$$

55. Solving by factoring:
$$x(14 - x) = 48$$
$$14x - x^2 = 48$$
$$-x^2 + 14x - 48 = 0$$
$$x^2 - 14x + 48 = 0$$
$$(x - 6)(x - 8) = 0$$
$$x = 6, 8$$

57. Solving by factoring:
$$(x + 5)^2 = 2x + 9$$
$$x^2 + 10x + 25 = 2x + 9$$
$$x^2 + 8x + 16 = 0$$
$$(x + 4)^2 = 0$$
$$x + 4 = 0$$
$$x = -4$$

59. Solving by factoring:
$$(y - 6)^2 = y - 4$$
$$y^2 - 12y + 36 = y - 4$$
$$y^2 - 13y + 40 = 0$$
$$(y - 5)(y - 8) = 0$$
$$y = 5, 8$$

61. Solving by factoring:
$$10^2 = (x + 2)^2 + x^2$$
$$100 = x^2 + 4x + 4 + x^2$$
$$100 = 2x^2 + 4x + 4$$
$$0 = 2x^2 + 4x - 96$$
$$0 = 2(x^2 + 2x - 48)$$
$$0 = 2(x + 8)(x - 6)$$
$$x = -8, 6$$

63. Solving by factoring:
$$2x^3 + 11x^2 + 12x = 0$$
$$x(2x^2 + 11x + 12) = 0$$
$$x(2x + 3)(x + 4) = 0$$
$$x = 0, -\tfrac{3}{2}, -4$$

65. Solving by factoring:
$$4y^3 - 2y^2 - 30y = 0$$
$$2y(2y^2 - y - 15) = 0$$
$$2y(2y + 5)(y - 3) = 0$$
$$y = 0, -\tfrac{5}{2}, 3$$

67. Solving by factoring:
$$8x^3 + 16x^2 = 10x$$
$$8x^3 + 16x^2 - 10x = 0$$
$$2x(4x^2 + 8x - 5) = 0$$
$$2x(2x - 1)(2x + 5) = 0$$
$$x = 0, \tfrac{1}{2}, -\tfrac{5}{2}$$

69. Solving by factoring:
$$20a^3 = -18a^2 + 18a$$
$$20a^3 + 18a^2 - 18a = 0$$
$$2a(10a^2 + 9a - 9) = 0$$
$$2a(5a - 3)(2a + 3) = 0$$
$$a = 0, \tfrac{3}{5}, -\tfrac{3}{2}$$

71. Solving by factoring:
$$16t^2 - 32t + 12 = 0$$
$$4(4t^2 - 8t + 3) = 0$$
$$4(2t - 1)(2t - 3) = 0$$
$$t = \tfrac{1}{2}, \tfrac{3}{2}$$

73. Solving the equation:
$$(a - 5)(a + 4) = -2a$$
$$a^2 - a - 20 = -2a$$
$$a^2 + a - 20 = 0$$
$$(a + 5)(a - 4) = 0$$
$$a = -5, 4$$

75. Solving the equation:
$$3x(x + 1) - 2x(x - 5) = -42$$
$$3x^2 + 3x - 2x^2 + 10x = -42$$
$$x^2 + 13x + 42 = 0$$
$$(x + 7)(x + 6) = 0$$
$$x = -7, -6$$

77.   Solving the equation:
$$2x(x+3) = x(x+2) - 3$$
$$2x^2 + 6x = x^2 + 2x - 3$$
$$x^2 + 4x + 3 = 0$$
$$(x+3)(x+1) = 0$$
$$x = -3, -1$$

79.   Solving the equation:
$$a(a-3) + 6 = 2a$$
$$a^2 - 3a + 6 = 2a$$
$$a^2 - 5a + 6 = 0$$
$$(a-2)(a-3) = 0$$
$$a = 2, 3$$

81.   Solving the equation:
$$15(x+20) + 15x = 2x(x+20)$$
$$15x + 300 + 15x = 2x^2 + 40x$$
$$30x + 300 = 2x^2 + 40x$$
$$0 = 2x^2 + 10x - 300$$
$$0 = 2(x^2 + 5x - 150)$$
$$0 = 2(x+15)(x-10)$$
$$x = -15, 10$$

83.   Solving the equation:

$$15 = a(a+2)$$
$$15 = a^2 + 2a$$
$$0 = a^2 + 2a - 15$$
$$0 = (a+5)(a-3)$$
$$a = -5, 3$$

85.   Solving by factoring:
$$x^3 + 3x^2 - 4x - 12 = 0$$
$$x^2(x+3) - 4(x+3) = 0$$
$$(x+3)(x^2 - 4) = 0$$
$$(x+3)(x+2)(x-2) = 0$$
$$x = -3, -2, 2$$

87.   Solving by factoring:
$$x^3 + x^2 - 16x - 16 = 0$$
$$x^2(x+1) - 16(x+1) = 0$$
$$(x+1)(x^2 - 16) = 0$$
$$(x+1)(x+4)(x-4) = 0$$
$$x = -1, -4, 4$$

89.   Writing the equation:
$$(x-3)(x-5) = 0$$
$$x^2 - 8x + 15 = 0$$

91.   **a.**   Writing the equation:
$$(x-3)(x-2) = 0$$
$$x^2 - 5x + 6 = 0$$

   b.   Writing the equation:
$$(x-1)(x-6) = 0$$
$$x^2 - 7x + 6 = 0$$

   c.   Writing the equation:
$$(x-3)(x+2) = 0$$
$$x^2 - x - 6 = 0$$

93.   Simplifying using the properties of exponents: $2^{-3} = \dfrac{1}{2^3} = \dfrac{1}{8}$

95.   Simplifying using the properties of exponents: $\dfrac{x^5}{x^{-3}} = x^{5-(-3)} = x^{5+3} = x^8$

97.   Simplifying using the properties of exponents: $\dfrac{\left(x^2\right)^3}{\left(x^{-3}\right)^4} = \dfrac{x^6}{x^{-12}} = x^{6-(-12)} = x^{6+12} = x^{18}$

99.   Writing in scientific notation: $0.0056 = 5.6 \times 10^{-3}$

101.   Writing in scientific notation: $5,670,000,000 = 5.67 \times 10^9$

103.   Let x and $x + 1$ represent the two consecutive integers. The equation is: $x(x+1) = 72$

105.   Let x and $x + 2$ represent the two consecutive odd integers. The equation is: $x(x+2) = 99$

107.   Let x and $x + 2$ represent the two consecutive even integers. The equation is: $x(x+2) = 5[x + (x+2)] - 10$

109. Let x represent the cost of the suit and $5x$ represent the cost of the bicycle. The equation is:

$$x + 5x = 90$$
$$6x = 90$$
$$x = 15$$
$$5x = 75$$

The suit costs $15 and the bicycle costs $75.

111. Let x represent the cost of the lot and $4x$ represent the cost of the house. The equation is:

$$x + 4x = 3000$$
$$5x = 3000$$
$$x = 600$$
$$4x = 2400$$

The lot cost $600 and the house cost $2,400.

6.8 Applications

1. Let x and $x + 2$ represent the two integers. The equation is:

$$x(x+2) = 80$$
$$x^2 + 2x = 80$$
$$x^2 + 2x - 80 = 0$$
$$(x+10)(x-8) = 0$$
$$x = -10, 8$$
$$x + 2 = -8, 10$$

The two numbers are either −10 and −8, or 8 and 10.

3. Let x and $x + 2$ represent the two integers. The equation is:

$$x(x+2) = 99$$
$$x^2 + 2x = 99$$
$$x^2 + 2x - 99 = 0$$
$$(x+11)(x-9) = 0$$
$$x = -11, 9$$
$$x + 2 = -9, 11$$

The two numbers are either −11 and −9, or 9 and 11.

5. Let x and $x + 2$ represent the two integers. The equation is:

$$x(x+2) = 5(x+x+2) - 10$$
$$x^2 + 2x = 5(2x+2) - 10$$
$$x^2 + 2x = 10x + 10 - 10$$
$$x^2 + 2x = 10x$$
$$x^2 - 8x = 0$$
$$x(x-8) = 0$$
$$x = 0, 8$$
$$x + 2 = 2, 10$$

The two numbers are either 0 and 2, or 8 and 10.

7. Let x and $14 - x$ represent the two numbers. The equation is:

$$x(14-x) = 48$$
$$14x - x^2 = 48$$
$$0 = x^2 - 14x + 48$$
$$0 = (x-8)(x-6)$$
$$x = 8, 6$$
$$14 - x = 6, 8$$

The two numbers are 6 and 8.

9.   Let x and $5x + 2$ represent the two numbers. The equation is:

$$x(5x+2) = 24$$
$$5x^2 + 2x = 24$$
$$5x^2 + 2x - 24 = 0$$
$$(5x+12)(x-2) = 0$$
$$x = -\frac{12}{5}, 2$$
$$5x + 2 = -10, 12$$

The two numbers are either $-\frac{12}{5}$ and -10, or 2 and 12.

11.   Let x and $4x$ represent the two numbers. The equation is:

$$x(4x) = 4(x+4x)$$
$$4x^2 = 4(5x)$$
$$4x^2 = 20x$$
$$4x^2 - 20x = 0$$
$$4x(x-5) = 0$$
$$x = 0, 5$$
$$4x = 0, 20$$

The two numbers are either 0 and 0, or 5 and 20.

13.   Let w represent the width and $w + 1$ represent the length. The equation is:

$$w(w+1) = 12$$
$$w^2 + w = 12$$
$$w^2 + w - 12 = 0$$
$$(w+4)(w-3) = 0$$
$$w = 3 \quad (w = -4 \text{ is impossible})$$
$$w + 1 = 4$$

The width is 3 inches and the length is 4 inches.

15.   Let b represent the base and $2b$ represent the height. The equation is:

$$\tfrac{1}{2}b(2b) = 9$$
$$b^2 = 9$$
$$b^2 - 9 = 0$$
$$(b+3)(b-3) = 0$$
$$b = 3 \quad (b = -3 \text{ is impossible})$$

The base is 3 inches.

17.   Let x and $x + 2$ represent the two legs. The equation is:

$$x^2 + (x+2)^2 = 10^2$$
$$x^2 + x^2 + 4x + 4 = 100$$
$$2x^2 + 4x + 4 = 100$$
$$2x^2 + 4x - 96 = 0$$
$$2(x^2 + 2x - 48) = 0$$
$$2(x+8)(x-6) = 0$$
$$x = 6 \quad (x = -8 \text{ is impossible})$$
$$x + 2 = 8$$

The legs are 6 inches and 8 inches.

19.   Let x represent the longer leg and $x + 1$ represent the hypotenuse. The equation is:

$$5^2 + x^2 = (x+1)^2$$
$$25 + x^2 = x^2 + 2x + 1$$
$$25 = 2x + 1$$
$$24 = 2x$$
$$x = 12$$

The longer leg is 12 meters.

21.   Setting $C = \$1,400$:

$$1400 = 400 + 700x - 100x^2$$
$$100x^2 - 700x + 1000 = 0$$
$$100\left(x^2 - 7x + 10\right) = 0$$
$$100(x-5)(x-2) = 0$$
$$x = 2, 5$$

The company can manufacture either 200 items or 500 items.

23.   Setting $C = \$2,200$:

$$2200 = 600 + 1000x - 100x^2$$
$$100x^2 - 1000x + 1600 = 0$$
$$100\left(x^2 - 10x + 16\right) = 0$$
$$100(x-2)(x-8) = 0$$
$$x = 2, 8$$

The company can manufacture either 200 videotapes or 800 videotapes.

25.   The revenue is given by: $R = xp = (1200 - 100p)p$. Setting $R = \$3,200$:

$$3200 = (1200 - 100p)p$$
$$3200 = 1200p - 100p^2$$
$$100p^2 - 1200p + 3200 = 0$$
$$100\left(p^2 - 12p + 32\right) = 0$$
$$100(p-4)(p-8) = 0$$
$$p = 4, 8$$

The company should sell the ribbons for either $4 or $8.

27.   The revenue is given by: $R = xp = (1700 - 100p)p$. Setting $R = \$7,000$:

$$7000 = (1700 - 100p)p$$
$$7000 = 1700p - 100p^2$$
$$100p^2 - 1700p + 7000 = 0$$
$$100\left(p^2 - 17p + 70\right) = 0$$
$$100(p-7)(p-10) = 0$$
$$p = 7, 10$$

The calculators should be sold for either $7 or $10.

29.   **a.**   Let x represent the distance from the base to the wall, and $2x + 2$ represent the height on the wall. Using the Pythagorean theorem:

$$x^2 + (2x+2)^2 = 13^2$$
$$x^2 + 4x^2 + 8x + 4 = 169$$
$$5x^2 + 8x - 165 = 0$$
$$(5x+33)(x-5) = 0$$
$$x = 5 \quad \left(x = -\tfrac{33}{5} \text{ is impossible}\right)$$

The base of the ladder is 5 feet from the wall.

b.   Since $2x + 2 = 2 \bullet 5 + 2 = 12$, the ladder reaches a height of 12 feet.

31. **a.** Finding when $h = 0$:
$$0 = -16t^2 + 396t + 100$$
$$16t^2 - 396t - 100 = 0$$
$$4t^2 - 99t - 25 = 0$$
$$(4t + 1)(t - 25) = 0$$
$$t = 25 \quad \left(t = -\tfrac{1}{4} \text{ is impossible}\right)$$

The bullet will land on the ground after 25 seconds.

b. Completing the table:

t (seconds)	h (feet)
0	100
5	1,680
10	2,460
15	2,440
20	1,620
25	0

33. Simplifying the expression: $\left(5x^3\right)^2 \left(2x^6\right)^3 = 25x^6 \cdot 8x^{18} = 200x^{24}$

35. Simplifying the expression: $\dfrac{x^4}{x^{-3}} = x^{4-(-3)} = x^{4+3} = x^7$

37. Simplifying the expression: $\left(2 \times 10^{-4}\right)\left(4 \times 10^5\right) = 8 \times 10^1 = 80$

39. Simplifying the expression: $20ab^2 - 16ab^2 + 6ab^2 = 10ab^2$

41. Multiplying using the distributive property:
$$2x^2\left(3x^2 + 3x - 1\right) = 2x^2\left(3x^2\right) + 2x^2\left(3x\right) - 2x^2\left(1\right) = 6x^4 + 6x^3 - 2x^2$$

43. Multiplying using the square of binomial formula: $(3y - 5)^2 = (3y)^2 - 2(3y)(5) + (5)^2 = 9y^2 - 30y + 25$

45. Multiplying using the difference of squares formula: $\left(2a^2 + 7\right)\left(2a^2 - 7\right) = \left(2a^2\right)^2 - (7)^2 = 4a^4 - 49$

Chapter 6 Review/Test

1. Factoring the polynomial: $10x - 20 = 10(x - 2)$
2. Factoring the polynomial: $4x^3 - 9x^2 = x^2\left(4x - 9\right)$

3. Factoring the polynomial: $5x - 5y = 5(x - y)$
4. Factoring the polynomial: $7x^3 + 2x = x\left(7x^2 + 2\right)$

5. Factoring the polynomial: $8x + 4 = 4(2x + 1)$

6. Factoring the polynomial: $2x^2 + 14x + 6 = 2\left(x^2 + 7x + 3\right)$

7. Factoring the polynomial: $24y^2 - 40y + 48 = 8\left(3y^2 - 5y + 6\right)$

8. Factoring the polynomial: $30xy^3 - 45x^3y^2 = 15xy^2\left(2y - 3x^2\right)$

9. Factoring the polynomial: $49a^3 - 14b^3 = 7\left(7a^3 - 2b^3\right)$

10. Factoring the polynomial: $6ab^2 + 18a^3b^3 - 24a^2b = 6ab\left(b + 3a^2b^2 - 4a\right)$

11. Factoring the polynomial: $xy + bx + ay + ab = x(y + b) + a(y + b) = (y + b)(x + a)$

12. Factoring the polynomial: $xy + 4x - 5y - 20 = x(y + 4) - 5(y + 4) = (y + 4)(x - 5)$

13. Factoring the polynomial: $2xy + 10x - 3y - 15 = 2x(y + 5) - 3(y + 5) = (y + 5)(2x - 3)$

14. Factoring the polynomial: $5x^2 - 4ax - 10bx + 8ab = x(5x - 4a) - 2b(5x - 4a) = (5x - 4a)(x - 2b)$

15. Factoring the polynomial: $y^2 + 9y + 14 = (y + 7)(y + 2)$

16. Factoring the polynomial: $w^2 + 15w + 50 = (w + 10)(w + 5)$

17. Factoring the polynomial: $a^2 - 14a + 48 = (a - 8)(a - 6)$

18. Factoring the polynomial: $r^2 - 18r + 72 = (r - 12)(r - 6)$

19. Factoring the polynomial: $y^2 + 20y + 99 = (y + 9)(y + 11)$

20. Factoring the polynomial: $y^2 + 8y + 12 = (y + 6)(y + 2)$

21. Factoring the polynomial: $2x^2 + 13x + 15 = (2x + 3)(x + 5)$

22. Factoring the polynomial: $4y^2 - 12y + 5 = (2y - 5)(2y - 1)$

23. Factoring the polynomial: $5y^2 + 11y + 6 = (5y + 6)(y + 1)$

24. Factoring the polynomial: $20a^2 - 27a + 9 = (5a - 3)(4a - 3)$

25. Factoring the polynomial: $6r^2 + 5rt - 6t^2 = (3r - 2t)(2r + 3t)$

26. Factoring the polynomial: $10x^2 - 29x - 21 = (5x + 3)(2x - 7)$

27. Factoring the polynomial: $n^2 - 81 = (n + 9)(n - 9)$

28. Factoring the polynomial: $4y^2 - 9 = (2y + 3)(2y - 3)$

29. The expression $x^2 + 49$ cannot be factored.

30. Factoring the polynomial: $36y^2 - 121x^2 = (6y + 11x)(6y - 11x)$

31. Factoring the polynomial: $64a^2 - 121b^2 = (8a + 11b)(8a - 11b)$

32. Factoring the polynomial: $64 - 9m^2 = (8 + 3m)(8 - 3m)$

33. Factoring the polynomial: $y^2 + 20y + 100 = (y + 10)(y + 10) = (y + 10)^2$

34. Factoring the polynomial: $m^2 - 16m + 64 = (m - 8)(m - 8) = (m - 8)^2$

35. Factoring the polynomial: $64t^2 + 16t + 1 = (8t + 1)(8t + 1) = (8t + 1)^2$

36. Factoring the polynomial: $16n^2 - 24n + 9 = (4n - 3)(4n - 3) = (4n - 3)^2$

37. Factoring the polynomial: $4r^2 - 12rt + 9t^2 = (2r - 3t)(2r - 3t) = (2r - 3t)^2$

38. Factoring the polynomial: $9m^2 + 30mn + 25n^2 = (3m + 5n)(3m + 5n) = (3m + 5n)^2$

39. Factoring the polynomial: $2x^2 + 20x + 48 = 2(x^2 + 10x + 24) = 2(x + 4)(x + 6)$

40. Factoring the polynomial: $a^3 - 10a^2 + 21a = a(a^2 - 10a + 21) = a(a - 7)(a - 3)$

41. Factoring the polynomial: $3m^3 - 18m^2 - 21m = 3m(m^2 - 6m - 7) = 3m(m - 7)(m + 1)$

42. Factoring the polynomial: $5y^4 + 10y^3 - 40y^2 = 5y^2(y^2 + 2y - 8) = 5y^2(y + 4)(y - 2)$

43. Factoring the polynomial: $8x^2 + 16x + 6 = 2(4x^2 + 8x + 3) = 2(2x + 1)(2x + 3)$

44. Factoring the polynomial: $3a^3 - 14a^2 - 5a = a(3a^2 - 14a - 5) = a(3a + 1)(a - 5)$

45. Factoring the polynomial: $20m^3 - 34m^2 + 6m = 2m(10m^2 - 17m + 3) = 2m(5m - 1)(2m - 3)$

46. Factoring the polynomial: $30x^2y - 55xy^2 + 15y^3 = 5y(6x^2 - 11xy + 3y^2) = 5y(3x - y)(2x - 3y)$

47.   Factoring the polynomial: $4x^2 + 40x + 100 = 4\left(x^2 + 10x + 25\right) = 4\left(x+5\right)\left(x+5\right) = 4\left(x+5\right)^2$

48.   Factoring the polynomial: $4x^3 + 12x^2 + 9x = x\left(4x^2 + 12x + 9\right) = x\left(2x+3\right)\left(2x+3\right) = x\left(2x+3\right)^2$

49.   Factoring the polynomial: $5x^2 - 45 = 5\left(x^2 - 9\right) = 5\left(x+3\right)\left(x-3\right)$

50.   Factoring the polynomial: $12x^3 - 27xy^2 = 3x\left(4x^2 - 9y^2\right) = 3x\left(2x+3y\right)\left(2x-3y\right)$

51.   Factoring the polynomial: $8a^3 - b^3 = \left(2a-b\right)\left(4a^2 + 2ab + b^2\right)$

52.   Factoring the polynomial: $27x^3 + 8y^3 = \left(3x+2y\right)\left(9x^2 - 6xy + 4y^2\right)$

53.   Factoring the polynomial: $125x^3 - 64y^3 = \left(5x - 4y\right)\left(25x^2 + 20xy + 16y^2\right)$

54.   Factoring the polynomial: $6a^3b + 33a^2b^2 + 15ab^3 = 3ab\left(2a^2 + 11ab + 5b^2\right) = 3ab\left(2a+b\right)\left(a+5b\right)$

55.   Factoring the polynomial: $x^5 - x^3 = x^3\left(x^2 - 1\right) = x^3\left(x+1\right)\left(x-1\right)$

56.   Factoring the polynomial: $4y^6 + 9y^4 = y^4\left(4y^2 + 9\right)$

57.   Factoring the polynomial: $12x^5 + 20x^4y - 8x^3y^2 = 4x^3\left(3x^2 + 5xy - 2y^2\right) = 4x^3\left(3x - y\right)\left(x + 2y\right)$

58.   Factoring the polynomial: $30a^4b + 35a^3b^2 - 15a^2b^3 = 5a^2b\left(6a^2 + 7ab - 3b^2\right) = 5a^2b\left(3a - b\right)\left(2a + 3b\right)$

59.   Factoring the polynomial: $18a^3b^2 + 3a^2b^3 - 6ab^4 = 3ab^2\left(6a^2 + ab - 2b^2\right) = 3ab^2\left(3a + 2b\right)\left(2a - b\right)$

60.   Setting each factor equal to 0:
$$x - 5 = 0 \qquad\qquad x + 2 = 0$$
$$x = 5 \qquad\qquad\qquad x = -2$$
The solutions are $-2, 5$.

61.   Setting each factor equal to 0:
$$2y + 5 = 0 \qquad\qquad 2y - 5 = 0$$
$$2y = -5 \qquad\qquad\quad 2y = 5$$
$$y = -\tfrac{5}{2} \qquad\qquad\quad y = \tfrac{5}{2}$$
The solutions are $-\tfrac{5}{2}, \tfrac{5}{2}$.

62.   Solving the equation by factoring:
$$m^2 + 3m = 10$$
$$m^2 + 3m - 10 = 0$$
$$\left(m+5\right)\left(m-2\right) = 0$$
$$m = -5, 2$$

63.   Solving the equation by factoring:
$$a^2 - 49 = 0$$
$$\left(a+7\right)\left(a-7\right) = 0$$
$$a = -7, 7$$

64.   Solving the equation by factoring:
$$m^2 - 9m = 0$$
$$m\left(m-9\right) = 0$$
$$m = 0, 9$$

65.   Solving the equation by factoring:
$$6y^2 = -13y - 6$$
$$6y^2 + 13y + 6 = 0$$
$$\left(3y+2\right)\left(2y+3\right) = 0$$
$$y = -\tfrac{2}{3}, -\tfrac{3}{2}$$

66.   Solving the equation by factoring:
$$9x^4 + 9x^3 = 10x^2$$
$$9x^4 + 9x^3 - 10x^2 = 0$$
$$x^2\left(9x^2 + 9x - 10\right) = 0$$
$$x^2\left(3x - 2\right)\left(3x + 5\right) = 0$$
$$x = 0, \tfrac{2}{3}, -\tfrac{5}{3}$$

67. Let x and $x + 2$ represent the two integers. The equation is:
$$x(x+2) = 120$$
$$x^2 + 2x = 120$$
$$x^2 + 2x - 120 = 0$$
$$(x+12)(x-10) = 0$$
$$x = -12, 10$$
$$x + 2 = -10, 12$$
The two integers are either −12 and −10, or 10 and 12.

68. Let x and $x + 1$ represent the two integers. The equation is:
$$x(x+1) = 110$$
$$x^2 + x = 110$$
$$x^2 + x - 110 = 0$$
$$(x+11)(x-10) = 0$$
$$x = -11, 10$$
$$x + 1 = -10, 11$$
The two integers are either −11 and −10, or 10 and 11.

69. Let x and $x + 2$ represent the two integers. The equation is:
$$x(x+2) = 3(x+x+2) - 1$$
$$x^2 + 2x = 3(2x+2) - 1$$
$$x^2 + 2x = 6x + 6 - 1$$
$$x^2 + 2x = 6x + 5$$
$$x^2 - 4x - 5 = 0$$
$$(x-5)(x+1) = 0$$
$$x = 5, -1$$
$$x + 2 = 7, 1$$
The two integers are either 5 and 7, or −1 and 1.

70. Let x and $20 - x$ represent the two numbers. The equation is:
$$x(20-x) = 75$$
$$20x - x^2 = 75$$
$$0 = x^2 - 20x + 75$$
$$0 = (x-15)(x-5)$$
$$x = 15, 5$$
$$20 - x = 5, 15$$
The two numbers are 5 and 15.

71. Let x and $2x - 1$ represent the two numbers. The equation is:
$$x(2x-1) = 66$$
$$2x^2 - x = 66$$
$$2x^2 - x - 66 = 0$$
$$(2x+11)(x-6) = 0$$
$$x = -\tfrac{11}{2}, 6$$
$$2x - 1 = -12, 11$$
The two numbers are either $-\frac{11}{2}$ and −12, or 6 and 11.

72.   Let b represent the base, and $8b$ represent the height. The equation is:

$$\frac{1}{2}(b)(8b) = 16$$
$$4b^2 = 16$$
$$4b^2 - 16 = 0$$
$$4(b^2 - 4) = 0$$
$$4(b+2)(b-2) = 0$$
$$b = 2 \quad (b = -2 \text{ is impossible})$$

The base is 2 inches.

Chapter 7
Rational Expressions

7.1 Reducing Rational Expressions to Lowest Terms

1. **a.** Simplifying: $\dfrac{5+1}{25-1} = \dfrac{6}{24} = \dfrac{1}{4}$

   b. Simplifying: $\dfrac{x+1}{x^2-1} = \dfrac{1(x+1)}{(x+1)(x-1)} = \dfrac{1}{x-1}$. The variable restriction is $x \neq -1,1$.

   c. Simplifying: $\dfrac{x^2-x}{x^2-1} = \dfrac{x(x-1)}{(x+1)(x-1)} = \dfrac{x}{x+1}$. The variable restriction is $x \neq -1,1$.

   d. Simplifying: $\dfrac{x^3-1}{x^2-1} = \dfrac{(x-1)(x^2+x+1)}{(x+1)(x-1)} = \dfrac{x^2+x+1}{x+1}$. The variable restriction is $x \neq -1,1$.

   e. Simplifying: $\dfrac{x^3-1}{x^3-x^2} = \dfrac{(x-1)(x^2+x+1)}{x^2(x-1)} = \dfrac{x^2+x+1}{x^2}$. The variable restriction is $x \neq 0,1$.

3. Reducing the rational expression: $\dfrac{a-3}{a^2-9} = \dfrac{1(a-3)}{(a+3)(a-3)} = \dfrac{1}{a+3}$. The variable restriction is $a \neq -3,3$.

5. Reducing the rational expression: $\dfrac{x+5}{x^2-25} = \dfrac{1(x+5)}{(x+5)(x-5)} = \dfrac{1}{x-5}$. The variable restriction is $x \neq -5,5$.

7. Reducing the rational expression: $\dfrac{2x^2-8}{4} = \dfrac{2(x^2-4)}{4} = \dfrac{2(x+2)(x-2)}{4} = \dfrac{(x+2)(x-2)}{2}$

   There are no variable restrictions.

9. Reducing the rational expression: $\dfrac{2x-10}{3x-6} = \dfrac{2(x-5)}{3(x-2)}$. The variable restriction is $x \neq 2$.

11. Reducing the rational expression: $\dfrac{10a+20}{5a+10} = \dfrac{10(a+2)}{5(a+2)} = \dfrac{2}{1} = 2$

13. Reducing the rational expression: $\dfrac{5x^2-5}{4x+4} = \dfrac{5(x^2-1)}{4(x+1)} = \dfrac{5(x+1)(x-1)}{4(x+1)} = \dfrac{5(x-1)}{4}$

15. Reducing the rational expression: $\dfrac{x-3}{x^2-6x+9} = \dfrac{1(x-3)}{(x-3)^2} = \dfrac{1}{x-3}$

17. Reducing the rational expression: $\dfrac{3x+15}{3x^2+24x+45}=\dfrac{3(x+5)}{3\left(x^2+8x+15\right)}=\dfrac{3(x+5)}{3(x+5)(x+3)}=\dfrac{1}{x+3}$

19. Reducing the rational expression: $\dfrac{a^2-3a}{a^3-8a^2+15a}=\dfrac{a(a-3)}{a\left(a^2-8a+15\right)}=\dfrac{a(a-3)}{a(a-3)(a-5)}=\dfrac{1}{a-5}$

21. Reducing the rational expression: $\dfrac{3x-2}{9x^2-4}=\dfrac{1(3x-2)}{(3x+2)(3x-2)}=\dfrac{1}{3x+2}$

23. Reducing the rational expression: $\dfrac{x^2+8x+15}{x^2+5x+6}=\dfrac{(x+5)(x+3)}{(x+2)(x+3)}=\dfrac{x+5}{x+2}$

25. Reducing the rational expression: $\dfrac{2m^3-2m^2-12m}{m^2-5m+6}=\dfrac{2m\left(m^2-m-6\right)}{(m-2)(m-3)}=\dfrac{2m(m-3)(m+2)}{(m-2)(m-3)}=\dfrac{2m(m+2)}{m-2}$

27. Reducing the rational expression: $\dfrac{x^3+3x^2-4x}{x^3-16x}=\dfrac{x\left(x^2+3x-4\right)}{x\left(x^2-16\right)}=\dfrac{x(x+4)(x-1)}{x(x+4)(x-4)}=\dfrac{x-1}{x-4}$

29. Reducing the rational expression: $\dfrac{4x^3-10x^2+6x}{2x^3+x^2-3x}=\dfrac{2x\left(2x^2-5x+3\right)}{x\left(2x^2+x-3\right)}=\dfrac{2x(2x-3)(x-1)}{x(2x+3)(x-1)}=\dfrac{2(2x-3)}{2x+3}$

31. Reducing the rational expression: $\dfrac{4x^2-12x+9}{4x^2-9}=\dfrac{(2x-3)^2}{(2x+3)(2x-3)}=\dfrac{2x-3}{2x+3}$

33. Reducing the rational expression: $\dfrac{x+3}{x^4-81}=\dfrac{x+3}{\left(x^2+9\right)\left(x^2-9\right)}=\dfrac{x+3}{\left(x^2+9\right)(x+3)(x-3)}=\dfrac{1}{\left(x^2+9\right)(x-3)}$

35. Reducing the rational expression: $\dfrac{3x^2+x-10}{x^4-16}=\dfrac{(3x-5)(x+2)}{\left(x^2+4\right)\left(x^2-4\right)}=\dfrac{(3x-5)(x+2)}{\left(x^2+4\right)(x+2)(x-2)}=\dfrac{3x-5}{\left(x^2+4\right)(x-2)}$

37. Reducing the rational expression: $\dfrac{42x^3-20x^2-48x}{6x^2-5x-4}=\dfrac{2x\left(21x^2-10x-24\right)}{(3x-4)(2x+1)}=\dfrac{2x(7x+6)(3x-4)}{(3x-4)(2x+1)}=\dfrac{2x(7x+6)}{2x+1}$

39. Reducing the rational expression: $\dfrac{x^3-y^3}{x^2-y^2}=\dfrac{(x-y)\left(x^2+xy+y^2\right)}{(x+y)(x-y)}=\dfrac{x^2+xy+y^2}{x+y}$

41. Reducing the rational expression: $\dfrac{x^3+8}{x^2-4}=\dfrac{(x+2)\left(x^2-2x+4\right)}{(x+2)(x-2)}=\dfrac{x^2-2x+4}{x-2}$

43. Reducing the rational expression: $\dfrac{x^3+8}{x^2+x-2}=\dfrac{(x+2)\left(x^2-2x+4\right)}{(x+2)(x-1)}=\dfrac{x^2-2x+4}{x-1}$

45. Reducing the rational expression: $\dfrac{xy+3x+2y+6}{xy+3x+5y+15}=\dfrac{x(y+3)+2(y+3)}{x(y+3)+5(y+3)}=\dfrac{(y+3)(x+2)}{(y+3)(x+5)}=\dfrac{x+2}{x+5}$

47. Reducing the rational expression: $\dfrac{x^2-3x+ax-3a}{x^2-3x+bx-3b}=\dfrac{x(x-3)+a(x-3)}{x(x-3)+b(x-3)}=\dfrac{(x-3)(x+a)}{(x-3)(x+b)}=\dfrac{x+a}{x+b}$

49. **a.** Adding: $\left(x^2 - 4x\right) + \left(4x - 16\right) = x^2 - 4x + 4x - 16 = x^2 - 16$

b. Subtracting: $\left(x^2 - 4x\right) - \left(4x - 16\right) = x^2 - 4x - 4x + 16 = x^2 - 8x + 16$

c. Multiplying: $\left(x^2 - 4x\right)\left(4x - 16\right) = 4x^3 - 16x^2 - 16x^2 + 64x = 4x^3 - 32x^2 + 64x$

d. Reducing: $\dfrac{x^2 - 4x}{4x - 16} = \dfrac{x(x-4)}{4(x-4)} = \dfrac{x}{4}$

51. Writing as a ratio: $\dfrac{8}{6} = \dfrac{4}{3}$

53. Writing as a ratio: $\dfrac{200}{250} = \dfrac{4}{5}$

55. Writing as a ratio: $\dfrac{32}{4} = \dfrac{8}{1}$

57. Completing the table:

Checks Written x	Total Cost $2.00 + 0.15x$	Cost per Check $\dfrac{2.00 + 0.15x}{x}$
0	$2.00	undefined
5	$2.75	$0.55
10	$3.50	$0.35
15	$4.25	$0.28
20	$5.00	$0.25

59. The average speed is: $\dfrac{122 \text{ miles}}{3 \text{ hours}} \approx 40.7 \text{ miles/hour}$

61. The average speed is: $\dfrac{785 \text{ feet}}{20 \text{ minutes}} = 39.25 \text{ feet/minute}$

63. Computing the strikeouts per inning:

Randy Johnson: $\dfrac{11.04 \text{ strikeouts}}{9 \text{ innings}} \approx 1.23 \text{ strikeouts/inning}$

Kerry Wood: $\dfrac{10.45 \text{ strikeouts}}{9 \text{ innings}} \approx 1.16 \text{ strikeouts/inning}$

Pedro Martinez: $\dfrac{10.40 \text{ strikeouts}}{9 \text{ innings}} \approx 1.16 \text{ strikeouts/inning}$

Hideo Nomo: $\dfrac{8.81 \text{ strikeouts}}{9 \text{ innings}} \approx 0.98 \text{ strikeouts/inning}$

65. Her average speed on level ground is: $\dfrac{20 \text{ minutes}}{2 \text{ miles}} = 10 \text{ minutes/mile}$, or $\dfrac{2 \text{ miles}}{20 \text{ minutes}} = 0.1 \text{ miles/minute}$

Her average speed downhill is: $\dfrac{40 \text{ minutes}}{6 \text{ miles}} = \dfrac{20}{3} \text{ minutes/mile}$, or $\dfrac{6 \text{ miles}}{40 \text{ minutes}} = \dfrac{3}{20} \text{ miles/minute}$

67. The average fuel consumption is: $\dfrac{168 \text{ miles}}{3.5 \text{ gallons}} = 48 \text{ miles/gallon}$

69. Substituting $x = 5$ and $y = 4$: $\dfrac{x^2 - y^2}{x - y} = \dfrac{5^2 - 4^2}{5 - 4} = \dfrac{25 - 16}{5 - 4} = \dfrac{9}{1} = 9$. The result is equal to $5 + 4 = 9$.

71. Completing the table:

x	$\dfrac{x-3}{3-x}$
-2	-1
-1	-1
0	-1
1	-1
2	-1

73. Completing the table:

x	$\dfrac{x-5}{x^2-25}$	$\dfrac{1}{x+5}$
0	$\frac{1}{5}$	$\frac{1}{5}$
2	$\frac{1}{7}$	$\frac{1}{7}$
-2	$\frac{1}{3}$	$\frac{1}{3}$
5	undefined	$\frac{1}{10}$
-5	undefined	undefined

Simplifying: $\dfrac{x-3}{3-x} = \dfrac{-(3-x)}{3-x} = -1$

75. Completing the table:

Stock	Price	Earnings per Share	P/E
Yahoo	37.80	1.07	35.33
Google	381.24	4.51	84.53
Disney	24.96	1.34	18.63
Nike	85.46	4.88	17.51
Ebay	40.96	0.73	56.11

The stocks Disney and Nike appear to be undervalued.

77. Simplifying the expression: $\dfrac{27x^5}{9x^2} - \dfrac{45x^8}{15x^5} = 3x^3 - 3x^3 = 0$

79. Simplifying the expression: $\dfrac{72a^3b^7}{9ab^5} + \dfrac{64a^5b^3}{8a^3b} = 8a^2b^2 + 8a^2b^2 = 16a^2b^2$

81. Dividing by the monomial: $\dfrac{38x^7 + 42x^5 - 84x^3}{2x^3} = \dfrac{38x^7}{2x^3} + \dfrac{42x^5}{2x^3} - \dfrac{84x^3}{2x^3} = 19x^4 + 21x^2 - 42$

83. Dividing by the monomial: $\dfrac{28a^5b^5 + 36ab^4 - 44a^4b}{4ab} = \dfrac{28a^5b^5}{4ab} + \dfrac{36ab^4}{4ab} - \dfrac{44a^4b}{4ab} = 7a^4b^4 + 9b^3 - 11a^3$

85. Simplifying: $\frac{3}{4} \cdot \frac{10}{21} = \frac{3 \cdot 2 \cdot 5}{2 \cdot 2 \cdot 3 \cdot 7} = \frac{5}{14}$

87. Simplifying: $\frac{4}{5} \div \frac{8}{9} = \frac{4}{5} \cdot \frac{9}{8} = \frac{2 \cdot 2 \cdot 9}{5 \cdot 2 \cdot 2 \cdot 2} = \frac{9}{10}$

89. Factoring: $x^2 - 9 = (x+3)(x-3)$

91. Factoring: $3x - 9 = 3(x-3)$

93. Factoring: $x^2 - x - 20 = (x-5)(x+4)$

95. Factoring: $a^2 + 5a = a(a+5)$

97. Simplifying: $\dfrac{a(a+5)(a-5)(a+4)}{a^2+5a} = \dfrac{a(a+5)(a-5)(a+4)}{a(a+5)} = (a-5)(a+4)$

99. Multiplying: $\dfrac{5,603}{11} \cdot \dfrac{1}{5,280} \cdot \dfrac{60}{1} \approx 5.8$

7.2 Multiplication and Division of Rational Expressions

1. Simplifying the expression: $\dfrac{x+y}{3} \cdot \dfrac{6}{x+y} = \dfrac{6(x+y)}{3(x+y)} = 2$

3. Simplifying the expression: $\dfrac{2x+10}{x^2} \cdot \dfrac{x^3}{4x+20} = \dfrac{2(x+5)}{x^2} \cdot \dfrac{x^3}{4(x+5)} = \dfrac{2x^3(x+5)}{4x^2(x+5)} = \dfrac{x}{2}$

5. Simplifying the expression: $\dfrac{9}{2a-8} \div \dfrac{3}{a-4} = \dfrac{9}{2a-8} \cdot \dfrac{a-4}{3} = \dfrac{9}{2(a-4)} \cdot \dfrac{a-4}{3} = \dfrac{9(a-4)}{6(a-4)} = \dfrac{3}{2}$

7. Simplifying the expression:

$$\dfrac{x+1}{x^2-9} \div \dfrac{2x+2}{x+3} = \dfrac{x+1}{x^2-9} \cdot \dfrac{x+3}{2x+2} = \dfrac{x+1}{(x+3)(x-3)} \cdot \dfrac{x+3}{2(x+1)} = \dfrac{(x+1)(x+3)}{2(x+3)(x-3)(x+1)} = \dfrac{1}{2(x-3)}$$

9. Simplifying the expression: $\dfrac{a^2+5a}{7a} \cdot \dfrac{4a^2}{a^2+4a} = \dfrac{a(a+5)}{7a} \cdot \dfrac{4a^2}{a(a+4)} = \dfrac{4a^3(a+5)}{7a^2(a+4)} \left(= \dfrac{4a(a+5)}{7(a+4)} \right)$

11. Simplifying the expression:

$$\dfrac{y^2-5y+6}{2y+4} \div \dfrac{2y-6}{y+2} = \dfrac{y^2-5y+6}{2y+4} \cdot \dfrac{y+2}{2y-6} = \dfrac{(y-2)(y-3)}{2(y+2)} \cdot \dfrac{y+2}{2(y-3)} = \dfrac{(y-2)(y-3)(y+2)}{4(y+2)(y-3)} = \dfrac{y-2}{4}$$

13. Simplifying the expression:

$$\dfrac{2x-8}{x^2-4} \cdot \dfrac{x^2+6x+8}{x-4} = \dfrac{2(x-4)}{(x+2)(x-2)} \cdot \dfrac{(x+4)(x+2)}{x-4} = \dfrac{2(x-4)(x+4)(x+2)}{(x+2)(x-2)(x-4)} = \dfrac{2(x+4)}{x-2}$$

15. Simplifying the expression:

$$\dfrac{x-1}{x^2-x-6} \cdot \dfrac{x^2+5x+6}{x^2-1} = \dfrac{x-1}{(x-3)(x+2)} \cdot \dfrac{(x+2)(x+3)}{(x+1)(x-1)} = \dfrac{(x-1)(x+2)(x+3)}{(x-3)(x+2)(x+1)(x-1)} = \dfrac{x+3}{(x-3)(x+1)}$$

17. Simplifying the expression:

$$\dfrac{a^2+10a+25}{a+5} \div \dfrac{a^2-25}{a-5} = \dfrac{a^2+10a+25}{a+5} \cdot \dfrac{a-5}{a^2-25} = \dfrac{(a+5)^2}{a+5} \cdot \dfrac{a-5}{(a+5)(a-5)} = \dfrac{(a+5)^2(a-5)}{(a+5)^2(a-5)} = 1$$

19. Simplifying the expression:

$$\dfrac{y^3-5y^2}{y^4+3y^3+2y^2} \div \dfrac{y^2-5y+6}{y^2-2y-3} = \dfrac{y^3-5y^2}{y^4+3y^3+2y^2} \cdot \dfrac{y^2-2y-3}{y^2-5y+6}$$

$$= \dfrac{y^2(y-5)}{y^2(y+2)(y+1)} \cdot \dfrac{(y-3)(y+1)}{(y-2)(y-3)}$$

$$= \dfrac{y^2(y-5)(y-3)(y+1)}{y^2(y+2)(y+1)(y-2)(y-3)}$$

$$= \dfrac{y-5}{(y+2)(y-2)}$$

21. Simplifying the expression:

$$\dfrac{2x^2+17x+21}{x^2+2x-35} \cdot \dfrac{x^3-125}{2x^2-7x-15} = \dfrac{(2x+3)(x+7)}{(x+7)(x-5)} \cdot \dfrac{(x-5)(x^2+5x+25)}{(2x+3)(x-5)}$$

$$= \dfrac{(2x+3)(x+7)(x-5)(x^2+5x+25)}{(x+7)(x-5)^2(2x+3)}$$

$$= \dfrac{x^2+5x+25}{x-5}$$

23.   Simplifying the expression:

$$\frac{2x^2+10x+12}{4x^2+24x+32} \cdot \frac{2x^2+18x+40}{x^2+8x+15} = \frac{2\left(x^2+5x+6\right)}{4\left(x^2+6x+8\right)} \cdot \frac{2\left(x^2+9x+20\right)}{x^2+8x+15}$$

$$= \frac{2(x+2)(x+3)}{4(x+4)(x+2)} \cdot \frac{2(x+5)(x+4)}{(x+5)(x+3)}$$

$$= \frac{4(x+2)(x+3)(x+4)(x+5)}{4(x+2)(x+3)(x+4)(x+5)}$$

$$= 1$$

25.   Simplifying the expression:

$$\frac{2a^2+7a+3}{a^2-16} \div \frac{4a^2+8a+3}{2a^2-5a-12} = \frac{2a^2+7a+3}{a^2-16} \cdot \frac{2a^2-5a-12}{4a^2+8a+3}$$

$$= \frac{(2a+1)(a+3)}{(a+4)(a-4)} \cdot \frac{(2a+3)(a-4)}{(2a+1)(2a+3)}$$

$$= \frac{(2a+1)(a+3)(2a+3)(a-4)}{(a+4)(a-4)(2a+1)(2a+3)}$$

$$= \frac{a+3}{a+4}$$

27.   Simplifying the expression:

$$\frac{4y^2-12y+9}{y^2-36} \div \frac{2y^2-5y+3}{y^2+5y-6} = \frac{4y^2-12y+9}{y^2-36} \cdot \frac{y^2+5y-6}{2y^2-5y+3}$$

$$= \frac{(2y-3)^2}{(y+6)(y-6)} \cdot \frac{(y+6)(y-1)}{(2y-3)(y-1)}$$

$$= \frac{(2y-3)^2(y+6)(y-1)}{(y+6)(y-6)(2y-3)(y-1)}$$

$$= \frac{2y-3}{y-6}$$

29.   Simplifying the expression:

$$\frac{x^2-1}{6x^2+42x+60} \cdot \frac{7x^2+17x+6}{x^3+1} \cdot \frac{6x+30}{7x^2-11x-6} = \frac{(x+1)(x-1)}{6(x+5)(x+2)} \cdot \frac{(7x+3)(x+2)}{(x+1)\left(x^2-x+1\right)} \cdot \frac{6(x+5)}{(7x+3)(x-2)}$$

$$= \frac{6(x+1)(x-1)(7x+3)(x+2)(x+5)}{6(x+5)(x+2)(x+1)(7x+3)(x-2)\left(x^2-x+1\right)}$$

$$= \frac{x-1}{(x-2)\left(x^2-x+1\right)}$$

31.   Simplifying the expression:

$$\frac{18x^3+21x^2-60x}{21x^2-25x-4} \cdot \frac{28x^2-17x-3}{16x^3+28x^2-30x} = \frac{3x\left(6x^2+7x-20\right)}{21x^2-25x-4} \cdot \frac{28x^2-17x-3}{2x\left(8x^2+14x-15\right)}$$

$$= \frac{3x(3x-4)(2x+5)}{(7x+1)(3x-4)} \cdot \frac{(7x+1)(4x-3)}{2x(4x-3)(2x+5)}$$

$$= \frac{3x(3x-4)(2x+5)(7x+1)(4x-3)}{2x(7x+1)(3x-4)(4x-3)(2x+5)}$$

$$= \frac{3}{2}$$

33. **a.**   Simplifying: $\dfrac{9-1}{27-1}=\dfrac{8}{26}=\dfrac{4}{13}$

     b.   Reducing: $\dfrac{x^2-1}{x^3-1}=\dfrac{(x+1)(x-1)}{(x-1)(x^2+x+1)}=\dfrac{x+1}{x^2+x+1}$

     c.   Multiplying: $\dfrac{x^2-1}{x^3-1}\cdot\dfrac{x-1}{x+1}=\dfrac{(x+1)(x-1)}{(x-1)(x^2+x+1)}\cdot\dfrac{x-1}{x+1}=\dfrac{x-1}{x^2+x+1}$

     d.   Dividing: $\dfrac{x^2-1}{x^3-1}\div\dfrac{x-1}{x^2+x+1}=\dfrac{(x+1)(x-1)}{(x-1)(x^2+x+1)}\cdot\dfrac{x^2+x+1}{x-1}=\dfrac{x+1}{x-1}$

35. Simplifying the expression: $(x^2-9)\left(\dfrac{2}{x+3}\right)=\dfrac{(x+3)(x-3)}{1}\cdot\dfrac{2}{x+3}=\dfrac{2(x+3)(x-3)}{x+3}=2(x-3)$

37. Simplifying the expression: $(x^2-x-6)\left(\dfrac{x+1}{x-3}\right)=\dfrac{(x-3)(x+2)}{1}\cdot\dfrac{x+1}{x-3}=\dfrac{(x-3)(x+2)(x+1)}{x-3}=(x+2)(x+1)$

39. Simplifying the expression: $(x^2-4x-5)\left(\dfrac{-2x}{x+1}\right)=\dfrac{(x-5)(x+1)}{1}\cdot\dfrac{-2x}{x+1}=\dfrac{-2x(x-5)(x+1)}{x+1}=-2x(x-5)$

41. Simplifying the expression:

$$\frac{x^2-9}{x^2-3x}\cdot\frac{2x+10}{xy+5x+3y+15}=\frac{(x+3)(x-3)}{x(x-3)}\cdot\frac{2(x+5)}{x(y+5)+3(y+5)}=\frac{2(x+3)(x-3)(x+5)}{x(x-3)(y+5)(x+3)}=\frac{2(x+5)}{x(y+5)}$$

43. Simplifying the expression:

$$\frac{2x^2+4x}{x^2-y^2}\cdot\frac{x^2+3x+xy+3y}{x^2+5x+6}=\frac{2x(x+2)}{(x+y)(x-y)}\cdot\frac{x(x+3)+y(x+3)}{(x+2)(x+3)}=\frac{2x(x+2)(x+3)(x+y)}{(x+y)(x-y)(x+2)(x+3)}=\frac{2x}{x-y}$$

45. Simplifying the expression:

$$\frac{x^3-3x^2+4x-12}{x^4-16}\cdot\frac{3x^2+5x-2}{3x^2-10x+3}=\frac{x^2(x-3)+4(x-3)}{(x^2+4)(x^2-4)}\cdot\frac{(3x-1)(x+2)}{(3x-1)(x-3)}$$

$$=\frac{(x-3)(x^2+4)}{(x^2+4)(x+2)(x-2)}\cdot\frac{(3x-1)(x+2)}{(3x-1)(x-3)}$$

$$=\frac{(x-3)(x^2+4)(3x-1)(x+2)}{(x^2+4)(x+2)(x-2)(3x-1)(x-3)}$$

$$=\frac{1}{x-2}$$

47. Simplifying the expression: $\left(1-\tfrac{1}{2}\right)\left(1-\tfrac{1}{3}\right)\left(1-\tfrac{1}{4}\right)\left(1-\tfrac{1}{5}\right)=\left(\tfrac{2}{2}-\tfrac{1}{2}\right)\left(\tfrac{3}{3}-\tfrac{1}{3}\right)\left(\tfrac{4}{4}-\tfrac{1}{4}\right)\left(\tfrac{5}{5}-\tfrac{1}{5}\right)=\tfrac{1}{2}\cdot\tfrac{2}{3}\cdot\tfrac{3}{4}\cdot\tfrac{4}{5}=\tfrac{1}{5}$

49. Simplifying the expression: $\left(1-\tfrac{1}{2}\right)\left(1-\tfrac{1}{3}\right)\left(1-\tfrac{1}{4}\right)\bullet\bullet\bullet\left(1-\tfrac{1}{99}\right)\left(1-\tfrac{1}{100}\right)=\tfrac{1}{2}\cdot\tfrac{2}{3}\cdot\tfrac{3}{4}\bullet\bullet\bullet\tfrac{98}{99}\cdot\tfrac{99}{100}=\tfrac{1}{100}$

51. Since 5,280 feet = 1 mile, the height is: $\dfrac{14,494\text{ feet}}{5,280\text{ feet/mile}}\approx 2.7$ miles

53. Converting to miles per hour: $\dfrac{1088\text{ feet}}{1\text{ second}}\cdot\dfrac{1\text{ mile}}{5280\text{ feet}}\cdot\dfrac{60\text{ seconds}}{1\text{ minute}}\cdot\dfrac{60\text{ minutes}}{1\text{ hour}}\approx 742$ miles/hour

55. Converting to miles per hour: $\dfrac{785\text{ feet}}{20\text{ minutes}}\cdot\dfrac{60\text{ minutes}}{1\text{ hour}}\cdot\dfrac{1\text{ mile}}{5280\text{ feet}}\approx 0.45$ miles/hour

57. Converting to miles per hour: $\dfrac{518\text{ feet}}{40\text{ seconds}}\cdot\dfrac{60\text{ seconds}}{1\text{ minute}}\cdot\dfrac{60\text{ minutes}}{1\text{ hour}}\cdot\dfrac{1\text{ mile}}{5280\text{ feet}}\approx 8.8$ miles/hour

59. **a.**   Finding his average speed: $\dfrac{10 \text{ kilometers}}{28\frac{1}{6}\text{ minutes}} \bullet \dfrac{60 \text{ minutes}}{1 \text{ hour}} \approx 21.3 \text{ kilometers/hour}$

   b.   Converting to miles per hour: $\dfrac{21.3 \text{ kilometers}}{1 \text{ hour}} \bullet \dfrac{6.2 \text{ miles}}{10 \text{ kilometers}} \approx 13.2 \text{ miles/hour}$

61.   Adding the fractions: $\frac{1}{2} + \frac{5}{2} = \frac{6}{2} = 3$ **63.**   Adding the fractions: $2 + \frac{3}{4} = \frac{2 \bullet 4}{1 \bullet 4} + \frac{3}{4} = \frac{8}{4} + \frac{3}{4} = \frac{11}{4}$

65.   Simplifying the expression: $\dfrac{10x^4}{2x^2} + \dfrac{12x^6}{3x^4} = 5x^2 + 4x^2 = 9x^2$

67.   Simplifying the expression: $\dfrac{12a^2 b^5}{3ab^3} + \dfrac{14a^4 b^7}{7a^3 b^5} = 4ab^2 + 2ab^2 = 6ab^2$

69.   Adding the fractions: $\frac{1}{5} + \frac{3}{5} = \frac{4}{5}$

71.   Adding the fractions: $\frac{1}{10} + \frac{3}{14} = \frac{1}{10} \bullet \frac{7}{7} + \frac{3}{14} \bullet \frac{5}{5} = \frac{7}{70} + \frac{15}{70} = \frac{22}{70} = \frac{11}{35}$

73.   Subtracting the fractions: $\frac{1}{10} - \frac{3}{14} = \frac{1}{10} \bullet \frac{7}{7} - \frac{3}{14} \bullet \frac{5}{5} = \frac{7}{70} - \frac{15}{70} = -\frac{8}{70} = -\frac{4}{35}$

75.   Multiplying: $2(x - 3) = 2x - 6$

77.   Multiplying: $(x + 4)(x - 5) = x^2 - 5x + 4x - 20 = x^2 - x - 20$

79.   Reducing the fraction: $\dfrac{x + 3}{x^2 - 9} = \dfrac{x + 3}{(x + 3)(x - 3)} = \dfrac{1}{x - 3}$

81.   Reducing the fraction: $\dfrac{x^2 - x - 30}{2(x + 5)(x - 5)} = \dfrac{(x + 5)(x - 6)}{2(x + 5)(x - 5)} = \dfrac{x - 6}{2(x - 5)}$

83.   Simplifying: $(x + 4)(x - 5) - 10 = x^2 - 5x + 4x - 20 - 10 = x^2 - x - 30$

7.3 Addition and Subtraction of Rational Expressions

1.   Combining the fractions: $\dfrac{3}{x} + \dfrac{4}{x} = \dfrac{7}{x}$ **3.**   Combining the fractions: $\dfrac{9}{a} - \dfrac{5}{a} = \dfrac{4}{a}$

5.   Combining the fractions: $\dfrac{1}{x + 1} + \dfrac{x}{x + 1} = \dfrac{1 + x}{x + 1} = 1$

7.   Combining the fractions: $\dfrac{y^2}{y - 1} - \dfrac{1}{y - 1} = \dfrac{y^2 - 1}{y - 1} = \dfrac{(y + 1)(y - 1)}{y - 1} = y + 1$

9.   Combining the fractions: $\dfrac{x^2}{x + 2} + \dfrac{4x + 4}{x + 2} = \dfrac{x^2 + 4x + 4}{x + 2} = \dfrac{(x + 2)^2}{x + 2} = x + 2$

11.   Combining the fractions: $\dfrac{x^2}{x - 2} - \dfrac{4x - 4}{x - 2} = \dfrac{x^2 - 4x + 4}{x - 2} = \dfrac{(x - 2)^2}{x - 2} = x - 2$

13.   Combining the fractions: $\dfrac{x + 2}{x + 6} - \dfrac{x - 4}{x + 6} = \dfrac{x + 2 - x + 4}{x + 6} = \dfrac{6}{x + 6}$

15.   Combining the fractions: $\dfrac{y}{2} - \dfrac{2}{y} = \dfrac{y \bullet y}{2 \bullet y} - \dfrac{2 \bullet 2}{y \bullet 2} = \dfrac{y^2}{2y} - \dfrac{4}{2y} = \dfrac{y^2 - 4}{2y} = \dfrac{(y + 2)(y - 2)}{2y}$

17.   Combining the fractions: $\dfrac{1}{2} + \dfrac{a}{3} = \dfrac{1 \bullet 3}{2 \bullet 3} + \dfrac{a \bullet 2}{3 \bullet 2} = \dfrac{3}{6} + \dfrac{2a}{6} = \dfrac{2a + 3}{6}$

19.   Combining the fractions: $\dfrac{x}{x + 1} + \dfrac{3}{4} = \dfrac{x \bullet 4}{(x + 1) \bullet 4} + \dfrac{3 \bullet (x + 1)}{4 \bullet (x + 1)} = \dfrac{4x}{4(x + 1)} + \dfrac{3x + 3}{4(x + 1)} = \dfrac{4x + 3x + 3}{4(x + 1)} = \dfrac{7x + 3}{4(x + 1)}$

21. Combining the fractions:

$$\frac{x+1}{x-2} - \frac{4x+7}{5x-10} = \frac{(x+1)\bullet 5}{(x-2)\bullet 5} - \frac{4x+7}{5(x-2)} = \frac{5x+5}{5(x-2)} - \frac{4x+7}{5(x-2)} = \frac{5x+5-4x-7}{5(x-2)} = \frac{x-2}{5(x-2)} = \frac{1}{5}$$

23. Combining the fractions:

$$\frac{4x-2}{3x+12} - \frac{x-2}{x+4} = \frac{4x-2}{3(x+4)} - \frac{(x-2)\bullet 3}{(x+4)\bullet 3} = \frac{4x-2}{3(x+4)} - \frac{3x-6}{3(x+4)} = \frac{4x-2-3x+6}{3(x+4)} = \frac{x+4}{3(x+4)} = \frac{1}{3}$$

25. Combining the fractions:

$$\frac{6}{x(x-2)} + \frac{3}{x} = \frac{6}{x(x-2)} + \frac{3\bullet(x-2)}{x\bullet(x-2)} = \frac{6}{x(x-2)} + \frac{3x-6}{x(x-2)} = \frac{6+3x-6}{x(x-2)} = \frac{3x}{x(x-2)} = \frac{3}{x-2}$$

27. Combining the fractions:

$$\frac{4}{a} - \frac{12}{a^2+3a} = \frac{4\bullet(a+3)}{a\bullet(a+3)} - \frac{12}{a(a+3)} = \frac{4a+12}{a(a+3)} - \frac{12}{a(a+3)} = \frac{4a+12-12}{a(a+3)} = \frac{4a}{a(a+3)} = \frac{4}{a+3}$$

29. Combining the fractions:

$$\frac{2}{x+5} - \frac{10}{x^2-25} = \frac{2\bullet(x-5)}{(x+5)\bullet(x-5)} - \frac{10}{(x+5)(x-5)}$$

$$= \frac{2x-10}{(x+5)(x-5)} - \frac{10}{(x+5)(x-5)}$$

$$= \frac{2x-10-10}{(x+5)(x-5)}$$

$$= \frac{2x-20}{(x+5)(x-5)}$$

$$= \frac{2(x-10)}{(x+5)(x-5)}$$

31. Combining the fractions:

$$\frac{x-4}{x-3} + \frac{6}{x^2-9} = \frac{(x-4)\bullet(x+3)}{(x-3)\bullet(x+3)} + \frac{6}{(x+3)(x-3)}$$

$$= \frac{x^2-x-12}{(x+3)(x-3)} + \frac{6}{(x+3)(x-3)}$$

$$= \frac{x^2-x-12+6}{(x+3)(x-3)}$$

$$= \frac{x^2-x-6}{(x+3)(x-3)}$$

$$= \frac{(x-3)(x+2)}{(x+3)(x-3)}$$

$$= \frac{x+2}{x+3}$$

33. Combining the fractions:

$$\frac{a-4}{a-3} + \frac{5}{a^2-a-6} = \frac{(a-4)\bullet(a+2)}{(a-3)\bullet(a+2)} + \frac{5}{(a-3)(a+2)}$$

$$= \frac{a^2-2a-8}{(a-3)(a+2)} + \frac{5}{(a-3)(a+2)}$$

$$= \frac{a^2-2a-8+5}{(a-3)(a+2)}$$

$$= \frac{a^2-2a-3}{(a-3)(a+2)}$$

$$= \frac{(a-3)(a+1)}{(a-3)(a+2)}$$

$$= \frac{a+1}{a+2}$$

35. Combining the fractions:

$$\frac{8}{x^2-16}-\frac{7}{x^2-x-12}=\frac{8}{(x+4)(x-4)}-\frac{7}{(x-4)(x+3)}$$

$$=\frac{8(x+3)}{(x+4)(x-4)(x+3)}-\frac{7(x+4)}{(x+4)(x-4)(x+3)}$$

$$=\frac{8x+24}{(x+4)(x-4)(x+3)}-\frac{7x+28}{(x+4)(x-4)(x+3)}$$

$$=\frac{8x+24-7x-28}{(x+4)(x-4)(x+3)}$$

$$=\frac{x-4}{(x+4)(x-4)(x+3)}$$

$$=\frac{1}{(x+4)(x+3)}$$

37. Combining the fractions:

$$\frac{4y}{y^2+6y+5}-\frac{3y}{y^2+5y+4}=\frac{4y}{(y+5)(y+1)}-\frac{3y}{(y+4)(y+1)}$$

$$=\frac{4y(y+4)}{(y+5)(y+1)(y+4)}-\frac{3y(y+5)}{(y+5)(y+1)(y+4)}$$

$$=\frac{4y^2+16y}{(y+5)(y+1)(y+4)}-\frac{3y^2+15y}{(y+5)(y+1)(y+4)}$$

$$=\frac{4y^2+16y-3y^2-15y}{(y+5)(y+1)(y+4)}$$

$$=\frac{y^2+y}{(y+5)(y+1)(y+4)}$$

$$=\frac{y(y+1)}{(y+5)(y+1)(y+4)}$$

$$=\frac{y}{(y+5)(y+4)}$$

39. Combining the fractions:

$$\frac{4x+1}{x^2+5x+4}-\frac{x+3}{x^2+4x+3}=\frac{4x+1}{(x+4)(x+1)}-\frac{x+3}{(x+3)(x+1)}$$

$$=\frac{(4x+1)(x+3)}{(x+4)(x+1)(x+3)}-\frac{(x+3)(x+4)}{(x+4)(x+1)(x+3)}$$

$$=\frac{4x^2+13x+3}{(x+4)(x+1)(x+3)}-\frac{x^2+7x+12}{(x+4)(x+1)(x+3)}$$

$$=\frac{4x^2+13x+3-x^2-7x-12}{(x+4)(x+1)(x+3)}$$

$$=\frac{3x^2+6x-9}{(x+4)(x+1)(x+3)}$$

$$=\frac{3(x+3)(x-1)}{(x+4)(x+1)(x+3)}$$

$$=\frac{3(x-1)}{(x+4)(x+1)}$$

41. Combining the fractions:

$$\frac{1}{x} + \frac{x}{3x+9} - \frac{3}{x^2+3x} = \frac{1}{x} + \frac{x}{3(x+3)} - \frac{3}{x(x+3)}$$

$$= \frac{1 \cdot 3(x+3)}{x \cdot 3(x+3)} + \frac{x \cdot x}{3(x+3) \cdot x} - \frac{3 \cdot 3}{x(x+3) \cdot 3}$$

$$= \frac{3x+9}{3x(x+3)} + \frac{x^2}{3x(x+3)} - \frac{9}{3x(x+3)}$$

$$= \frac{3x+9+x^2-9}{3x(x+3)}$$

$$= \frac{x^2+3x}{3x(x+3)}$$

$$= \frac{x(x+3)}{3x(x+3)}$$

$$= \frac{1}{3}$$

43. **a.** Multiplying: $\dfrac{4}{9} \cdot \dfrac{1}{6} = \dfrac{2 \cdot 2}{3 \cdot 3} \cdot \dfrac{1}{2 \cdot 3} = \dfrac{2}{3 \cdot 3 \cdot 3} = \dfrac{2}{27}$

b. Dividing: $\dfrac{4}{9} \div \dfrac{1}{6} = \dfrac{4}{9} \cdot \dfrac{6}{1} = \dfrac{2 \cdot 2}{3 \cdot 3} \cdot \dfrac{2 \cdot 3}{1} = \dfrac{2 \cdot 2 \cdot 2}{3} = \dfrac{8}{3}$

c. Adding: $\dfrac{4}{9} + \dfrac{1}{6} = \dfrac{4}{9} \cdot \dfrac{2}{2} + \dfrac{1}{6} \cdot \dfrac{3}{3} = \dfrac{8}{18} + \dfrac{3}{18} = \dfrac{11}{18}$

d. Multiplying: $\dfrac{x+2}{x-2} \cdot \dfrac{3x+10}{x^2-4} = \dfrac{x+2}{x-2} \cdot \dfrac{3x+10}{(x+2)(x-2)} = \dfrac{3x+10}{(x-2)^2}$

e. Dividing: $\dfrac{x+2}{x-2} \div \dfrac{3x+10}{x^2-4} = \dfrac{x+2}{x-2} \cdot \dfrac{(x+2)(x-2)}{3x+10} = \dfrac{(x+2)^2}{3x+10}$

f. Subtracting:

$$\frac{x+2}{x-2} - \frac{3x+10}{x^2-4} = \frac{x+2}{x-2} \cdot \frac{x+2}{x+2} - \frac{3x+10}{(x+2)(x-2)}$$

$$= \frac{x^2+4x+4}{(x+2)(x-2)} - \frac{3x+10}{(x+2)(x-2)}$$

$$= \frac{x^2+x-6}{(x+2)(x-2)}$$

$$= \frac{(x+3)(x-2)}{(x+2)(x-2)}$$

$$= \frac{x+3}{x+2}$$

45. Completing the table:

Number x	Reciprocal $\dfrac{1}{x}$	Sum $1+\dfrac{1}{x}$	Sum $\dfrac{x+1}{x}$
1	1	2	2
2	$\dfrac{1}{2}$	$\dfrac{3}{2}$	$\dfrac{3}{2}$
3	$\dfrac{1}{3}$	$\dfrac{4}{3}$	$\dfrac{4}{3}$
4	$\dfrac{1}{4}$	$\dfrac{5}{4}$	$\dfrac{5}{4}$

47. Completing the table:

x	$x+\dfrac{4}{x}$	$\dfrac{x^2+4}{x}$	$x+4$
1	5	5	5
2	4	4	6
3	$\dfrac{13}{3}$	$\dfrac{13}{3}$	7
4	5	5	8

49. Combining the fractions: $1 + \dfrac{1}{x+2} = \dfrac{1 \bullet (x+2)}{1 \bullet (x+2)} + \dfrac{1}{x+2} = \dfrac{x+2}{x+2} + \dfrac{1}{x+2} = \dfrac{x+2+1}{x+2} = \dfrac{x+3}{x+2}$

51. Combining the fractions: $1 - \dfrac{1}{x+3} = \dfrac{1 \bullet (x+3)}{1 \bullet (x+3)} - \dfrac{1}{x+3} = \dfrac{x+3}{x+3} - \dfrac{1}{x+3} = \dfrac{x+3-1}{x+3} = \dfrac{x+2}{x+3}$

53. Solving the equation:
$$2x + 3(x-3) = 6$$
$$2x + 3x - 9 = 6$$
$$5x - 9 = 6$$
$$5x = 15$$
$$x = 3$$

55. Solving the equation:
$$x - 3(x+3) = x - 3$$
$$x - 3x - 9 = x - 3$$
$$-2x - 9 = x - 3$$
$$-3x - 9 = -3$$
$$-3x = 6$$
$$x = -2$$

57. Solving the equation:
$$7 - 2(3x+1) = 4x + 3$$
$$7 - 6x - 2 = 4x + 3$$
$$-6x + 5 = 4x + 3$$
$$-10x + 5 = 3$$
$$-10x = -2$$
$$x = \tfrac{1}{5}$$

59. Solving the quadratic equation:

$$x^2 + 5x + 6 = 0$$
$$(x+2)(x+3) = 0$$
$$x = -2, -3$$

61. Solving the quadratic equation:
$$x^2 - x = 6$$
$$x^2 - x - 6 = 0$$
$$(x-3)(x+2) = 0$$
$$x = -2, 3$$

63. Solving the quadratic equation:

$$x^2 - 5x = 0$$
$$x(x-5) = 0$$
$$x = 0, 5$$

65. Simplifying: $6\left(\tfrac{1}{2}\right) = 3$

67. Simplifying: $\dfrac{0}{5} = 0$

69. Simplifying: $\dfrac{5}{0}$ is undefined

71. Simplifying: $1 - \tfrac{5}{2} = \tfrac{2}{2} - \tfrac{5}{2} = -\tfrac{3}{2}$

73. Simplifying: $6\left(\dfrac{x}{3} + \dfrac{5}{2}\right) = 6 \bullet \dfrac{x}{3} + 6 \bullet \dfrac{5}{2} = 2x + 15$

75. Simplifying: $x^2\left(1 - \dfrac{5}{x}\right) = x^2 \bullet 1 - x^2 \bullet \dfrac{5}{x} = x^2 - 5x$

77. Solving the equation:
$$2x + 15 = 3$$
$$2x = -12$$
$$x = -6$$

79. Solving the equation:
$$-2x - 9 = x - 3$$
$$-3x - 9 = -3$$
$$-3x = 6$$
$$x = -2$$

7.4 Equations Involving Rational Expressions

1. Multiplying both sides of the equation by 6:

$$6\left(\dfrac{x}{3} + \dfrac{1}{2}\right) = 6\left(-\dfrac{1}{2}\right)$$
$$2x + 3 = -3$$
$$2x = -6$$
$$x = -3$$

Since $x = -3$ checks in the original equation, the solution is $x = -3$.

3. Multiplying both sides of the equation by $5a$:

$$5a\left(\dfrac{4}{a}\right) = 5a\left(\dfrac{1}{5}\right)$$
$$20 = a$$

Since $a = 20$ checks in the original equation, the solution is $a = 20$.

5. Multiplying both sides of the equation by x:
$$x\left(\frac{3}{x}+1\right)=x\left(\frac{2}{x}\right)$$
$$3+x=2$$
$$x=-1$$
Since $x=-1$ checks in the original equation, the solution is $x=-1$.

7. Multiplying both sides of the equation by $5a$:
$$5a\left(\frac{3}{a}-\frac{2}{a}\right)=5a\left(\frac{1}{5}\right)$$
$$15-10=a$$
$$a=5$$
Since $a=5$ checks in the original equation, the solution is $a=5$.

9. Multiplying both sides of the equation by $2x$:
$$2x\left(\frac{3}{x}+2\right)=2x\left(\frac{1}{2}\right)$$
$$6+4x=x$$
$$6=-3x$$
$$x=-2$$
Since $x=-2$ checks in the original equation, the solution is $x=-2$.

11. Multiplying both sides of the equation by $4y$:
$$4y\left(\frac{1}{y}-\frac{1}{2}\right)=4y\left(-\frac{1}{4}\right)$$
$$4-2y=-y$$
$$4=y$$
Since $y=4$ checks in the original equation, the solution is $y=4$.

13. Multiplying both sides of the equation by x^2:
$$x^2\left(1-\frac{8}{x}\right)=x^2\left(-\frac{15}{x^2}\right)$$
$$x^2-8x=-15$$
$$x^2-8x+15=0$$
$$(x-3)(x-5)=0$$
$$x=3,5$$
Both $x=3$ and $x=5$ check in the original equation.

15. Multiplying both sides of the equation by $2x$:
$$2x\left(\frac{x}{2}-\frac{4}{x}\right)=2x\left(-\frac{7}{2}\right)$$
$$x^2-8=-7x$$
$$x^2+7x-8=0$$
$$(x+8)(x-1)=0$$
$$x=-8,1$$
Both $x=-8$ and $x=1$ check in the original equation.

17. Multiplying both sides of the equation by 6:
$$6\left(\frac{x-3}{2}+\frac{2x}{3}\right)=6\left(\frac{5}{6}\right)$$
$$3(x-3)+2(2x)=5$$
$$3x-9+4x=5$$
$$7x-9=5$$
$$7x=14$$
$$x=2$$
Since $x=2$ checks in the original equation, the solution is $x=2$.

19. Multiplying both sides of the equation by 12:
$$12\left(\frac{x+1}{3}+\frac{x-3}{4}\right)=12\left(\frac{1}{6}\right)$$
$$4(x+1)+3(x-3)=2$$
$$4x+4+3x-9=2$$
$$7x-5=2$$
$$7x=7$$
$$x=1$$
Since $x=1$ checks in the original equation, the solution is $x=1$.

21.   Multiplying both sides of the equation by $5(x+2)$:

$$5(x+2) \cdot \frac{6}{x+2} = 5(x+2) \cdot \tfrac{3}{5}$$
$$30 = 3x + 6$$
$$24 = 3x$$
$$x = 8$$

Since $x = 8$ checks in the original equation, the solution is $x = 8$.

23.   Multiplying both sides of the equation by $(y-2)(y-3)$:

$$(y-2)(y-3) \cdot \frac{3}{y-2} = (y-2)(y-3) \cdot \frac{2}{y-3}$$
$$3(y-3) = 2(y-2)$$
$$3y - 9 = 2y - 4$$
$$y = 5$$

Since $y = 5$ checks in the original equation, the solution is $y = 5$.

25.   Multiplying both sides of the equation by $3(x-2)$:

$$3(x-2)\left(\frac{x}{x-2} + \tfrac{2}{3}\right) = 3(x-2)\left(\frac{2}{x-2}\right)$$
$$3x + 2(x-2) = 6$$
$$3x + 2x - 4 = 6$$
$$5x - 4 = 6$$
$$5x = 10$$
$$x = 2$$

Since $x = 2$ does not check in the original equation, there is no solution $(\varnothing)$.

27.   Multiplying both sides of the equation by $2(x-2)$:

$$2(x-2)\left(\frac{x}{x-2} + \tfrac{3}{2}\right) = 2(x-2) \cdot \frac{9}{2(x-2)}$$
$$2x + 3(x-2) = 9$$
$$2x + 3x - 6 = 9$$
$$5x - 6 = 9$$
$$5x = 15$$
$$x = 3$$

Since $x = 3$ checks in the original equation, the solution is $x = 3$.

29.   Multiplying both sides of the equation by $x^2 + 5x + 6 = (x+2)(x+3)$:

$$(x+2)(x+3)\left(\frac{5}{x+2} + \frac{1}{x+3}\right) = (x+2)(x+3) \cdot \frac{-1}{(x+2)(x+3)}$$
$$5(x+3) + 1(x+2) = -1$$
$$5x + 15 + x + 2 = -1$$
$$6x + 17 = -1$$
$$6x = -18$$
$$x = -3$$

Since $x = -3$ does not check in the original equation, there is no solution $(\varnothing)$.

31.   Multiplying both sides of the equation by $x^2 - 4 = (x+2)(x-2)$:

$$(x+2)(x-2)\left(\frac{8}{x^2-4}+\frac{3}{x+2}\right)=(x+2)(x-2)\cdot\frac{1}{x-2}$$
$$8+3(x-2)=1(x+2)$$
$$8+3x-6=x+2$$
$$3x+2=x+2$$
$$2x=0$$
$$x=0$$

Since $x = 0$ checks in the original equation, the solution is $x = 0$.

33.   Multiplying both sides of the equation by $2(a-3)$:

$$2(a-3)\left(\frac{a}{2}+\frac{3}{a-3}\right)=2(a-3)\cdot\frac{a}{a-3}$$
$$a(a-3)+6=2a$$
$$a^2-3a+6=2a$$
$$a^2-5a+6=0$$
$$(a-2)(a-3)=0$$
$$a=2,3$$

Since $a = 3$ does not check in the original equation, the solution is $a = 2$.

35.   Since $y^2 - 4 = (y+2)(y-2)$ and $y^2 + 2y = y(y+2)$, the LCD is $y(y+2)(y-2)$. Multiplying by the LCD:

$$y(y+2)(y-2)\cdot\frac{6}{(y+2)(y-2)}=y(y+2)(y-2)\cdot\frac{4}{y(y+2)}$$
$$6y=4(y-2)$$
$$6y=4y-8$$
$$2y=-8$$
$$y=-4$$

Since $y = -4$ checks in the original equation, the solution is $y = -4$.

37.   Since $a^2 - 9 = (a+3)(a-3)$ and $a^2 + a - 12 = (a+4)(a-3)$, the LCD is $(a+3)(a-3)(a+4)$. Multiplying by the LCD:

$$(a+3)(a-3)(a+4)\cdot\frac{2}{(a+3)(a-3)}=(a+3)(a-3)(a+4)\cdot\frac{3}{(a+4)(a-3)}$$
$$2(a+4)=3(a+3)$$
$$2a+8=3a+9$$
$$-a+8=9$$
$$-a=1$$
$$a=-1$$

Since $a = -1$ checks in the original equation, the solution is $a = -1$.

39.   Multiplying both sides of the equation by $x^2 - 4x - 5 = (x-5)(x+1)$:

$$(x-5)(x+1)\left(\frac{3x}{x-5}-\frac{2x}{x+1}\right)=(x-5)(x+1)\cdot\frac{-42}{(x-5)(x+1)}$$
$$3x(x+1)-2x(x-5)=-42$$
$$3x^2+3x-2x^2+10x=-42$$
$$x^2+13x+42=0$$
$$(x+7)(x+6)=0$$
$$x=-7,-6$$

Both $x = -7$ and $x = -6$ check in the original equation.

41.   Multiplying both sides of the equation by $x^2 + 5x + 6 = (x+2)(x+3)$:

$$(x+2)(x+3)\left(\frac{2x}{x+2}\right) = (x+2)(x+3)\left(\frac{x}{x+3} - \frac{3}{x^2+5x+6}\right)$$
$$2x(x+3) = x(x+2) - 3$$
$$2x^2 + 6x = x^2 + 2x - 3$$
$$x^2 + 4x + 3 = 0$$
$$(x+3)(x+1) = 0$$
$$x = -3, -1$$

Since $x = -3$ does not check in the original equation, the solution is $x = -1$.

43.   **a.**   Solving the equation:

$$5x - 1 = 0$$
$$5x = 1$$
$$x = \tfrac{1}{5}$$

b.   Solving the equation:

$$\frac{5}{x} - 1 = 0$$
$$x\left(\frac{5}{x} - 1\right) = x(0)$$
$$5 - x = 0$$
$$x = 5$$

c.   Solving the equation:

$$\frac{x}{5} - 1 = \frac{2}{3}$$
$$15\left(\frac{x}{5} - 1\right) = 15\left(\frac{2}{3}\right)$$
$$3x - 15 = 10$$
$$3x = 25$$
$$x = \frac{25}{3}$$

d.   Solving the equation:

$$\frac{5}{x} - 1 = \frac{2}{3}$$
$$3x\left(\frac{5}{x} - 1\right) = 3x\left(\frac{2}{3}\right)$$
$$15 - 3x = 2x$$
$$15 = 5x$$
$$x = 3$$

e.   Solving the equation:

$$\frac{5}{x^2} + 5 = \frac{26}{x}$$
$$x^2\left(\frac{5}{x^2} + 5\right) = x^2\left(\frac{26}{x}\right)$$
$$5 + 5x^2 = 26x$$
$$5x^2 - 26x + 5 = 0$$
$$(5x-1)(x-5) = 0$$
$$x = \tfrac{1}{5}, 5$$

45.   **a.**   Dividing: $\dfrac{7}{a^2 - 5a - 6} \div \dfrac{a+2}{a+1} = \dfrac{7}{(a-6)(a+1)} \cdot \dfrac{a+1}{a+2} = \dfrac{7}{(a-6)(a+2)}$

b.   Adding:

$$\frac{7}{a^2 - 5a - 6} + \frac{a+2}{a+1} = \frac{7}{(a-6)(a+1)} + \frac{a+2}{a+1} \cdot \frac{a-6}{a-6}$$
$$= \frac{7}{(a-6)(a+1)} + \frac{a^2 - 4a - 12}{(a-6)(a+1)}$$
$$= \frac{a^2 - 4a - 5}{(a-6)(a+1)}$$
$$= \frac{(a-5)(a+1)}{(a-6)(a+1)}$$
$$= \frac{a-5}{a-6}$$

c.   Solving the equation:

$$\frac{7}{a^2 - 5a - 6} + \frac{a+2}{a+1} = 2$$

$$\frac{7}{(a-6)(a+1)} + \frac{a+2}{a+1} = 2$$

$$(a-6)(a+1)\left(\frac{7}{(a-6)(a+1)} + \frac{a+2}{a+1}\right) = 2(a-6)(a+1)$$

$$7 + (a-6)(a+2) = 2(a-6)(a+1)$$

$$7 + a^2 - 4a - 12 = 2a^2 - 10a - 12$$

$$0 = a^2 - 6a - 7$$

$$0 = (a-7)(a+1)$$

$$a = -1, 7$$

Upon checking, only $a = 7$ checks in the original equation.

47.   Let x represent the number. The equation is:

$$2(x-3) - 5 = 3$$

$$2x - 6 - 5 = 3$$

$$2x - 11 = 3$$

$$2x = 14$$

$$x = 7$$

The number is 7.

49.   Let w represent the width, and $2w + 5$ represent the length. Using the perimeter formula:

$$2w + 2(2w + 5) = 34$$

$$2w + 4w + 10 = 34$$

$$6w + 10 = 34$$

$$6w = 24$$

$$w = 4$$

$$2w + 5 = 13$$

The length is 13 inches and the width is 4 inches.

51.   Let x and $x + 2$ represent the two integers. The equation is:

$$x(x+2) = 48$$

$$x^2 + 2x = 48$$

$$x^2 + 2x - 48 = 0$$

$$(x+8)(x-6) = 0$$

$$x = -8, 6$$

$$x + 2 = -6, 8$$

The two integers are either –8 and –6, or 6 and 8.

53.   Let x and $x + 2$ represent the two legs. The equation is:

$$x^2 + (x+2)^2 = 10^2$$

$$x^2 + x^2 + 4x + 4 = 100$$

$$2x^2 + 4x - 96 = 0$$

$$x^2 + 2x - 48 = 0$$

$$(x+8)(x-6) = 0$$

$$x = 6 \quad (x = -8 \text{ is impossible})$$

$$x + 2 = 8$$

The legs are 6 inches and 8 inches.

55.   Multiplying both sides of the equation by $2x$:

$$2x\left(\frac{1}{x}+\frac{1}{2x}\right)=2x\left(\frac{9}{2}\right)$$
$$2+1=9x$$
$$9x=3$$
$$x=\frac{1}{3}$$

57.   Multiplying both sides of the equation by $30x$:

$$30x\left(\frac{1}{10}-\frac{1}{15}\right)=30x\left(\frac{1}{x}\right)$$
$$3x-2x=30$$
$$x=30$$

59.   Evaluating when $x=-6$: $y=-\dfrac{6}{-6}=1$ **61.**   Evaluating when $x=2$: $y=-\dfrac{6}{2}=-3$

7.5 Applications

1.   Let x and $3x$ represent the two numbers. The equation is:

$$\frac{1}{x}+\frac{1}{3x}=\frac{16}{3}$$
$$3x\left(\frac{1}{x}+\frac{1}{3x}\right)=3x\left(\frac{16}{3}\right)$$
$$3+1=16x$$
$$16x=4$$
$$x=\frac{1}{4}$$
$$3x=\frac{3}{4}$$

The numbers are $\frac{1}{4}$ and $\frac{3}{4}$.

3.   Let x represent the number. The equation is:

$$x+\frac{1}{x}=\frac{13}{6}$$
$$6x\left(x+\frac{1}{x}\right)=6x\left(\frac{13}{6}\right)$$
$$6x^2+6=13x$$
$$6x^2-13x+6=0$$
$$(3x-2)(2x-3)=0$$
$$x=\frac{2}{3},\frac{3}{2}$$

The number is either $\frac{2}{3}$ or $\frac{3}{2}$.

5.   Let x represent the number. The equation is:

$$\frac{7+x}{9+x}=\frac{5}{7}$$
$$7(9+x)\cdot\frac{7+x}{9+x}=7(9+x)\cdot\frac{5}{7}$$
$$7(7+x)=5(9+x)$$
$$49+7x=45+5x$$
$$49+2x=45$$
$$2x=-4$$
$$x=-2$$

The number is -2.

7. Let x and $x + 2$ represent the two integers. The equation is:

$$\frac{1}{x} + \frac{1}{x+2} = \frac{5}{12}$$

$$12x(x+2)\left(\frac{1}{x} + \frac{1}{x+2}\right) = 12x(x+2)\left(\frac{5}{12}\right)$$

$$12(x+2) + 12x = 5x(x+2)$$

$$12x + 24 + 12x = 5x^2 + 10x$$

$$0 = 5x^2 - 14x - 24$$

$$(5x+6)(x-4) = 0$$

$$x = 4 \quad \left(x = -\frac{6}{5} \text{ is impossible}\right)$$

$$x + 2 = 6$$

The integers are 4 and 6.

9. Let x represent the rate of the boat in still water:

	d	r	t
Upstream	26	$x-3$	$\frac{26}{x-3}$
Downstream	38	$x+3$	$\frac{38}{x+3}$

The equation is:

$$\frac{26}{x-3} = \frac{38}{x+3}$$

$$(x+3)(x-3) \cdot \frac{26}{x-3} = (x+3)(x-3) \cdot \frac{38}{x+3}$$

$$26(x+3) = 38(x-3)$$

$$26x + 78 = 38x - 114$$

$$-12x + 78 = -114$$

$$-12x = -192$$

$$x = 16$$

The speed of the boat in still water is 16 mph.

11. Let x represent the plane speed in still air:

	d	r	t
Against Wind	140	$x-20$	$\frac{140}{x-20}$
With Wind	160	$x+20$	$\frac{160}{x+20}$

The equation is:

$$\frac{140}{x-20} = \frac{160}{x+20}$$

$$(x+20)(x-20) \cdot \frac{140}{x-20} = (x+20)(x-20) \cdot \frac{160}{x+20}$$

$$140(x+20) = 160(x-20)$$

$$140x + 2800 = 160x - 3200$$

$$-20x + 2800 = -3200$$

$$-20x = -6000$$

$$x = 300$$

The plane speed in still air is 300 mph.

13. Let x and $x + 20$ represent the rates of each plane:

	d	r	t
Plane 1	285	$x+20$	$\frac{285}{x+20}$
Plane 2	255	x	$\frac{255}{x}$

The equation is:

$$\frac{285}{x+20} = \frac{255}{x}$$

$$x(x+20) \cdot \frac{285}{x+20} = x(x+20) \cdot \frac{255}{x}$$

$$285x = 255(x+20)$$

$$285x = 255x + 5100$$

$$30x = 5100$$

$$x = 170$$

$$x + 20 = 190$$

The plane speeds are 170 mph and 190 mph.

15. Let x represent her rate downhill:

	d	r	t
Level Ground	2	$x-3$	$\dfrac{2}{x-3}$
Downhill	6	x	$\dfrac{6}{x}$

The equation is:

$$\frac{2}{x-3}+\frac{6}{x}=1$$

$$x(x-3)\left(\frac{2}{x-3}+\frac{6}{x}\right)=x(x-3)\cdot 1$$

$$2x+6(x-3)=x(x-3)$$

$$2x+6x-18=x^2-3x$$

$$8x-18=x^2-3x$$

$$0=x^2-11x+18$$

$$0=(x-2)(x-9)$$

$$x=9\quad (x=2\text{ is impossible})$$

Tina runs 9 mph on the downhill part of the course.

17. Let x represent her rate on level ground:

	d	r	t
Level Ground	4	x	$\dfrac{4}{x}$
Downhill	5	$x+2$	$\dfrac{5}{x+2}$

The equation is:

$$\frac{4}{x}+\frac{5}{x+2}=1$$

$$x(x+2)\left(\frac{4}{x}+\frac{5}{x+2}\right)=x(x+2)\cdot 1$$

$$4(x+2)+5x=x(x+2)$$

$$4x+8+5x=x^2+2x$$

$$9x+8=x^2+2x$$

$$0=x^2-7x-8$$

$$0=(x-8)(x+1)$$

$$x=8\quad (x=-1\text{ is impossible})$$

Jerri jogs 8 mph on level ground.

19. Let t represent the time to fill the pool with both pipes left open. The equation is:

$$\frac{1}{12}-\frac{1}{15}=\frac{1}{t}$$

$$60t\left(\frac{1}{12}-\frac{1}{15}\right)=60t\cdot\frac{1}{t}$$

$$5t-4t=60$$

$$t=60$$

It will take 60 hours to fill the pool with both pipes left open.

21.   Let t represent the time to fill the bathtub with both faucets open. The equation is:

$$\frac{1}{10} + \frac{1}{12} = \frac{1}{t}$$

$$60t\left(\frac{1}{10} + \frac{1}{12}\right) = 60t \cdot \frac{1}{t}$$

$$6t + 5t = 60$$

$$11t = 60$$

$$t = \frac{60}{11} = 5\frac{5}{11}$$

It will take $5\frac{5}{11}$ minutes to fill the tub with both faucets open.

23.   Let t represent the time to fill the sink with both the faucet and the drain left open. The equation is:

$$\frac{1}{3} - \frac{1}{4} = \frac{1}{t}$$

$$12t\left(\frac{1}{3} - \frac{1}{4}\right) = 12t \cdot \frac{1}{t}$$

$$4t - 3t = 12$$

$$t = 12$$

It will take 12 minutes for the sink to overflow with both the faucet and drain left open.

25.   Sketching the graph:

27.   Sketching the graph:

29.   Sketching the graph:

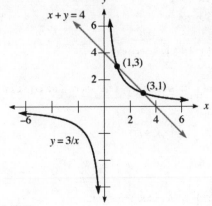

The intersection points are (1,3) and (3,1).

31.   Factoring the polynomial: $15a^3b^3 - 20a^2b - 35ab^2 = 5ab\left(3a^2b^2 - 4a - 7b\right)$

33.   Factoring the polynomial: $x^2 - 4x - 12 = (x-6)(x+2)$

35.   Factoring the polynomial: $x^4 - 16 = \left(x^2 + 4\right)\left(x^2 - 4\right) = \left(x^2 + 4\right)(x+2)(x-2)$

37.   Factoring the polynomial: $5x^3 - 25x^2 - 30x = 5x(x^2 - 5x - 6) = 5x(x - 6)(x + 1)$

39.   Solving the equation by factoring:

$$x^2 - 6x = 0$$
$$x(x - 6) = 0$$
$$x = 0, 6$$

41.   Solving the equation by factoring:

$$x(x + 2) = 80$$
$$x^2 + 2x = 80$$
$$x^2 + 2x - 80 = 0$$
$$(x + 10)(x - 8) = 0$$
$$x = -10, 8$$

43.   Let x and $x + 3$ represent the two integers. The equation is:

$$x^2 + (x + 3)^2 = 15^2$$
$$x^2 + x^2 + 6x + 9 = 225$$
$$2x^2 + 6x - 216 = 0$$
$$x^2 + 3x - 108 = 0$$
$$(x + 12)(x - 9) = 0$$
$$x = 9 \qquad (x = -12 \text{ is impossible})$$
$$x + 3 = 12$$

The two legs are 9 inches and 12 inches.

45.   Simplifying: $\frac{1}{3} \div \frac{3}{4} = \frac{1}{3} \bullet \frac{4}{3} = \frac{4}{9}$

47.   Simplifying: $1 + \frac{2}{3} = \frac{3}{3} + \frac{2}{3} = \frac{5}{3}$

49.   Simplifying: $y^7 \bullet \frac{3x^5}{y^4} = \frac{3x^5 y^7}{y^4} = 3x^5 y^3$

51.   Simplifying: $\frac{3x^5}{y^4} \bullet \frac{y^7}{6x^2} = \frac{3x^5 y^7}{6x^2 y^4} = \frac{x^3 y^3}{2}$

53.   Factoring; $xy^2 + y = y(xy + 1)$

55.   Reducing the fraction: $\frac{3x^5 y^3}{6x^2} = \frac{x^3 y^3}{2}$

57.   Reducing the fraction: $\dfrac{x^2 - 9}{x^2 - 5x + 6} = \dfrac{(x + 3)(x - 3)}{(x - 2)(x - 3)} = \dfrac{x + 3}{x - 2}$

7.6 Complex Fractions

1.   Simplifying the complex fraction: $\dfrac{\frac{3}{4}}{\frac{7}{8}} = \dfrac{\frac{3}{4} \bullet 8}{\frac{7}{8} \bullet 8} = \dfrac{6}{1} = 6$

3.   Simplifying the complex fraction: $\dfrac{\frac{2}{3}}{4} = \dfrac{\frac{2}{3} \bullet 3}{4 \bullet 3} = \dfrac{2}{12} = \dfrac{1}{6}$

5.   Simplifying the complex fraction: $\dfrac{\frac{x^2}{y}}{\frac{x}{y^3}} = \dfrac{\frac{x^2}{y} \bullet y^3}{\frac{x}{y^3} \bullet y^3} = \dfrac{x^2 y^2}{x} = xy^2$

7.   Simplifying the complex fraction: $\dfrac{\frac{4x^3}{y^6}}{\frac{8x^2}{y^7}} = \dfrac{\frac{4x^3}{y^6} \bullet y^7}{\frac{8x^2}{y^7} \bullet y^7} = \dfrac{4x^3 y}{8x^2} = \dfrac{xy}{2}$

9.   Simplifying the complex fraction: $\dfrac{y + \frac{1}{x}}{x + \frac{1}{y}} = \dfrac{\left(y + \frac{1}{x}\right) \bullet xy}{\left(x + \frac{1}{y}\right) \bullet xy} = \dfrac{xy^2 + y}{x^2 y + x} = \dfrac{y(xy + 1)}{x(xy + 1)} = \dfrac{y}{x}$

11.   Simplifying the complex fraction: $\dfrac{1+\dfrac{1}{a}}{1-\dfrac{1}{a}} = \dfrac{\left(1+\dfrac{1}{a}\right)\bullet a}{\left(1-\dfrac{1}{a}\right)\bullet a} = \dfrac{a+1}{a-1}$

13.   Simplifying the complex fraction: $\dfrac{\dfrac{x+1}{x^2-9}}{\dfrac{2}{x+3}} = \dfrac{\dfrac{x+1}{(x+3)(x-3)}\bullet(x+3)(x-3)}{\dfrac{2}{x+3}\bullet(x+3)(x-3)} = \dfrac{x+1}{2(x-3)}$

15.   Simplifying the complex fraction: $\dfrac{\dfrac{1}{a+2}}{\dfrac{1}{a^2-a-6}} = \dfrac{\dfrac{1}{a+2}\bullet(a-3)(a+2)}{\dfrac{1}{(a-3)(a+2)}\bullet(a-3)(a+2)} = \dfrac{a-3}{1} = a-3$

17.   Simplifying the complex fraction: $\dfrac{1-\dfrac{9}{y^2}}{1-\dfrac{1}{y}-\dfrac{6}{y^2}} = \dfrac{\left(1-\dfrac{9}{y^2}\right)\bullet y^2}{\left(1-\dfrac{1}{y}-\dfrac{6}{y^2}\right)\bullet y^2} = \dfrac{y^2-9}{y^2-y-6} = \dfrac{(y+3)(y-3)}{(y+2)(y-3)} = \dfrac{y+3}{y+2}$

19.   Simplifying the complex fraction: $\dfrac{\dfrac{1}{y}+\dfrac{1}{x}}{\dfrac{1}{xy}} = \dfrac{\left(\dfrac{1}{y}+\dfrac{1}{x}\right)\bullet xy}{\left(\dfrac{1}{xy}\right)\bullet xy} = \dfrac{x+y}{1} = x+y$

21.   Simplifying the complex fraction: $\dfrac{1-\dfrac{1}{a^2}}{1-\dfrac{1}{a}} = \dfrac{\left(1-\dfrac{1}{a^2}\right)\bullet a^2}{\left(1-\dfrac{1}{a}\right)\bullet a^2} = \dfrac{a^2-1}{a^2-a} = \dfrac{(a+1)(a-1)}{a(a-1)} = \dfrac{a+1}{a}$

23.   Simplifying the complex fraction: $\dfrac{\dfrac{1}{10x}-\dfrac{y}{10x^2}}{\dfrac{1}{10}-\dfrac{y}{10x}} = \dfrac{\left(\dfrac{1}{10x}-\dfrac{y}{10x^2}\right)\bullet10x^2}{\left(\dfrac{1}{10}-\dfrac{y}{10x}\right)\bullet10x^2} = \dfrac{x-y}{x^2-xy} = \dfrac{1(x-y)}{x(x-y)} = \dfrac{1}{x}$

25.   Simplifying the complex fraction: $\dfrac{\dfrac{1}{a+1}+2}{\dfrac{1}{a+1}+3} = \dfrac{\left(\dfrac{1}{a+1}+2\right)\bullet(a+1)}{\left(\dfrac{1}{a+1}+3\right)\bullet(a+1)} = \dfrac{1+2(a+1)}{1+3(a+1)} = \dfrac{1+2a+2}{1+3a+3} = \dfrac{2a+3}{3a+4}$

27.   Simplifying each parenthesis first:

$1-\dfrac{1}{x} = \dfrac{x}{x}-\dfrac{1}{x} = \dfrac{x-1}{x}$

$1-\dfrac{1}{x+1} = \dfrac{x+1}{x+1}-\dfrac{1}{x+1} = \dfrac{x}{x+1}$

$1-\dfrac{1}{x+2} = \dfrac{x+2}{x+2}-\dfrac{1}{x+2} = \dfrac{x+1}{x+2}$

Now performing the multiplication: $\left(1-\dfrac{1}{x}\right)\left(1-\dfrac{1}{x+1}\right)\left(1-\dfrac{1}{x+2}\right) = \dfrac{x-1}{x}\bullet\dfrac{x}{x+1}\bullet\dfrac{x+1}{x+2} = \dfrac{x-1}{x+2}$

29.   Simplifying each parenthesis first:

$$1 + \frac{1}{x+3} = \frac{x+3}{x+3} + \frac{1}{x+3} = \frac{x+4}{x+3}$$

$$1 + \frac{1}{x+2} = \frac{x+2}{x+2} + \frac{1}{x+2} = \frac{x+3}{x+2}$$

$$1 + \frac{1}{x+1} = \frac{x+1}{x+1} + \frac{1}{x+1} = \frac{x+2}{x+1}$$

Now performing the multiplication: $\left(1 + \frac{1}{x+3}\right)\left(1 + \frac{1}{x+2}\right)\left(1 + \frac{1}{x+1}\right) = \frac{x+4}{x+3} \cdot \frac{x+3}{x+2} \cdot \frac{x+2}{x+1} = \frac{x+4}{x+1}$

31.   Simplifying each term in the sequence:

$$2 + \frac{1}{2+1} = 2 + \frac{1}{3} = \frac{6}{3} + \frac{1}{3} = \frac{7}{3}$$

$$2 + \frac{1}{2 + \frac{1}{2+1}} = 2 + \frac{1}{\frac{7}{3}} = 2 + \frac{3}{7} = \frac{14}{7} + \frac{3}{7} = \frac{17}{7}$$

$$2 + \frac{2}{2 + \frac{1}{2 + \frac{1}{2+1}}} = 2 + \frac{1}{\frac{17}{7}} = 2 + \frac{7}{17} = \frac{34}{17} + \frac{7}{17} = \frac{41}{17}$$

33.   Completing the table:

Number x	Reciprocal $\frac{1}{x}$	Quotient $\frac{x}{1/x}$	Square x^2
1	1	1	1
2	$\frac{1}{2}$	4	4
3	$\frac{1}{3}$	9	9
4	$\frac{1}{4}$	16	16

35.   Completing the table:

Number x	Reciprocal $\frac{1}{x}$	Sum $1 + \frac{1}{x}$	Quotient $\frac{1 + \frac{1}{x}}{\frac{1}{x}}$
1	1	2	2
2	$\frac{1}{2}$	$\frac{3}{2}$	3
3	$\frac{1}{3}$	$\frac{4}{3}$	4
4	$\frac{1}{4}$	$\frac{5}{4}$	5

37.   Solving the inequality:

$$2x + 3 < 5$$
$$2x + 3 - 3 < 5 - 3$$
$$2x < 2$$
$$\tfrac{1}{2}(2x) < \tfrac{1}{2}(2)$$
$$x < 1$$

39.   Solving the inequality:

$$-3x \leq 21$$
$$-\tfrac{1}{3}(-3x) \geq -\tfrac{1}{3}(21)$$
$$x \geq -7$$

41.   Solving the inequality:

$$-2x + 8 > -4$$
$$-2x + 8 - 8 > -4 - 8$$
$$-2x > -12$$
$$-\tfrac{1}{2}(-2x) < -\tfrac{1}{2}(-12)$$
$$x < 6$$

43.   Solving the inequality:

$$4 - 2(x+1) \geq -2$$
$$4 - 2x - 2 \geq -2$$
$$-2x + 2 \geq -2$$
$$-2x + 2 - 2 \geq -2 - 2$$
$$-2x \geq -4$$
$$-\tfrac{1}{2}(-2x) \leq -\tfrac{1}{2}(-4)$$
$$x \leq 2$$

45.   Solving the equation:

$$21 = 6x$$
$$x = \tfrac{21}{6} = \tfrac{7}{2}$$

47.   Solving the equation:

$$x^2 + x = 6$$
$$x^2 + x - 6 = 0$$
$$(x+3)(x-2) = 0$$
$$x = -3, 2$$

7.7 Proportions

1. Solving the proportion:
$$\frac{x}{2} = \frac{6}{12}$$
$$12x = 12$$
$$x = 1$$

3. Solving the proportion:
$$\frac{2}{5} = \frac{4}{x}$$
$$2x = 20$$
$$x = 10$$

5. Solving the proportion:
$$\frac{10}{20} = \frac{20}{x}$$
$$10x = 400$$
$$x = 40$$

7. Solving the proportion:
$$\frac{a}{3} = \frac{5}{12}$$
$$12a = 15$$
$$a = \frac{15}{12} = \frac{5}{4}$$

9. Solving the proportion:
$$\frac{2}{x} = \frac{6}{7}$$
$$6x = 14$$
$$x = \frac{14}{6} = \frac{7}{3}$$

11. Solving the proportion:
$$\frac{x+1}{3} = \frac{4}{x}$$
$$x^2 + x = 12$$
$$x^2 + x - 12 = 0$$
$$(x+4)(x-3) = 0$$
$$x = -4, 3$$

13. Solving the proportion:
$$\frac{x}{2} = \frac{8}{x}$$
$$x^2 = 16$$
$$x^2 - 16 = 0$$
$$(x+4)(x-4) = 0$$
$$x = -4, 4$$

15. Solving the proportion:
$$\frac{4}{a+2} = \frac{a}{2}$$
$$a^2 + 2a = 8$$
$$a^2 + 2a - 8 = 0$$
$$(a+4)(a-2) = 0$$
$$a = -4, 2$$

17. Solving the proportion:
$$\frac{1}{x} = \frac{x-5}{6}$$
$$x^2 - 5x = 6$$
$$x^2 - 5x - 6 = 0$$
$$(x-6)(x+1) = 0$$
$$x = -1, 6$$

19. Comparing hits to games, the proportion is:
$$\frac{6}{18} = \frac{x}{45}$$
$$18x = 270$$
$$x = 15$$
He will get 15 hits in 45 games.

21. Comparing ml alcohol to ml water, the proportion is:
$$\frac{12}{16} = \frac{x}{28}$$
$$16x = 336$$
$$x = 21$$
The solution will have 21 ml of alcohol.

23. Comparing grams of fat to total grams, the proportion is:
$$\frac{13}{100} = \frac{x}{350}$$
$$100x = 4550$$
$$x = 45.5$$
There are 45.5 grams of fat in 350 grams of ice cream.

25. Comparing inches on the map to actual miles, the proportion is:
$$\frac{3.5}{100} = \frac{x}{420}$$
$$100x = 1470$$
$$x = 14.7$$
They are 14.7 inches apart on the map.

27.   Comparing inches on the map to actual miles, the proportion is:

$$\frac{0.5}{5} = \frac{1.25}{x}$$
$$0.5x = 6.25$$
$$x = 12.5$$

The actual distance is 12.5 miles.

29.   Comparing driving time to distance:

$$\frac{46}{34} = \frac{x}{12.5}$$
$$34x = 575$$
$$x \approx 17$$

The driving time is approximately 17 minutes.

31.   Comparing miles to hours, the proportion is:

$$\frac{245}{5} = \frac{x}{7}$$
$$5x = 1715$$
$$x = 343$$

He will travel 343 miles.

33.   Answers will vary.

35.   Reducing the fraction: $\dfrac{xy+5x+3y+15}{x^2+ax+3x+3a} = \dfrac{x(y+5)+3(y+5)}{x(x+a)+3(x+a)} = \dfrac{(y+5)(x+3)}{(x+a)(x+3)} = \dfrac{y+5}{x+a}$

37.   Dividing the fractions:

$$\frac{3x+6}{x^2+4x+3} \div \frac{x^2+x-2}{x^2+2x-3} = \frac{3x+6}{x^2+4x+3} \bullet \frac{x^2+2x-3}{x^2+x-2}$$
$$= \frac{3(x+2)}{(x+3)(x+1)} \bullet \frac{(x+3)(x-1)}{(x+2)(x-1)}$$
$$= \frac{3(x+2)(x+3)(x-1)}{(x+3)(x+1)(x+2)(x-1)}$$
$$= \frac{3}{x+1}$$

39.   Subtracting the fractions:

$$\frac{2}{x^2-1} - \frac{5}{x^2+3x-4} = \frac{2}{(x+1)(x-1)} - \frac{5}{(x+4)(x-1)}$$
$$= \frac{2 \bullet (x+4)}{(x+1)(x-1)(x+4)} - \frac{5 \bullet (x+1)}{(x+4)(x-1)(x+1)}$$
$$= \frac{2x+8}{(x+1)(x-1)(x+4)} - \frac{5x+5}{(x+1)(x-1)(x+4)}$$
$$= \frac{2x+8-5x-5}{(x+1)(x-1)(x+4)}$$
$$= \frac{-3x+3}{(x+1)(x-1)(x+4)}$$
$$= \frac{-3(x-1)}{(x+1)(x-1)(x+4)}$$
$$= \frac{-3}{(x+1)(x+4)}$$

41.   Evaluating when $x = 3$: $y = 5(3) = 15$

43.   Evaluating when $x = 5$: $y = \dfrac{20}{5} = 4$

45.   Substituting $y = 72$:

$$72 = 2x^2$$
$$2x^2 - 72 = 0$$
$$2\left(x^2 - 36\right) = 0$$
$$2(x+6)(x-6) = 0$$
$$x = -6, 6$$

47. Substituting $y = 72$ and $x = 4$:

$$72 = K \bullet 4$$
$$K = 18$$

49. Substituting $y = 45$ and $x = 3$:

$$45 = K(3)^2$$
$$45 = 9K$$
$$K = 5$$

7.8 Variation

1. The variation equation is $y = Kx$. Finding K:

$$10 = K \bullet 5$$
$$K = 2$$

So $y = 2x$. Substituting $x = 4$: $y = 2 \bullet 4 = 8$

3. The variation equation is $y = Kx$. Finding K:

$$39 = K \bullet 3$$
$$K = 13$$

So $y = 13x$. Substituting $x = 10$: $y = 13 \bullet 10 = 130$

5. The variation equation is $y = Kx$. Finding K:

$$-24 = K \bullet 4$$
$$K = -6$$

So $y = -6x$. Substituting $y = -30$:

$$-6x = -30$$
$$x = 5$$

7. The variation equation is $y = Kx$. Finding K:

$$-7 = K \bullet (-1)$$
$$K = 7$$

So $y = 7x$. Substituting $y = -21$:

$$7x = -21$$
$$x = -3$$

9. The variation equation is $y = Kx^2$. Finding K:

$$75 = K \bullet 5^2$$
$$75 = 25K$$
$$K = 3$$

So $y = 3x^2$. Substituting $x = 1$: $y = 3 \bullet 1^2 = 3 \bullet 1 = 3$

11. The variation equation is $y = Kx^2$. Finding K:

$$48 = K \bullet 4^2$$
$$48 = 16K$$
$$K = 3$$

So $y = 3x^2$. Substituting $x = 9$: $y = 3 \bullet 9^2 = 3 \bullet 81 = 243$

13. The variation equation is $y = \dfrac{K}{x}$. Finding K:

$$5 = \frac{K}{2}$$
$$K = 10$$

So $y = \dfrac{10}{x}$. Substituting $x = 5$: $y = \dfrac{10}{5} = 2$

15. The variation equation is $y = \dfrac{K}{x}$. Finding K:

$$2 = \frac{K}{1}$$
$$K = 2$$

So $y = \dfrac{2}{x}$. Substituting $x = 4$: $y = \dfrac{2}{4} = \dfrac{1}{2}$

17. The variation equation is $y = \dfrac{K}{x}$. Finding K:

$$5 = \frac{K}{3}$$
$$K = 15$$

So $y = \dfrac{15}{x}$. Substituting $y = 15$:

$$\frac{15}{x} = 15$$
$$15 = 15x$$
$$x = 1$$

19. The variation equation is $y = \dfrac{K}{x}$. Finding K:

$$10 = \frac{K}{10}$$
$$K = 100$$

So $y = \dfrac{100}{x}$. Substituting $y = 100$:

$$\frac{100}{x} = 20$$
$$100 = 20x$$
$$x = 5$$

21. The variation equation is $y = \dfrac{K}{x^2}$. Finding K:

$$4 = \frac{K}{5^2}$$
$$4 = \frac{K}{25}$$
$$K = 100$$

So $y = \dfrac{100}{x^2}$. Substituting $x = 2$: $y = \dfrac{100}{2^2} = \dfrac{100}{4} = 25$

23. The variation equation is $y = \dfrac{K}{x^2}$. Finding K:

$$4 = \frac{K}{3^2}$$
$$4 = \frac{K}{9}$$
$$K = 36$$

So $y = \dfrac{36}{x^2}$. Substituting $x = 2$: $y = \dfrac{36}{2^2} = \dfrac{36}{4} = 9$

25. The variation equation is $t = Kd$. Finding K:

$$42 = K \bullet 2$$
$$K = 21$$

So $t = 21d$. Substituting $d = 4$: $t = 21 \bullet 4 = 84$ pounds

27. The variation equation is $P = KI^2$. Finding K:

$$30 = K \bullet 2^2$$
$$30 = 4K$$
$$K = \frac{15}{2}$$

So $P = \dfrac{15}{2}I^2$. Substituting $I = 7$: $P = \dfrac{15}{2} \bullet 7^2 = \dfrac{15}{2} \bullet 49 = 367.5$

29. The variation equation is $M = Kh$. Finding K:

$$157 = K \bullet 20$$
$$K = 7.85$$

So $M = 7.85h$. Substituting $h = 30$: $M = 7.85 \bullet 30 = \$235.50$

31. The variation equation is $F = \dfrac{K}{d^2}$. Finding K:

$$150 = \frac{K}{4000^2}$$
$$150 = \frac{K}{1.6 \times 10^7}$$
$$K = 2.4 \times 10^9$$

So $F = \dfrac{2.4 \times 10^9}{d^2}$. Substituting $d = 5000$: $F = \dfrac{2.4 \times 10^9}{(5000)^2} = \dfrac{2.4 \times 10^9}{2.5 \times 10^7} = 96$ pounds

33. The variation equation is $I = \dfrac{K}{R}$. Finding K:

$$30 = \frac{K}{2}$$
$$K = 60$$

So $I = \dfrac{60}{R}$. Substituting $R = 5$: $I = \dfrac{60}{5} = 12$ amps

35. Adding the two equations:

$$5x = 10$$
$$x = 2$$

Substituting into the first equation:

$$2(2) + y = 3$$
$$4 + y = 3$$
$$y = -1$$

The solution is $(2, -1)$.

37.   Multiplying the second equation by –4:
$$4x - 5y = 1$$
$$-4x + 8y = 8$$
Adding the two equations:
$$3y = 9$$
$$y = 3$$
Substituting into the second equation:
$$x - 2(3) = -2$$
$$x - 6 = -2$$
$$x = 4$$
The solution is $(4, 3)$.

39.   Substituting into the first equation:
$$5x + 2(3x - 2) = 7$$
$$5x + 6x - 4 = 7$$
$$11x - 4 = 7$$
$$11x = 11$$
$$x = 1$$
Substituting into the second equation: $y = 3(1) - 2 = 3 - 2 = 1$. The solution is $(1, 1)$.

41.   Substituting into the first equation:
$$2(2y + 1) - 3y = 4$$
$$4y + 2 - 3y = 4$$
$$y + 2 = 4$$
$$y = 2$$
Substituting into the second equation: $x = 2(2) + 1 = 4 + 1 = 5$. The solution is $(5, 2)$.

Chapter 7 Review/Test

1.   Reducing the rational expression: $\dfrac{7}{14x - 28} = \dfrac{7}{14(x - 2)} = \dfrac{1}{2(x - 2)}$. The variable restriction is $x \neq 2$.

2.   Reducing the rational expression: $\dfrac{a + 6}{a^2 - 36} = \dfrac{1(a + 6)}{(a + 6)(a - 6)} = \dfrac{1}{a - 6}$. The variable restriction is $a \neq -6, 6$.

3.   Reducing the rational expression: $\dfrac{8x - 4}{4x + 12} = \dfrac{4(2x - 1)}{4(x + 3)} = \dfrac{2x - 1}{x + 3}$. The variable restriction is $x \neq -3$.

4.   Reducing the rational expression: $\dfrac{x + 4}{x^2 + 8x + 16} = \dfrac{1(x + 4)}{(x + 4)^2} = \dfrac{1}{x + 4}$. The variable restriction is $x \neq -4$.

5.   Reducing the rational expression: $\dfrac{3x^3 + 16x^2 - 12x}{2x^3 + 9x^2 - 18x} = \dfrac{x(3x^2 + 16x - 12)}{x(2x^2 + 9x - 18)} = \dfrac{x(3x - 2)(x + 6)}{x(2x - 3)(x + 6)} = \dfrac{3x - 2}{2x - 3}$

The variable restriction is $x \neq -6, 0, \frac{3}{2}$.

6.   Reducing the rational expression: $\dfrac{x + 2}{x^4 - 16} = \dfrac{x + 2}{(x^2 + 4)(x^2 - 4)} = \dfrac{x + 2}{(x^2 + 4)(x + 2)(x - 2)} = \dfrac{1}{(x^2 + 4)(x - 2)}$

The variable restriction is $x \neq -2, 2$.

7.   Reducing the rational expression: $\dfrac{x^2 + 5x - 14}{x + 7} = \dfrac{(x + 7)(x - 2)}{x + 7} = x - 2$. The variable restriction is $x \neq -7$.

8.   Reducing the rational expression: $\dfrac{a^2 + 16a + 64}{a + 8} = \dfrac{(a + 8)^2}{a + 8} = a + 8$. The variable restriction is $a \neq -8$.

9.   Reducing the rational expression: $\dfrac{xy+bx+ay+ab}{xy+5x+ay+5a} = \dfrac{x(y+b)+a(y+b)}{x(y+5)+a(y+5)} = \dfrac{(y+b)(x+a)}{(y+5)(x+a)} = \dfrac{y+b}{y+5}$

The variable restriction is $y \ne -5, x \ne -a$.

10.   Performing the operations: $\dfrac{3x+9}{x^2} \cdot \dfrac{x^3}{6x+18} = \dfrac{3(x+3)}{x^2} \cdot \dfrac{x^3}{6(x+3)} = \dfrac{3x^3(x+3)}{6x^2(x+3)} = \dfrac{x}{2}$

11.   Performing the operations:

$$\dfrac{x^2+8x+16}{x^2+x-12} \div \dfrac{x^2-16}{x^2-x-6} = \dfrac{x^2+8x+16}{x^2+x-12} \cdot \dfrac{x^2-x-6}{x^2-16}$$

$$= \dfrac{(x+4)^2}{(x+4)(x-3)} \cdot \dfrac{(x+2)(x-3)}{(x+4)(x-4)}$$

$$= \dfrac{(x+4)^2(x+2)(x-3)}{(x+4)^2(x-3)(x-4)}$$

$$= \dfrac{x+2}{x-4}$$

12.   Performing the operations: $\left(a^2-4a-12\right)\left(\dfrac{a-6}{a+2}\right) = \dfrac{(a-6)(a+2)}{1} \cdot \dfrac{a-6}{a+2} = \dfrac{(a-6)^2(a+2)}{a+2} = (a-6)^2$

13.   Performing the operations:

$$\dfrac{3x^2-2x-1}{x^2+6x+8} \div \dfrac{3x^2+13x+4}{x^2+8x+16} = \dfrac{3x^2-2x-1}{x^2+6x+8} \cdot \dfrac{x^2+8x+16}{3x^2+13x+4}$$

$$= \dfrac{(3x+1)(x-1)}{(x+4)(x+2)} \cdot \dfrac{(x+4)^2}{(3x+1)(x+4)}$$

$$= \dfrac{(x+4)^2(3x+1)(x-1)}{(x+4)^2(x+2)(3x+1)}$$

$$= \dfrac{x-1}{x+2}$$

14.   Performing the operations: $\dfrac{2x}{2x+3} + \dfrac{3}{2x+3} = \dfrac{2x+3}{2x+3} = 1$

15.   Performing the operations: $\dfrac{x^2}{x-9} - \dfrac{18x-81}{x-9} = \dfrac{x^2-18x+81}{x-9} = \dfrac{(x-9)^2}{x-9} = x-9$

16.   Performing the operations: $\dfrac{a+4}{a+8} - \dfrac{a-9}{a+8} = \dfrac{a+4-a+9}{a+8} = \dfrac{13}{a+8}$

17.   Performing the operations: $\dfrac{x}{x+9} + \dfrac{5}{x} = \dfrac{x \cdot x}{(x+9) \cdot x} + \dfrac{5 \cdot (x+9)}{x \cdot (x+9)} = \dfrac{x^2}{x(x+9)} + \dfrac{5x+45}{x(x+9)} = \dfrac{x^2+5x+45}{x(x+9)}$

18.   Performing the operations: $\dfrac{5}{4x+20} + \dfrac{x}{x+5} = \dfrac{5}{4(x+5)} + \dfrac{x \cdot 4}{(x+5) \cdot 4} = \dfrac{5}{4(x+5)} + \dfrac{4x}{4(x+5)} = \dfrac{4x+5}{4(x+5)}$

19.   Performing the operations:

$$\frac{3}{x^2-36}-\frac{2}{x^2-4x-12}=\frac{3}{(x+6)(x-6)}-\frac{2}{(x-6)(x+2)}$$

$$=\frac{3(x+2)}{(x+6)(x-6)(x+2)}-\frac{2(x+6)}{(x+6)(x-6)(x+2)}$$

$$=\frac{3x+6}{(x+6)(x-6)(x+2)}-\frac{2x+12}{(x+6)(x-6)(x+2)}$$

$$=\frac{3x+6-2x-12}{(x+6)(x-6)(x+2)}$$

$$=\frac{x-6}{(x+6)(x-6)(x+2)}$$

$$=\frac{1}{(x+6)(x+2)}$$

20.   Performing the operations:

$$\frac{3a}{a^2+8a+15}-\frac{2}{a+5}=\frac{3a}{(a+5)(a+3)}-\frac{2(a+3)}{(a+5)(a+3)}$$

$$=\frac{3a}{(a+5)(a+3)}-\frac{2a+6}{(a+5)(a+3)}$$

$$=\frac{3a-2a-6}{(a+5)(a+3)}$$

$$=\frac{a-6}{(a+5)(a+3)}$$

21.   Multiplying both sides of the equation by $2x$:

$$2x\left(\frac{3}{x}+\frac{1}{2}\right)=2x\left(\frac{5}{x}\right)$$
$$6+x=10$$
$$x=4$$

Since $x=4$ checks in the original equation, the solution is $x=4$.

22.   Multiplying both side of the equation by $2(a-3)$:

$$2(a-3)\bullet\frac{a}{a-3}=2(a-3)\bullet\tfrac{3}{2}$$
$$2a=3(a-3)$$
$$2a=3a-9$$
$$-a=-9$$
$$a=9$$

Since $a=9$ checks in the original equation, the solution is $a=9$.

23.   Multiplying both sides of the equation by x^2:

$$x^2\left(1-\frac{7}{x}\right)=x^2\left(\frac{-6}{x^2}\right)$$
$$x^2-7x=-6$$
$$x^2-7x+6=0$$
$$(x-6)(x-1)=0$$
$$x=1,6$$

Both $x=1$ and $x=6$ check in the original equation.

24. Multiplying both side of the equation by $x^2 + 4x - 12 = (x+6)(x-2)$:

$$(x+6)(x-2)\left(\frac{3}{x+6} - \frac{1}{x-2}\right) = (x+6)(x-2) \cdot \frac{-8}{(x+6)(x-2)}$$
$$3(x-2) - 1(x+6) = -8$$
$$3x - 6 - x - 6 = -8$$
$$2x - 12 = -8$$
$$2x = 4$$
$$x = 2$$

Since $x = 2$ does not check in the original equation, there is no solution.

25. Since $y^2 - 16 = (y+4)(y-4)$ and $y^2 + 4y = y(y+4)$, multiply each side of the equation by $y(y+4)(y-4)$:

$$y(y+4)(y-4) \cdot \frac{2}{(y+4)(y-4)} = y(y+4)(y-4) \cdot \frac{10}{y(y+4)}$$
$$2y = 10(y-4)$$
$$2y = 10y - 40$$
$$-8y = -40$$
$$y = 5$$

Since $y = 5$ checks in the original equation, the solution is $y = 5$.

26. Let x represent the number. The equation is:

$$x + 7\left(\frac{1}{x}\right) = \frac{16}{3}$$
$$3x\left(x + \frac{7}{x}\right) = 3x \cdot \frac{16}{3}$$
$$3x^2 + 21 = 16x$$
$$3x^2 - 16x + 21 = 0$$
$$(3x - 7)(x - 3) = 0$$
$$x = 3, \frac{7}{3}$$

The number is either 3 or $\frac{7}{3}$.

27. Let x represent the speed of the boat in still water. Completing the table:

	d	r	t
Upstream	48	$x-3$	$\dfrac{48}{x-3}$
Downstream	72	$x+3$	$\dfrac{72}{x+3}$

The equation is:

$$\frac{48}{x-3} = \frac{72}{x+3}$$
$$(x+3)(x-3) \cdot \frac{48}{x-3} = (x+3)(x-3) \cdot \frac{72}{x+3}$$
$$48(x+3) = 72(x-3)$$
$$48x + 144 = 72x - 216$$
$$-24x + 144 = -216$$
$$-24x = -360$$
$$x = 15$$

The speed of the boat in still water is 15 mph.

28. Let t represent the time to fill the pool with both pipes left open. The equation is:

$$\frac{1}{21} - \frac{1}{28} = \frac{1}{t}$$

$$84t \cdot \left(\frac{1}{21} - \frac{1}{28}\right) = 84t \cdot \frac{1}{t}$$

$$4t - 3t = 84$$

$$t = 84$$

It will take 84 hours to fill the pool with both pipes left open.

29. Simplifying the complex fraction: $\dfrac{\dfrac{x+4}{x^2-16}}{\dfrac{2}{x-4}} = \dfrac{\dfrac{x+4}{(x+4)(x-4)}}{\dfrac{2}{x-4}} = \dfrac{\dfrac{1}{x-4} \cdot (x-4)}{\dfrac{2}{x-4} \cdot (x-4)} = \dfrac{1}{2}$

30. Simplifying the complex fraction: $\dfrac{1 - \dfrac{9}{y^2}}{1 + \dfrac{4}{y} - \dfrac{21}{y^2}} = \dfrac{\left(1 - \dfrac{9}{y^2}\right) \cdot y^2}{\left(1 + \dfrac{4}{y} - \dfrac{21}{y^2}\right) \cdot y^2} = \dfrac{y^2 - 9}{y^2 + 4y - 21} = \dfrac{(y+3)(y-3)}{(y+7)(y-3)} = \dfrac{y+3}{y+7}$

31. Simplifying the complex fraction: $\dfrac{\dfrac{1}{a-2} + 4}{\dfrac{1}{a-2} + 1} = \dfrac{\left(\dfrac{1}{a-2} + 4\right)(a-2)}{\left(\dfrac{1}{a-2} + 1\right)(a-2)} = \dfrac{1 + 4(a-2)}{1 + 1(a-2)} = \dfrac{1 + 4a - 8}{1 + a - 2} = \dfrac{4a-7}{a-1}$

32. Writing as a fraction: $\dfrac{40}{100} = \dfrac{2}{5}$

33. Writing as a fraction: $\dfrac{40 \text{ seconds}}{3 \text{ minutes}} = \dfrac{40 \text{ seconds}}{180 \text{ seconds}} = \dfrac{2}{9}$

34. Solving the proportion:

$$\frac{x}{9} = \frac{4}{3}$$

$$3x = 36$$

$$x = 12$$

35. Solving the proportion:

$$\frac{a}{3} = \frac{12}{a}$$

$$a^2 = 36$$

$$a^2 - 36 = 0$$

$$(a+6)(a-6) = 0$$

$$a = -6, 6$$

36. Solving the proportion:

$$\frac{8}{x-2} = \frac{x}{6}$$

$$x^2 - 2x = 48$$

$$x^2 - 2x - 48 = 0$$

$$(x+6)(x-8) = 0$$

$$x = -6, 8$$

37. The variation equation is $y = Kx$. Finding K:

$$-20 = K \cdot 4$$

$$K = -5$$

So $y = -5x$. Substituting $x = 7$: $y = -5 \cdot 7 = -35$

38. The variation equation is $y = \dfrac{K}{x}$. Finding K:

$$3 = \frac{K}{2}$$

$$K = 6$$

So $y = \dfrac{6}{x}$. Substituting $x = 12$: $y = \dfrac{6}{12} = \dfrac{1}{2}$

Chapter 8
Roots and Radicals

8.1 Definitions and Common Roots

1. Finding the root: $\sqrt{9} = 3$

3. Finding the root: $-\sqrt{9} = -3$

5. Finding the root: $\sqrt{-25}$ is not a real number

7. Finding the root: $-\sqrt{144} = -12$

9. Finding the root: $\sqrt{625} = 25$

11. Finding the root: $\sqrt{-49}$ is not a real number

13. Finding the root: $-\sqrt{64} = -8$

15. Finding the root: $-\sqrt{100} = -10$

17. Finding the root: $\sqrt{1225} = 35$

19. Finding the root: $\sqrt[4]{1} = 1$

21. Finding the root: $\sqrt[3]{-8} = -2$

23. Finding the root: $-\sqrt[3]{125} = -5$

25. Finding the root: $\sqrt[3]{-1} = -1$

27. Finding the root: $\sqrt[3]{-27} = -3$

29. Finding the root: $-\sqrt[4]{16} = -2$

31. Simplifying the expression: $\sqrt{x^2} = x$

33. Simplifying the expression: $\sqrt{9x^2} = 3x$

35. Simplifying the expression: $\sqrt{x^2 y^2} = xy$

37. Simplifying the expression: $\sqrt{(a+b)^2} = a+b$

39. Simplifying the expression: $\sqrt{49x^2 y^2} = 7xy$

41. Simplifying the expression: $\sqrt[3]{x^3} = x$

43. Simplifying the expression: $\sqrt[3]{8x^3} = 2x$

45. Simplifying the expression: $\sqrt{x^4} = x^2$

47. Simplifying the expression: $\sqrt{36a^6} = 6a^3$

49. Simplifying the expression: $\sqrt{25a^8 b^4} = 5a^4 b^2$

51. Simplifying the expression: $\sqrt[3]{x^6} = x^2$

53. Simplifying the expression: $\sqrt[3]{27a^{12}} = 3a^4$

55. **a.** Simplifying the expression: $\sqrt[4]{16} = 2$

   b. Simplifying the expression: $\sqrt[4]{x^4} = x$

   c. Simplifying the expression: $\sqrt[4]{x^8} = x^2$

   d. Simplifying the expression: $\sqrt[4]{16x^8 y^{12}} = 2x^2 y^3$

57. **a.** Simplifying the expression: $\sqrt{9} + \sqrt{16} = 3 + 4 = 7$

   b. Simplifying the expression: $\sqrt{9+16} = \sqrt{25} = 5$

   c. Simplifying the expression: $\sqrt{9 \bullet 16} = \sqrt{144} = 12$

   d. Simplifying the expression: $\sqrt{9} \sqrt{16} = 3 \bullet 4 = 12$

59. **a.** Simplifying the expression: $\sqrt{25} - \sqrt{16} = 5 - 4 = 1$

   b. Simplifying the expression: $\sqrt{25 - 16} = \sqrt{9} = 3$

   c. Simplifying the expression: $\sqrt{25} \sqrt{16} = 5 \bullet 4 = 20$

   d. Simplifying the expression: $\sqrt{25 \bullet 16} = \sqrt{400} = 20$

61. **a.** Approximating the expression: $\dfrac{1+\sqrt{5}}{2} \approx \dfrac{1+2.236}{2} = \dfrac{3.236}{2} = 1.618$

   b. Approximating the expression: $\dfrac{1-\sqrt{5}}{2} \approx \dfrac{1-2.236}{2} = \dfrac{-1.236}{2} = -0.618$

   c. Approximating the expression: $\dfrac{1+\sqrt{5}}{2} + \dfrac{1-\sqrt{5}}{2} \approx 1.618 - 0.618 = 1$

63. **a.** Evaluating the root: $\sqrt{9} = 3$ **b.** Evaluating the root: $\sqrt{900} = 30$

   c. Evaluating the root: $\sqrt{0.09} = 0.3$

65. Simplifying each expression:

$$\dfrac{5+\sqrt{49}}{2} = \dfrac{5+7}{2} = \dfrac{12}{2} = 6 \qquad\qquad \dfrac{5-\sqrt{49}}{2} = \dfrac{5-7}{2} = \dfrac{-2}{2} = -1$$

67. Simplifying each expression:

$$\dfrac{2+\sqrt{16}}{2} = \dfrac{2+4}{2} = \dfrac{6}{2} = 3 \qquad\qquad \dfrac{2-\sqrt{16}}{2} = \dfrac{2-4}{2} = \dfrac{-2}{2} = -1$$

69. Simplifying the expression: $\sqrt{x^2 + 6x + 9} = \sqrt{(x+3)^2} = x + 3$

71. Completing the table:

Length L (feet)	Time T (seconds)
1	1.11
2	1.57
3	1.92
4	2.22
5	2.48
6	2.72

73. Using the Pythagorean Theorem:

$$x^2 = 3^2 + 4^2 = 9 + 16 = 25$$
$$x = \sqrt{25} = 5$$

75. Using the Pythagorean Theorem:

$$x^2 = 5^2 + 10^2 = 25 + 100 = 125$$
$$x = \sqrt{125} \approx 11.2$$

77. Let l represent the length of the wire. Using the Pythagorean Theorem:

$$l^2 = 24^2 + 18^2 = 576 + 324 = 900$$
$$l = \sqrt{900} = 30$$

The length of the wire is 30 feet.

79. Let x represent the length of the log. Using the Pythagorean theorem:

$$x^2 = 5^2 + 12^2 = 25 + 144 = 169$$
$$x = \sqrt{169} = 13$$

The log must be 13 feet to just barely reach.

81. Reducing the rational expression: $\dfrac{x^2 - 16}{x+4} = \dfrac{(x+4)(x-4)}{x+4} = x - 4$

83. Reducing the rational expression: $\dfrac{10a + 20}{5a^2 - 20} = \dfrac{10(a+2)}{5(a^2 - 4)} = \dfrac{10(a+2)}{5(a+2)(a-2)} = \dfrac{2}{a-2}$

85. Reducing the rational expression: $\dfrac{2x^2 - 5x - 3}{x^2 - 3x} = \dfrac{(2x+1)(x-3)}{x(x-3)} = \dfrac{2x+1}{x}$

87. Reducing the rational expression: $\dfrac{xy + 3x + 2y + 6}{xy + 3x + ay + 3a} = \dfrac{x(y+3) + 2(y+3)}{x(y+3) + a(y+3)} = \dfrac{(y+3)(x+2)}{(y+3)(x+a)} = \dfrac{x+2}{x+a}$

89. Simplifying: $3 \bullet \sqrt{16} = 3 \bullet 4 = 12$ **91.** Factoring: $75 = 25 \bullet 3$

93. Factoring: $50 = 25 \bullet 2$ **95.** Factoring: $40 = 4 \bullet 10$

97. Factoring: $x^5 = x^4 \bullet x$

99. Factoring: $12x^2 = 4x^2 \bullet 3$

101. Factoring: $50x^3y^2 = 25x^2y^2 \bullet 2x$

8.2 Properties of Radicals

1. Simplifying the radical expression: $\sqrt{8} = \sqrt{4 \bullet 2} = \sqrt{4}\sqrt{2} = 2\sqrt{2}$

3. Simplifying the radical expression: $\sqrt{12} = \sqrt{4 \bullet 3} = \sqrt{4}\sqrt{3} = 2\sqrt{3}$

5. Simplifying the radical expression: $\sqrt[3]{24} = \sqrt[3]{8 \bullet 3} = \sqrt[3]{8}\sqrt[3]{3} = 2\sqrt[3]{3}$

7. Simplifying the radical expression: $\sqrt{50x^2} = \sqrt{25x^2 \bullet 2} = \sqrt{25x^2}\sqrt{2} = 5x\sqrt{2}$

9. Simplifying the radical expression: $\sqrt{45a^2b^2} = \sqrt{9a^2b^2 \bullet 5} = \sqrt{9a^2b^2}\sqrt{5} = 3ab\sqrt{5}$

11. Simplifying the radical expression: $\sqrt[3]{54x^3} = \sqrt[3]{27x^3 \bullet 2} = \sqrt[3]{27x^3}\sqrt[3]{2} = 3x\sqrt[3]{2}$

13. Simplifying the radical expression: $\sqrt{32x^4} = \sqrt{16x^4 \bullet 2} = \sqrt{16x^4}\sqrt{2} = 4x^2\sqrt{2}$

15. Simplifying the radical expression: $5\sqrt{80} = 5\sqrt{16 \bullet 5} = 5\sqrt{16}\sqrt{5} = 5 \bullet 4\sqrt{5} = 20\sqrt{5}$

17. Simplifying the radical expression: $\frac{1}{2}\sqrt{28x^3} = \frac{1}{2}\sqrt{4x^2 \bullet 7x} = \frac{1}{2}\sqrt{4x^2}\sqrt{7x} = \frac{1}{2} \bullet 2x\sqrt{7x} = x\sqrt{7x}$

19. Simplifying the radical expression: $x\sqrt[3]{8x^4} = x\sqrt[3]{8x^3 \bullet x} = x\sqrt[3]{8x^3}\sqrt[3]{x} = x \bullet 2x\sqrt[3]{x} = 2x^2\sqrt[3]{x}$

21. Simplifying the radical expression: $2a\sqrt[3]{27a^5} = 2a\sqrt[3]{27a^3 \bullet a^2} = 2a\sqrt[3]{27a^3}\sqrt[3]{a^2} = 2a \bullet 3a\sqrt[3]{a^2} = 6a^2\sqrt[3]{a^2}$

23. Simplifying the radical expression: $\frac{4}{3}\sqrt{45a^3} = \frac{4}{3}\sqrt{9a^2 \bullet 5a} = \frac{4}{3}\sqrt{9a^2}\sqrt{5a} = \frac{4}{3} \bullet 3a\sqrt{5a} = 4a\sqrt{5a}$

25. Simplifying the radical expression: $3\sqrt{50xy^2} = 3\sqrt{25y^2 \bullet 2x} = 3\sqrt{25y^2}\sqrt{2x} = 3 \bullet 5y\sqrt{2x} = 15y\sqrt{2x}$

27. Simplifying the radical expression: $7\sqrt{12x^2y} = 7\sqrt{4x^2 \bullet 3y} = 7\sqrt{4x^2}\sqrt{3y} = 7 \bullet 2x\sqrt{3y} = 14x\sqrt{3y}$

29. Simplifying the radical expression: $\sqrt{\dfrac{16}{25}} = \dfrac{\sqrt{16}}{\sqrt{25}} = \dfrac{4}{5}$ **31.** Simplifying the radical expression: $\sqrt{\dfrac{4}{9}} = \dfrac{\sqrt{4}}{\sqrt{9}} = \dfrac{2}{3}$

33. Simplifying the radical expression: $\sqrt[3]{\dfrac{8}{27}} = \dfrac{\sqrt[3]{8}}{\sqrt[3]{27}} = \dfrac{2}{3}$ **35.** Simplifying the radical expression: $\sqrt[4]{\dfrac{16}{81}} = \dfrac{\sqrt[4]{16}}{\sqrt[4]{81}} = \dfrac{2}{3}$

37. Simplifying the radical expression: $\sqrt{\dfrac{100x^2}{25}} = \sqrt{4x^2} = 2x$

39. Simplifying the radical expression: $\sqrt{\dfrac{81a^2b^2}{9}} = \sqrt{9a^2b^2} = 3ab$

41. Simplifying the radical expression: $\sqrt[3]{\dfrac{27x^3}{8y^3}} = \dfrac{\sqrt[3]{27x^3}}{\sqrt[3]{8y^3}} = \dfrac{3x}{2y}$

43. Simplifying the radical expression: $\sqrt{\dfrac{50}{9}} = \dfrac{\sqrt{50}}{\sqrt{9}} = \dfrac{\sqrt{25 \bullet 2}}{3} = \dfrac{5\sqrt{2}}{3}$

45. Simplifying the radical expression: $\sqrt{\dfrac{75}{25}} = \sqrt{3}$

47. Simplifying the radical expression: $\sqrt{\dfrac{128}{49}} = \dfrac{\sqrt{128}}{\sqrt{49}} = \dfrac{\sqrt{64 \bullet 2}}{7} = \dfrac{8\sqrt{2}}{7}$

49. Simplifying the radical expression: $\sqrt{\dfrac{288x}{25}} = \dfrac{\sqrt{288x}}{\sqrt{25}} = \dfrac{\sqrt{144 \bullet 2x}}{5} = \dfrac{12\sqrt{2x}}{5}$

51. Simplifying the radical expression: $\sqrt{\dfrac{54a^2}{25}} = \dfrac{\sqrt{54a^2}}{\sqrt{25}} = \dfrac{\sqrt{9a^2 \bullet 6}}{5} = \dfrac{3a\sqrt{6}}{5}$

53. Simplifying the radical expression: $\dfrac{3\sqrt{50}}{2} = \dfrac{3\sqrt{25 \cdot 2}}{2} = \dfrac{3 \cdot 5\sqrt{2}}{2} = \dfrac{15\sqrt{2}}{2}$

55. Simplifying the radical expression: $\dfrac{7\sqrt{28y^2}}{3} = \dfrac{7\sqrt{4y^2 \cdot 7}}{3} = \dfrac{7 \cdot 2y\sqrt{7}}{3} = \dfrac{14y\sqrt{7}}{3}$

57. Simplifying the radical expression: $\dfrac{5\sqrt{72a^2b^2}}{\sqrt{36}} = \dfrac{5\sqrt{36a^2b^2 \cdot 2}}{6} = \dfrac{5 \cdot 6ab\sqrt{2}}{6} = \dfrac{30ab\sqrt{2}}{6} = 5ab\sqrt{2}$

59. Simplifying the radical expression: $\dfrac{6\sqrt{8x^2y}}{\sqrt{4}} = \dfrac{6\sqrt{4x^2 \cdot 2y}}{2} = \dfrac{6 \cdot 2x\sqrt{2y}}{2} = \dfrac{12x\sqrt{2y}}{2} = 6x\sqrt{2y}$

61. **a.** Substituting the values: $\sqrt{b^2 - 4ac} = \sqrt{(4)^2 - 4(2)(-3)} = \sqrt{16 + 24} = \sqrt{40} = \sqrt{4 \cdot 10} = 2\sqrt{10}$

     b. Substituting the values: $\sqrt{b^2 - 4ac} = \sqrt{(1)^2 - 4(1)(-6)} = \sqrt{1 + 24} = \sqrt{25} = 5$

     c. Substituting the values: $\sqrt{b^2 - 4ac} = \sqrt{(1)^2 - 4(1)(-11)} = \sqrt{1 + 44} = \sqrt{45} = \sqrt{9 \cdot 5} = 3\sqrt{5}$

     d. Substituting the values: $\sqrt{b^2 - 4ac} = \sqrt{(6)^2 - 4(3)(2)} = \sqrt{36 - 24} = \sqrt{12} = \sqrt{4 \cdot 3} = 2\sqrt{3}$

63. **a.** Simplifying: $\sqrt{32x^{10}y^5} = \sqrt{16x^{10}y^4 \cdot 2y} = \sqrt{16x^{10}y^4}\,\sqrt{2y} = 4x^5y^2\sqrt{2y}$

     b. Simplifying: $\sqrt[3]{32x^{10}y^5} = \sqrt[3]{8x^9y^3 \cdot 4xy^2} = \sqrt[3]{8x^9y^3}\,\sqrt[3]{4xy^2} = 2x^3y\sqrt[3]{4xy^2}$

     c. Simplifying: $\sqrt[4]{32x^{10}y^5} = \sqrt[4]{16x^8y^4 \cdot 2x^2y} = \sqrt[4]{16x^8y^4}\,\sqrt[4]{2x^2y} = 2x^2y\sqrt[4]{2x^2y}$

     d. Simplifying: $\sqrt[5]{32x^{10}y^5} = 2x^2y$

65. **a.** Simplifying: $\sqrt{4} = 2$ **b.** Simplifying: $\sqrt{0.04} = 0.2$

     c. Simplifying: $\sqrt{400} = 20$ **d.** Simplifying: $\sqrt{0.0004} = 0.02$

67. Completing the table:

x	$\sqrt{x}$	$2\sqrt{x}$	$\sqrt{4x}$
1	1	2	2
2	1.414	2.828	2.828
3	1.732	3.464	3.464
4	2	4	4

69. Completing the table:

x	$\sqrt{x}$	$3\sqrt{x}$	$\sqrt{9x}$
1	1	3	3
2	1.414	4.243	4.243
3	1.732	5.196	5.196
4	2	6	6

71. Performing the operations: $\dfrac{8x}{x^2 - 5x} \cdot \dfrac{x^2 - 25}{4x^2 + 4x} = \dfrac{8x}{x(x-5)} \cdot \dfrac{(x+5)(x-5)}{4x(x+1)} = \dfrac{8x(x+5)(x-5)}{4x^2(x-5)(x+1)} = \dfrac{2(x+5)}{x(x+1)}$

73. Performing the operations:

$$\dfrac{x^2 + 3x - 4}{3x^2 + 7x - 20} \div \dfrac{x^2 - 2x + 1}{3x^2 - 2x - 5} = \dfrac{x^2 + 3x - 4}{3x^2 + 7x - 20} \cdot \dfrac{3x^2 - 2x - 5}{x^2 - 2x + 1}$$

$$= \dfrac{(x+4)(x-1)}{(3x-5)(x+4)} \cdot \dfrac{(3x-5)(x+1)}{(x-1)^2}$$

$$= \dfrac{(x+4)(x-1)(3x-5)(x+1)}{(3x-5)(x+4)(x-1)^2}$$

$$= \dfrac{x+1}{x-1}$$

75. Performing the operations: $\left(x^2 - 36\right)\left(\dfrac{x+3}{x-6}\right) = \dfrac{(x+6)(x-6)}{1} \cdot \dfrac{x+3}{x-6} = \dfrac{(x+6)(x-6)(x+3)}{x-6} = (x+6)(x+3)$

77. Simplifying: $\sqrt{4x^3y^2} = \sqrt{4x^2y^2 \cdot x} = 2xy\sqrt{x}$ **79.** Simplifying: $\dfrac{6}{2}\sqrt{16} = 3 \cdot 4 = 12$

81. Simplifying: $\dfrac{\sqrt{2}}{\sqrt{4}} = \dfrac{\sqrt{2}}{2}$ **83.** Simplifying: $\dfrac{\sqrt[3]{18}}{\sqrt[3]{27}} = \dfrac{\sqrt[3]{18}}{3}$

85. Multiplying: $\dfrac{\sqrt{2}}{\sqrt{3}} \cdot \dfrac{\sqrt{3}}{\sqrt{3}} = \dfrac{\sqrt{6}}{\sqrt{9}} = \dfrac{\sqrt{6}}{3}$ **87.** Multiplying: $\sqrt[3]{3} \cdot \sqrt[3]{9} = \sqrt[3]{27} = 3$

8.3 Simplified Form for Radicals

1. Simplifying the radical expression: $\sqrt{\dfrac{1}{2}} = \dfrac{\sqrt{1}}{\sqrt{2}} \cdot \dfrac{\sqrt{2}}{\sqrt{2}} = \dfrac{\sqrt{2}}{\sqrt{4}} = \dfrac{\sqrt{2}}{2}$

3. Simplifying the radical expression: $\sqrt{\dfrac{1}{3}} = \dfrac{\sqrt{1}}{\sqrt{3}} \cdot \dfrac{\sqrt{3}}{\sqrt{3}} = \dfrac{\sqrt{3}}{\sqrt{9}} = \dfrac{\sqrt{3}}{3}$

5. Simplifying the radical expression: $\sqrt{\dfrac{2}{5}} = \dfrac{\sqrt{2}}{\sqrt{5}} \cdot \dfrac{\sqrt{5}}{\sqrt{5}} = \dfrac{\sqrt{10}}{\sqrt{25}} = \dfrac{\sqrt{10}}{5}$

7. Simplifying the radical expression: $\sqrt{\dfrac{3}{2}} = \dfrac{\sqrt{3}}{\sqrt{2}} \cdot \dfrac{\sqrt{2}}{\sqrt{2}} = \dfrac{\sqrt{6}}{\sqrt{4}} = \dfrac{\sqrt{6}}{2}$

9. Simplifying the radical expression: $\sqrt{\dfrac{20}{3}} = \dfrac{\sqrt{20}}{\sqrt{3}} \cdot \dfrac{\sqrt{3}}{\sqrt{3}} = \dfrac{\sqrt{60}}{\sqrt{9}} = \dfrac{\sqrt{4 \cdot 15}}{3} = \dfrac{2\sqrt{15}}{3}$

11. Simplifying the radical expression: $\sqrt{\dfrac{45}{6}} = \sqrt{\dfrac{15}{2}} = \dfrac{\sqrt{15}}{\sqrt{2}} \cdot \dfrac{\sqrt{2}}{\sqrt{2}} = \dfrac{\sqrt{30}}{\sqrt{4}} = \dfrac{\sqrt{30}}{2}$

13. Simplifying the radical expression: $\sqrt{\dfrac{20}{5}} = \sqrt{4} = 2$ **15.** Simplifying the radical expression: $\dfrac{\sqrt{21}}{\sqrt{3}} = \sqrt{\dfrac{21}{3}} = \sqrt{7}$

17. Simplifying the radical expression: $\dfrac{\sqrt{35}}{\sqrt{7}} = \sqrt{\dfrac{35}{7}} = \sqrt{5}$

19. Simplifying the radical expression: $\dfrac{10\sqrt{15}}{5\sqrt{3}} = \dfrac{10}{5} \cdot \sqrt{\dfrac{15}{3}} = 2\sqrt{5}$

21. Simplifying the radical expression: $\dfrac{6\sqrt{21}}{3\sqrt{7}} = \dfrac{6}{3} \cdot \sqrt{\dfrac{21}{7}} = 2\sqrt{3}$

23. Simplifying the radical expression: $\dfrac{6\sqrt{35}}{12\sqrt{5}} = \dfrac{6}{12} \cdot \sqrt{\dfrac{35}{5}} = \dfrac{1}{2}\sqrt{7} = \dfrac{\sqrt{7}}{2}$

25. Simplifying the radical expression: $\sqrt{\dfrac{4x^2y^2}{2}} = \sqrt{2x^2y^2} = xy\sqrt{2}$

27. Simplifying the radical expression: $\sqrt{\dfrac{5x^2y}{3}} = \dfrac{\sqrt{5x^2y}}{\sqrt{3}} \cdot \dfrac{\sqrt{3}}{\sqrt{3}} = \dfrac{\sqrt{15x^2y}}{\sqrt{9}} = \dfrac{x\sqrt{15y}}{3}$

29. Simplifying the radical expression: $\sqrt{\dfrac{16a^4}{5}} = \dfrac{\sqrt{16a^4}}{\sqrt{5}} \cdot \dfrac{\sqrt{5}}{\sqrt{5}} = \dfrac{4a^2\sqrt{5}}{\sqrt{25}} = \dfrac{4a^2\sqrt{5}}{5}$

31. Simplifying the radical expression: $\sqrt{\dfrac{72a^5}{5}} = \dfrac{\sqrt{72a^5}}{\sqrt{5}} \cdot \dfrac{\sqrt{5}}{\sqrt{5}} = \dfrac{\sqrt{360a^5}}{\sqrt{25}} = \dfrac{\sqrt{36a^4 \cdot 10a}}{5} = \dfrac{6a^2\sqrt{10a}}{5}$

33. Simplifying the radical expression: $\sqrt{\dfrac{20x^2y^3}{3}} = \dfrac{\sqrt{20x^2y^3}}{\sqrt{3}} \cdot \dfrac{\sqrt{3}}{\sqrt{3}} = \dfrac{\sqrt{60x^2y^3}}{\sqrt{9}} = \dfrac{\sqrt{4x^2y^2 \cdot 15y}}{3} = \dfrac{2xy\sqrt{15y}}{3}$

35.   Simplifying the radical expression: $\dfrac{2\sqrt{20x^2y^3}}{3} = \dfrac{2\sqrt{4x^2y^2 \cdot 5y}}{3} = \dfrac{2 \cdot 2xy\sqrt{5y}}{3} = \dfrac{4xy\sqrt{5y}}{3}$

37.   Simplifying the radical expression: $\dfrac{6\sqrt{54a^2b^3}}{5} = \dfrac{6\sqrt{9a^2b^2 \cdot 6b}}{5} = \dfrac{6 \cdot 3ab\sqrt{6b}}{5} = \dfrac{18ab\sqrt{6b}}{5}$

39.   Simplifying the radical expression: $\dfrac{3\sqrt{72x^4}}{\sqrt{2x}} = 3\sqrt{\dfrac{72x^4}{2x}} = 3\sqrt{36x^3} = 3\sqrt{36x^2 \cdot x} = 3 \cdot 6x\sqrt{x} = 18x\sqrt{x}$

41.   Simplifying the radical expression: $\sqrt[3]{\dfrac{1}{2}} = \dfrac{\sqrt[3]{1}}{\sqrt[3]{2}} \cdot \dfrac{\sqrt[3]{4}}{\sqrt[3]{4}} = \dfrac{\sqrt[3]{4}}{\sqrt[3]{8}} = \dfrac{\sqrt[3]{4}}{2}$

43.   Simplifying the radical expression: $\sqrt[3]{\dfrac{1}{9}} = \dfrac{\sqrt[3]{1}}{\sqrt[3]{9}} \cdot \dfrac{\sqrt[3]{3}}{\sqrt[3]{3}} = \dfrac{\sqrt[3]{3}}{\sqrt[3]{27}} = \dfrac{\sqrt[3]{3}}{3}$

45.   **a.**   Simplifying the radical expression: $\dfrac{6}{\sqrt{\pi}} = \dfrac{6}{\sqrt{\pi}} \cdot \dfrac{\sqrt{\pi}}{\sqrt{\pi}} = \dfrac{6\sqrt{\pi}}{\pi}$

b.   Simplifying the radical expression: $\sqrt{\dfrac{A}{\pi}} = \dfrac{\sqrt{A}}{\sqrt{\pi}} \cdot \dfrac{\sqrt{\pi}}{\sqrt{\pi}} = \dfrac{\sqrt{A\pi}}{\pi}$

c.   Simplifying the radical expression: $\sqrt[3]{\dfrac{3V}{4\pi}} = \dfrac{\sqrt[3]{3V}}{\sqrt[3]{4\pi}} \cdot \dfrac{\sqrt[3]{2\pi^2}}{\sqrt[3]{2\pi^2}} = \dfrac{\sqrt[3]{6V\pi^2}}{\sqrt[3]{8\pi^3}} = \dfrac{\sqrt[3]{6V\pi^2}}{2\pi}$

d.   Simplifying the radical expression: $\dfrac{2}{\sqrt[3]{2\pi}} = \dfrac{2}{\sqrt[3]{2\pi}} \cdot \dfrac{\sqrt[3]{4\pi^2}}{\sqrt[3]{4\pi^2}} = \dfrac{2\sqrt[3]{4\pi^2}}{\sqrt[3]{8\pi^3}} = \dfrac{2\sqrt[3]{4\pi^2}}{2\pi} = \dfrac{\sqrt[3]{4\pi^2}}{\pi}$

47.   Completing the table:

x	$\sqrt{x}$	$\dfrac{1}{\sqrt{x}}$	$\dfrac{\sqrt{x}}{x}$
1	1	1	1
2	1.414	0.707	0.707
3	1.732	0.577	0.577
4	2	0.5	0.5
5	2.236	0.447	0.447
6	2.449	0.408	0.408

49.   Completing the table:

x	$\sqrt{x^2}$	$\sqrt{x^3}$	$x\sqrt{x}$
1	1	1	1
2	2	2.828	2.828
3	3	5.196	5.196
4	4	8	8
5	5	11.180	11.180
6	6	14.697	14.697

51.   Drawing the figure:

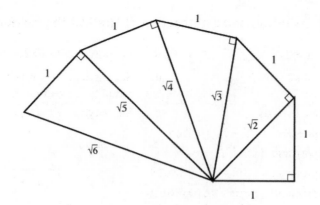

53.   Simplifying the terms in the sequence:

$$\sqrt{1^2+1}=\sqrt{1+1}=\sqrt{2}$$

$$\sqrt{\left(\sqrt{2}\right)^2+1}=\sqrt{2+1}=\sqrt{3}$$

$$\sqrt{\left(\sqrt{3}\right)^2+1}=\sqrt{3+1}=\sqrt{4}=2$$

55.   Combining the terms: $3x+7x=(3+7)x=10x$ **57.**   Combining the terms: $15x+8x=(15+8)x=23x$

59.   Combining the terms: $7a-3a+6a=(7-3+6)a=10a$

61.   Combining the rational expressions: $\dfrac{x^2}{x+5}+\dfrac{10x+25}{x+5}=\dfrac{x^2+10x+25}{x+5}=\dfrac{(x+5)^2}{x+5}=x+5$

63.   Combining the rational expressions: $\dfrac{a}{3}+\dfrac{2}{5}=\dfrac{a\bullet 5}{3\bullet 5}+\dfrac{2\bullet 3}{5\bullet 3}=\dfrac{5a}{15}+\dfrac{6}{15}=\dfrac{5a+6}{15}$

65.   Combining the rational expressions:

$$\dfrac{6}{a^2-9}-\dfrac{5}{a^2-a-6}=\dfrac{6}{(a+3)(a-3)}-\dfrac{5}{(a-3)(a+2)}$$

$$=\dfrac{6\bullet(a+2)}{(a+3)(a-3)(a+2)}-\dfrac{5\bullet(a+3)}{(a+3)(a-3)(a+2)}$$

$$=\dfrac{6a+12}{(a+3)(a-3)(a+2)}-\dfrac{5a+15}{(a+3)(a-3)(a+2)}$$

$$=\dfrac{6a+12-5a-15}{(a+3)(a-3)(a+2)}$$

$$=\dfrac{a-3}{(a+3)(a-3)(a+2)}$$

$$=\dfrac{1}{(a+3)(a+2)}$$

67.   Combining the terms: $15x+8x=(15+8)x=23x$

69.   Combining the terms: $25y+3y-y=(25+3-1)y=27y$

71.   Combining the terms: $2ab+5ab=(2+5)ab=7ab$ **73.**   Combining the terms: $2xy-9xy+50x=-7xy+50x$

75.   Simplifying: $\dfrac{6+2x}{4}=\dfrac{2(3+x)}{4}=\dfrac{3+x}{2}$

8.4 Addition and Subtraction of Radical Expressions

1.   **a.**   Simplifying: $3x+4x=7x$ **b.**   Simplifying: $3y+4y=7y$

   c.   Simplifying: $3\sqrt{2}+4\sqrt{2}=7\sqrt{2}$ **d.**   Simplifying: $3\sqrt{5}+4\sqrt{5}=7\sqrt{5}$

3.   **a.**   Simplifying: $x+6x=7x$ **b.**   Simplifying: $t+6t=7t$

   c.   Simplifying: $\sqrt{3}+6\sqrt{3}=7\sqrt{3}$ **d.**   Simplifying: $\sqrt{x}+6\sqrt{x}=7\sqrt{x}$

5.   Simplifying the radical expression: $9\sqrt{5}-7\sqrt{5}=2\sqrt{5}$

7.   Simplifying the radical expression: $5\sqrt{5}+\sqrt{5}=6\sqrt{5}$

9.   Simplifying the radical expression: $14\sqrt{13}-\sqrt{13}=13\sqrt{13}$

11.   Simplifying the radical expression: $-3\sqrt{10}+9\sqrt{10}=6\sqrt{10}$

13.   Simplifying the radical expression: $\frac{5}{8}\sqrt{5}-\frac{3}{7}\sqrt{5}=\frac{35}{56}\sqrt{5}-\frac{24}{56}\sqrt{5}=\frac{11}{56}\sqrt{5}$

15.   Simplifying the radical expression: $\sqrt{8}+2\sqrt{2}=\sqrt{4\bullet 2}+2\sqrt{2}=2\sqrt{2}+2\sqrt{2}=4\sqrt{2}$

17.   Simplifying the radical expression: $3\sqrt{3}-\sqrt{27}=3\sqrt{3}-\sqrt{9\bullet 3}=3\sqrt{3}-3\sqrt{3}=0$

19. Simplifying the radical expression:
$$5\sqrt{12} - 10\sqrt{48} = 5\sqrt{4\cdot 3} - 10\sqrt{16\cdot 3} = 5\cdot 2\sqrt{3} - 10\cdot 4\sqrt{3} = 10\sqrt{3} - 40\sqrt{3} = -30\sqrt{3}$$

21. Simplifying the radical expression: $-\sqrt{75} - \sqrt{3} = -\sqrt{25\cdot 3} - \sqrt{3} = -5\sqrt{3} - \sqrt{3} = -6\sqrt{3}$

23. Simplifying the radical expression: $\frac{1}{5}\sqrt{75} - \frac{1}{2}\sqrt{12} = \frac{1}{5}\sqrt{25\cdot 3} - \frac{1}{2}\sqrt{4\cdot 3} = \frac{1}{5}\cdot 5\sqrt{3} - \frac{1}{2}\cdot 2\sqrt{3} = \sqrt{3} - \sqrt{3} = 0$

25. Simplifying the radical expression: $\frac{3}{4}\sqrt{8} + \frac{3}{10}\sqrt{75} = \frac{3}{4}\sqrt{4\cdot 2} + \frac{3}{10}\sqrt{25\cdot 3} = \frac{3}{4}\cdot 2\sqrt{2} + \frac{3}{10}\cdot 5\sqrt{3} = \frac{3}{2}\sqrt{2} + \frac{3}{2}\sqrt{3}$

27. Simplifying the radical expression: $\sqrt{27} - 2\sqrt{12} + \sqrt{3} = \sqrt{9\cdot 3} - 2\sqrt{4\cdot 3} + \sqrt{3} = 3\sqrt{3} - 4\sqrt{3} + \sqrt{3} = 0$

29. Simplifying the radical expression:
$$\begin{aligned}
\frac{5}{6}\sqrt{72} - \frac{3}{8}\sqrt{8} + \frac{3}{10}\sqrt{50} &= \frac{5}{6}\sqrt{36\cdot 2} - \frac{3}{8}\sqrt{4\cdot 2} + \frac{3}{10}\sqrt{25\cdot 2} \\
&= \frac{5}{6}\cdot 6\sqrt{2} - \frac{3}{8}\cdot 2\sqrt{2} + \frac{3}{10}\cdot 5\sqrt{2} \\
&= 5\sqrt{2} - \frac{3}{4}\sqrt{2} + \frac{3}{2}\sqrt{2} \\
&= \frac{20}{4}\sqrt{2} - \frac{3}{4}\sqrt{2} + \frac{6}{4}\sqrt{2} \\
&= \frac{23}{4}\sqrt{2}
\end{aligned}$$

31. Simplifying the radical expression:
$$\begin{aligned}
6\sqrt{48} - 2\sqrt{12} + 5\sqrt{27} &= 6\sqrt{16\cdot 3} - 2\sqrt{4\cdot 3} + 5\sqrt{9\cdot 3} \\
&= 6\cdot 4\sqrt{3} - 2\cdot 2\sqrt{3} + 5\cdot 3\sqrt{3} \\
&= 24\sqrt{3} - 4\sqrt{3} + 15\sqrt{3} \\
&= 35\sqrt{3}
\end{aligned}$$

33. Simplifying the radical expression: $9 + 6\sqrt{2} + 2 - 18 - 6\sqrt{2} = (9 + 2 - 18) + \left(6\sqrt{2} - 6\sqrt{2}\right) = -7$

35. Simplifying the radical expression: $4 + 4\sqrt{5} + 5 - 8 - 4\sqrt{5} - 1 = (4 + 5 - 8 - 1) + \left(4\sqrt{5} - 4\sqrt{5}\right) = 0$

37. Simplifying the radical expression: $\sqrt{x^3} + x\sqrt{x} = \sqrt{x^2\cdot x} + x\sqrt{x} = x\sqrt{x} + x\sqrt{x} = 2x\sqrt{x}$

39. Simplifying the radical expression: $5\sqrt{3a^2} - a\sqrt{3} = 5a\sqrt{3} - a\sqrt{3} = 4a\sqrt{3}$

41. Simplifying the radical expression:
$$5\sqrt{8x^3} + x\sqrt{50x} = 5\sqrt{4x^2\cdot 2x} + x\sqrt{25\cdot 2x} = 5\cdot 2x\sqrt{2x} + x\cdot 5\sqrt{2x} = 10x\sqrt{2x} + 5x\sqrt{2x} = 15x\sqrt{2x}$$

43. Simplifying the radical expression:
$$3\sqrt{75x^3 y} - 2x\sqrt{3xy} = 3\sqrt{25x^2\cdot 3xy} - 2x\sqrt{3xy} = 3\cdot 5x\sqrt{3xy} - 2x\sqrt{3xy} = 15x\sqrt{3xy} - 2x\sqrt{3xy} = 13x\sqrt{3xy}$$

45. Simplifying the radical expression: $\sqrt{20ab^2} - b\sqrt{45a} = \sqrt{4b^2\cdot 5a} - b\sqrt{9\cdot 5a} = 2b\sqrt{5a} - 3b\sqrt{5a} = -b\sqrt{5a}$

47. Simplifying the radical expression:
$$9\sqrt{18x^3} - 2x\sqrt{48x} = 9\sqrt{9x^2\cdot 2x} - 2x\sqrt{16\cdot 3x} = 9\cdot 3x\sqrt{2x} - 2x\cdot 4\sqrt{3x} = 27x\sqrt{2x} - 8x\sqrt{3x}$$

49. Simplifying the radical expression:
$$\begin{aligned}
7\sqrt{50x^2 y} + 8x\sqrt{8y} - 7\sqrt{32x^2 y} &= 7\sqrt{25x^2\cdot 2y} + 8x\sqrt{4\cdot 2y} - 7\sqrt{16x^2\cdot 2y} \\
&= 7\cdot 5x\sqrt{2y} + 8x\cdot 2\sqrt{2y} - 7\cdot 4x\sqrt{2y} \\
&= 35x\sqrt{2y} + 16x\sqrt{2y} - 28x\sqrt{2y} \\
&= 23x\sqrt{2y}
\end{aligned}$$

51. Simplifying the expression: $\dfrac{6 + 2\sqrt{2}}{2} = \dfrac{2\left(3 + \sqrt{2}\right)}{2} = 3 + \sqrt{2}$

53. Simplifying the expression: $\dfrac{9 - 3\sqrt{3}}{3} = \dfrac{3\left(3 - \sqrt{3}\right)}{3} = 3 - \sqrt{3}$

55. Simplifying the expression: $\dfrac{8 - \sqrt{24}}{6} = \dfrac{8 - \sqrt{4\cdot 6}}{6} = \dfrac{8 - 2\sqrt{6}}{6} = \dfrac{2\left(4 - \sqrt{6}\right)}{6} = \dfrac{4 - \sqrt{6}}{3}$

57. Simplifying the expression: $\dfrac{6+\sqrt{8}}{2}=\dfrac{6+\sqrt{4\cdot 2}}{2}=\dfrac{6+2\sqrt{2}}{2}=\dfrac{2\left(3+\sqrt{2}\right)}{2}=3+\sqrt{2}$

59. Simplifying the expression: $\dfrac{-10+\sqrt{50}}{10}=\dfrac{-10+\sqrt{25\cdot 2}}{10}=\dfrac{-10+5\sqrt{2}}{10}=\dfrac{5\left(-2+\sqrt{2}\right)}{10}=\dfrac{-2+\sqrt{2}}{2}$

61. **a.** Approximating: $\dfrac{3+\sqrt{12}}{2}=\dfrac{3+\sqrt{4\cdot 3}}{2}=\dfrac{3+2\sqrt{3}}{2}\approx\dfrac{3+2\left(1.732\right)}{2}=\dfrac{6.464}{2}=3.232$

    b. Approximating: $\dfrac{3-\sqrt{12}}{2}=\dfrac{3-\sqrt{4\cdot 3}}{2}=\dfrac{3-2\sqrt{3}}{2}\approx\dfrac{3-2\left(1.732\right)}{2}=\dfrac{-0.464}{2}=-0.232$

    c. Approximating: $\dfrac{3+\sqrt{12}}{2}+\dfrac{3-\sqrt{12}}{2}=3.232-0.232=3$

63. Completing the table:

x	$\sqrt{x^2+9}$	$x+3$
1	3.162	4
2	3.606	5
3	4.243	6
4	5	7
5	5.831	8
6	6.708	9

65. Completing the table:

x	$\sqrt{x+3}$	$\sqrt{x}+\sqrt{3}$
1	2	2.732
2	2.236	3.146
3	2.449	3.464
4	2.646	3.732
5	2.828	3.968
6	3	4.182

67. The correct statement is: $4\sqrt{3}+5\sqrt{3}=9\sqrt{3}$

69. Multiplying the expressions: $\left(3x+y\right)^2=\left(3x\right)^2+2\left(3x\right)\left(y\right)+y^2=9x^2+6xy+y^2$

71. Multiplying the expressions: $\left(3x-4y\right)\left(3x+4y\right)=\left(3x\right)^2-\left(4y\right)^2=9x^2-16y^2$

73. Multiplying each side of the equation by 6:

$$6\left(\frac{x}{3}-\frac{1}{2}\right)=6\left(\frac{5}{2}\right)$$
$$2x-3=15$$
$$2x=18$$
$$x=9$$

Since $x=9$ checks in the original equation, the solution is $x=9$.

75. Multiplying each side of the equation by x^2:

$$x^2\left(1-\frac{5}{x}\right)=x^2\left(\frac{-6}{x^2}\right)$$
$$x^2-5x=-6$$
$$x^2-5x+6=0$$
$$\left(x-2\right)\left(x-3\right)=0$$
$$x=2,3$$

Both $x=2$ and $x=3$ check in the original equation.

77. Multiplying each side of the equation by $2(a-4)$:

$$2(a-4)\left(\frac{a}{a-4}-\frac{a}{2}\right)=2(a-4)\cdot\frac{4}{a-4}$$
$$2a-a(a-4)=8$$
$$2a-a^2+4a=8$$
$$-a^2+6a-8=0$$
$$a^2-6a+8=0$$
$$(a-4)(a-2)=0$$
$$a=2,4$$

Since $a=4$ does not check in the original equation, the solution is $a=2$.

79. Multiplying: $\sqrt{5}\sqrt{2}=\sqrt{10}$

81. Multiplying: $\sqrt{5}\sqrt{5}=\sqrt{25}=5$

83. Multiplying: $\sqrt{x}\sqrt{x}=\sqrt{x^2}=x$

85. Multiplying: $\sqrt{5}\sqrt{7}=\sqrt{35}$

87. Combining like terms: $5+7\sqrt{5}+2\sqrt{5}+14=19+9\sqrt{5}$

89. Combining like terms: $x-7\sqrt{x}+3\sqrt{x}-21=x-4\sqrt{x}-21$

8.5 Multiplication and Division of Radicals

1. Multiplying the radicals: $\sqrt{3}\sqrt{2}=\sqrt{6}$

3. Multiplying the radicals: $\sqrt{6}\sqrt{2}=\sqrt{12}=\sqrt{4\cdot3}=2\sqrt{3}$

5. Multiplying the radicals: $\left(2\sqrt{3}\right)\left(5\sqrt{7}\right)=10\sqrt{21}$

7. Multiplying the radicals: $\left(4\sqrt{3}\right)\left(2\sqrt{6}\right)=8\sqrt{18}=8\sqrt{9\cdot2}=8\cdot3\sqrt{2}=24\sqrt{2}$

9. Multiplying the radicals: $\left(2\sqrt{2}\right)^2=\left(2\sqrt{2}\right)\left(2\sqrt{2}\right)=4\sqrt{4}=4\cdot2=8$

11. Multiplying the radicals: $\left(-2\sqrt{6}\right)^2=\left(-2\sqrt{6}\right)\left(-2\sqrt{6}\right)=4\sqrt{36}=4\cdot6=24$

13. Multiplying the radicals: $\left(-1+5\sqrt{2}+1\right)^2=\left(5\sqrt{2}\right)^2=\left(5\sqrt{2}\right)\left(5\sqrt{2}\right)=25\sqrt{4}=25\cdot2=50$

15. Multiplying the radicals: $\left[2\left(-3+\sqrt{2}\right)+6\right]^2=\left(-6+2\sqrt{2}+6\right)=\left(2\sqrt{2}\right)^2=\left(2\sqrt{2}\right)\left(2\sqrt{2}\right)=4\sqrt{4}=4\cdot2=8$

17. Multiplying the radicals: $\sqrt{2}\left(\sqrt{3}-1\right)=\sqrt{6}-\sqrt{2}$

19. Multiplying the radicals: $\sqrt{2}\left(\sqrt{3}+\sqrt{2}\right)=\sqrt{6}+\sqrt{4}=\sqrt{6}+2$

21. Multiplying the radicals: $\sqrt{3}\left(2\sqrt{2}+\sqrt{3}\right)=2\sqrt{6}+\sqrt{9}=2\sqrt{6}+3$

23. Multiplying the radicals: $2\sqrt{3}\left(\sqrt{2}+\sqrt{5}\right)=2\sqrt{6}+2\sqrt{15}$

25. Simplifying the expression: $\left(\sqrt{2}+1\right)^2=\left(\sqrt{2}\right)^2+2\left(\sqrt{2}\right)(1)+(1)^2=2+2\sqrt{2}+1=3+2\sqrt{2}$

27. Simplifying the expression: $\left(\sqrt{x}+3\right)^2=\left(\sqrt{x}\right)^2+2\left(\sqrt{x}\right)(3)+(3)^2=x+6\sqrt{x}+9$

29. Simplifying the expression: $\left(\sqrt{a}-\frac{1}{2}\right)^2=\left(\sqrt{a}\right)^2-2\left(\sqrt{a}\right)\left(\frac{1}{2}\right)+\left(\frac{1}{2}\right)^2=a-\sqrt{a}+\frac{1}{4}$

31. Simplifying the expression: $\left(\sqrt{5}+3\right)\left(\sqrt{5}+2\right)=5+3\sqrt{5}+2\sqrt{5}+6=11+5\sqrt{5}$

33. Simplifying the expression: $\left(\sqrt{3}+\frac{1}{2}\right)\left(\sqrt{2}+\frac{1}{3}\right)=\sqrt{6}+\frac{1}{2}\sqrt{2}+\frac{1}{3}\sqrt{3}+\frac{1}{6}$

35. Simplifying the expression: $\left(\sqrt{x}+6\right)\left(\sqrt{x}-6\right)=\left(\sqrt{x}\right)^2-(6)^2=x-36$

37.   Simplifying the expression: $\left(\sqrt{a}+\frac{1}{3}\right)\left(\sqrt{a}+\frac{2}{3}\right)=a+\frac{1}{3}\sqrt{a}+\frac{2}{3}\sqrt{a}+\frac{2}{9}=a+\sqrt{a}+\frac{2}{9}$

39.   Simplifying the expression: $\left(\sqrt{5}-2\right)\left(\sqrt{5}+2\right)=\left(\sqrt{5}\right)^2-(2)^2=5-4=1$

41.   Simplifying the expression: $\left(2\sqrt{7}+3\right)\left(3\sqrt{7}-4\right)=6\sqrt{49}+9\sqrt{7}-8\sqrt{7}-12=42+\sqrt{7}-12=30+\sqrt{7}$

43.   Simplifying the expression:
$$\left(3+\sqrt{2}\right)^2-6\left(3+\sqrt{2}\right)=\left(3+\sqrt{2}\right)\left(3+\sqrt{2}\right)-6\left(3+\sqrt{2}\right)=9+3\sqrt{2}+3\sqrt{2}+2-18-6\sqrt{2}=-7$$

45.   Simplifying the expression:
$$\left(2+\sqrt{5}\right)^2-4\left(2+\sqrt{5}\right)=\left(2+\sqrt{5}\right)\left(2+\sqrt{5}\right)-4\left(2+\sqrt{5}\right)=4+2\sqrt{5}+2\sqrt{5}+5-8-4\sqrt{5}=1$$

47.   Simplifying the expression:
$$\left(7-\sqrt{5}\right)^2-14\left(7-\sqrt{5}\right)+44=\left(7-\sqrt{5}\right)\left(7-\sqrt{5}\right)-14\left(7-\sqrt{5}\right)+44$$
$$=49-7\sqrt{5}-7\sqrt{5}+5-98+14\sqrt{5}+44$$
$$=0$$

49.   Rationalizing the denominator: $\dfrac{\sqrt{3}}{\sqrt{5}-\sqrt{2}}\cdot\dfrac{\sqrt{5}+\sqrt{2}}{\sqrt{5}+\sqrt{2}}=\dfrac{\sqrt{15}+\sqrt{6}}{5-2}=\dfrac{\sqrt{15}+\sqrt{6}}{3}$

51.   Rationalizing the denominator: $\dfrac{\sqrt{5}}{\sqrt{5}+\sqrt{2}}\cdot\dfrac{\sqrt{5}-\sqrt{2}}{\sqrt{5}-\sqrt{2}}=\dfrac{\sqrt{25}-\sqrt{10}}{5-2}=\dfrac{5-\sqrt{10}}{3}$

53.   Rationalizing the denominator: $\dfrac{8}{3-\sqrt{5}}\cdot\dfrac{3+\sqrt{5}}{3+\sqrt{5}}=\dfrac{8\left(3+\sqrt{5}\right)}{9-5}=\dfrac{8\left(3+\sqrt{5}\right)}{4}=2\left(3+\sqrt{5}\right)=6+2\sqrt{5}$

55.   Rationalizing the denominator: $\dfrac{\sqrt{3}+\sqrt{2}}{\sqrt{3}-\sqrt{2}}\cdot\dfrac{\sqrt{3}+\sqrt{2}}{\sqrt{3}+\sqrt{2}}=\dfrac{3+\sqrt{6}+\sqrt{6}+2}{3-2}=\dfrac{5+2\sqrt{6}}{1}=5+2\sqrt{6}$

57.   **a.**   Subtracting: $\left(\sqrt{7}+\sqrt{3}\right)-\left(\sqrt{7}-\sqrt{3}\right)=\sqrt{7}+\sqrt{3}-\sqrt{7}+\sqrt{3}=2\sqrt{3}$

    b.   Multiplying: $\left(\sqrt{7}+\sqrt{3}\right)\left(\sqrt{7}-\sqrt{3}\right)=7-\sqrt{21}+\sqrt{21}-3=4$

    c.   Squaring: $\left(\sqrt{7}-\sqrt{3}\right)^2=\left(\sqrt{7}-\sqrt{3}\right)\left(\sqrt{7}-\sqrt{3}\right)=7-\sqrt{21}-\sqrt{21}+3=10-2\sqrt{21}$

    d.   Dividing: $\dfrac{\sqrt{7}-\sqrt{3}}{\sqrt{7}+\sqrt{3}}\cdot\dfrac{\sqrt{7}-\sqrt{3}}{\sqrt{7}-\sqrt{3}}=\dfrac{7-\sqrt{21}-\sqrt{21}+3}{7-3}=\dfrac{10-2\sqrt{21}}{4}=\dfrac{2\left(5-\sqrt{21}\right)}{4}=\dfrac{5-\sqrt{21}}{2}$

59.   **a.**   Adding: $\left(\sqrt{x}+2\right)+\left(\sqrt{x}-2\right)=\sqrt{x}+2+\sqrt{x}-2=2\sqrt{x}$

    b.   Multiplying: $\left(\sqrt{x}+2\right)\left(\sqrt{x}-2\right)=x+2\sqrt{x}-2\sqrt{x}-4=x-4$

    c.   Squaring: $\left(\sqrt{x}+2\right)^2=\left(\sqrt{x}+2\right)\left(\sqrt{x}+2\right)=x+2\sqrt{x}+2\sqrt{x}+4=x+4\sqrt{x}+4$

    d.   Dividing: $\dfrac{\sqrt{x}+2}{\sqrt{x}-2}\cdot\dfrac{\sqrt{x}+2}{\sqrt{x}+2}=\dfrac{x+2\sqrt{x}+2\sqrt{x}+4}{x-4}=\dfrac{x+4\sqrt{x}+4}{x-4}$

61.   **a.**   Adding: $\left(5+\sqrt{2}\right)+\left(5-\sqrt{2}\right)=5+\sqrt{2}+5-\sqrt{2}=10$

    b.   Multiplying: $\left(5+\sqrt{2}\right)\left(5-\sqrt{2}\right)=25+5\sqrt{2}-5\sqrt{2}-2=23$

    c.   Squaring: $\left(5+\sqrt{2}\right)^2=\left(5+\sqrt{2}\right)\left(5+\sqrt{2}\right)=25+5\sqrt{2}+5\sqrt{2}+2=27+10\sqrt{2}$

    d.   Dividing: $\dfrac{5+\sqrt{2}}{5-\sqrt{2}}\cdot\dfrac{5+\sqrt{2}}{5+\sqrt{2}}=\dfrac{25+5\sqrt{2}+5\sqrt{2}+2}{25-2}=\dfrac{27+10\sqrt{2}}{23}$

63. **a.** Adding: $\sqrt{2}+\left(\sqrt{6}+\sqrt{2}\right)=\sqrt{2}+\sqrt{6}+\sqrt{2}=2\sqrt{2}+\sqrt{6}$

     b. Multiplying: $\sqrt{2}\left(\sqrt{6}+\sqrt{2}\right)=\sqrt{12}+\sqrt{4}=2\sqrt{3}+2$

     c. Dividing: $\dfrac{\sqrt{6}+\sqrt{2}}{\sqrt{2}}\cdot\dfrac{\sqrt{2}}{\sqrt{2}}=\dfrac{\sqrt{12}+\sqrt{4}}{2}=\dfrac{2\sqrt{3}+2}{2}=1+\sqrt{3}$

     d. Dividing: $\dfrac{\sqrt{2}}{\sqrt{6}+\sqrt{2}}\cdot\dfrac{\sqrt{6}-\sqrt{2}}{\sqrt{6}-\sqrt{2}}=\dfrac{\sqrt{12}-\sqrt{4}}{6-2}=\dfrac{2\sqrt{3}-2}{4}=\dfrac{\sqrt{3}-1}{2}$

65. **a.** Adding: $\dfrac{1+\sqrt{5}}{2}+\dfrac{1-\sqrt{5}}{2}=\dfrac{2}{2}=1$

     b. Multiplying: $\left(\dfrac{1+\sqrt{5}}{2}\right)\left(\dfrac{1-\sqrt{5}}{2}\right)=\dfrac{1+\sqrt{5}-\sqrt{5}-5}{4}=\dfrac{-4}{4}=-1$

67. The correct statement is: $2\left(3\sqrt{5}\right)=6\sqrt{5}$

69. The correct statement is: $\left(\sqrt{3}+7\right)^{2}=\left(\sqrt{3}\right)^{2}+2\left(\sqrt{3}\right)(7)+(7)^{2}=3+14\sqrt{3}+49=52+14\sqrt{3}$

71. Solving the equation:
$$x^{2}+5x-6=0$$
$$(x+6)(x-1)=0$$
$$x=-6,1$$

73. Solving the equation:
$$x^{2}-3x=0$$
$$x(x-3)=0$$
$$x=0,3$$

75. Solving the proportion:
$$\frac{x}{3}=\frac{27}{x}$$
$$x^{2}=81$$
$$x^{2}-81=0$$
$$(x+9)(x-9)=0$$
$$x=-9,9$$

77. Solving the proportion:
$$\frac{x}{5}=\frac{3}{x+2}$$
$$x^{2}+2x=15$$
$$x^{2}+2x-15=0$$
$$(x+5)(x-3)=0$$
$$x=-5,3$$

79. Simplifying: $7^{2}=7\bullet 7=49$

81. Simplifying: $(-9)^{2}=(-9)\bullet(-9)=81$

83. Simplifying: $\left(\sqrt{x+1}\right)^{2}=x+1$

85. Simplifying: $\left(\sqrt{2x-3}\right)^{2}=2x-3$

87. Simplifying: $(x+3)^{2}=x^{2}+2(x)(3)+3^{2}=x^{2}+6x+9$

89. Solving the equation:
$$3a-2=4$$
$$3a=6$$
$$a=2$$

91. Solving the equation:
$$x+15=x^{2}+6x+9$$
$$0=x^{2}+5x-6$$
$$0=(x+6)(x-1)$$
$$x=-6,1$$

93.   a.   Substituting $x = -6$:
$$\sqrt{-6+15} = \sqrt{9} = 3$$
$$-6 + 3 = -3$$
$x = -6$ is not a solution.

b.   Substituting $x = 1$:
$$\sqrt{1+15} = \sqrt{16} = 4$$
$$1 + 3 = 4$$
$x = 1$ is a solution.

95.   The values are given in the table:

x	$y = 3\sqrt{x}$
-4	undefined
-1	undefined
0	$3\sqrt{0} = 0$
1	$3\sqrt{1} = 3$
4	$3\sqrt{4} = 6$
9	$3\sqrt{9} = 9$
16	$3\sqrt{16} = 12$

8.6 Equations Involving Radicals

1.   Solving the equation by squaring:
$$\sqrt{x+1} = 2$$
$$\left(\sqrt{x+1}\right)^2 = (2)^2$$
$$x + 1 = 4$$
$$x = 3$$
This value checks in the original equation.

3.   Solving the equation by squaring:
$$\sqrt{x+5} = 7$$
$$\left(\sqrt{x+5}\right)^2 = (7)^2$$
$$x + 5 = 49$$
$$x = 44$$
This value checks in the original equation.

5.   Solving the equation by squaring:
$$\sqrt{x-9} = -6$$
$$\left(\sqrt{x-9}\right)^2 = (-6)^2$$
$$x - 9 = 36$$
$$x = 45$$
Since $\sqrt{45-9} = \sqrt{36} = 6 \neq -6$, there is no solution to the equation $(\varnothing)$.

7.   Solving the equation by squaring:
$$\sqrt{x-5} = -4$$
$$\left(\sqrt{x-5}\right)^2 = (-4)^2$$
$$x - 5 = 16$$
$$x = 21$$
Since $\sqrt{21-5} = \sqrt{16} = 4 \neq -4$, there is no solution to the equation $(\varnothing)$.

9.   Solving the equation by squaring:
$$\sqrt{x-8} = 0$$
$$\left(\sqrt{x-8}\right)^2 = (0)^2$$
$$x - 8 = 0$$
$$x = 8$$
This value checks in the original equation.

11.   Solving the equation by squaring:
$$\sqrt{2x+1} = 3$$
$$\left(\sqrt{2x+1}\right)^2 = (3)^2$$
$$2x + 1 = 9$$
$$2x = 8$$
$$x = 4$$
This value checks in the original equation.

13.   Solving the equation by squaring:
$$\sqrt{2x-3} = -5$$
$$\left(\sqrt{2x-3}\right)^2 = (-5)^2$$
$$2x-3 = 25$$
$$2x = 28$$
$$x = 14$$
Since $\sqrt{2 \bullet 14 - 3} = \sqrt{28-3} = \sqrt{25} = 5 \neq -5$, there is no solution to the equation $(\varnothing)$.

15.   Solving the equation by squaring:
$$\sqrt{3x+6} = 2$$
$$\left(\sqrt{3x+6}\right)^2 = (2)^2$$
$$3x+6 = 4$$
$$3x = -2$$
$$x = -\frac{2}{3}$$

This value checks in the original equation.

17.   Solving the equation by squaring:
$$2\sqrt{x} = 10$$
$$\sqrt{x} = 5$$
$$\left(\sqrt{x}\right)^2 = (5)^2$$
$$x = 25$$

This value checks in the original equation.

19.   Solving the equation by squaring:
$$3\sqrt{a} = 6$$
$$\sqrt{a} = 2$$
$$\left(\sqrt{a}\right)^2 = (2)^2$$
$$a = 4$$

This value checks in the original equation.

21.   Solving the equation by squaring:
$$\sqrt{3x+4} - 3 = 2$$
$$\sqrt{3x+4} = 5$$
$$\left(\sqrt{3x+4}\right)^2 = (5)^2$$
$$3x+4 = 25$$
$$3x = 21$$
$$x = 7$$

This value checks in the original equation.

23.   Solving the equation by squaring:
$$\sqrt{5y-4} - 2 = 4$$
$$\sqrt{5y-4} = 6$$
$$\left(\sqrt{5y-4}\right)^2 = (6)^2$$
$$5y-4 = 36$$
$$5y = 40$$
$$y = 8$$

This value checks in the original equation.

25.   Solving the equation by squaring:
$$\sqrt{2x+1} + 5 = 2$$
$$\sqrt{2x+1} = -3$$
$$\left(\sqrt{2x+1}\right)^2 = (-3)^2$$
$$2x+1 = 9$$
$$2x = 8$$
$$x = 4$$
Since $\sqrt{2 \bullet 4 + 1} + 5 = \sqrt{9} + 5 = 3 + 5 = 8 \neq 2$, there is no solution to the equation $(\varnothing)$.

27.   Solving the equation by squaring:

$$\sqrt{x+3} = x-3$$
$$\left(\sqrt{x+3}\right)^2 = (x-3)^2$$
$$x+3 = x^2 - 6x + 9$$
$$0 = x^2 - 7x + 6$$
$$0 = (x-6)(x-1)$$
$$x = 6, 1$$

Since $x=1$ does not check in the original equation, the solution is $x=6$.

29.   Solving the equation by squaring:

$$\sqrt{a+2} = a+2$$
$$\left(\sqrt{a+2}\right)^2 = (a+2)^2$$
$$a+2 = a^2 + 4a + 4$$
$$0 = a^2 + 3a + 2$$
$$0 = (a+2)(a+1)$$
$$a = -2, -1$$

Both values check in the original equation.

31.   Solving the equation by squaring:

$$\sqrt{2x+9} = x+5$$
$$\left(\sqrt{2x+9}\right)^2 = (x+5)^2$$
$$2x+9 = x^2 + 10x + 25$$
$$0 = x^2 + 8x + 16$$
$$0 = (x+4)^2$$
$$x = -4$$

This value checks in the original equation.

33.   a.   Solving the equation:

$$\sqrt{y} - 4 = 6$$
$$\sqrt{y} = 10$$
$$\left(\sqrt{y}\right)^2 = (10)^2$$
$$y = 100$$

This value checks in the original equation.

b.   Solving the equation:

$$\sqrt{y-4} = 6$$
$$\left(\sqrt{y-4}\right)^2 = (6)^2$$
$$y - 4 = 36$$
$$y = 40$$

This value checks in the original equation.

c.   Solving the equation:

$$\sqrt{y-4} = -6$$
$$\left(\sqrt{y-4}\right)^2 = (-6)^2$$
$$y - 4 = 36$$
$$y = 40$$

This value does not check in the original equation, so there is no solution $(\varnothing)$.

d.   Solving the equation:

$$\sqrt{y-4} = y-6$$
$$\left(\sqrt{y-4}\right)^2 = (y-6)^2$$
$$y - 4 = y^2 - 12y + 36$$
$$0 = y^2 - 13y + 40$$
$$0 = (y-8)(y-5)$$
$$y = 5, 8$$

Since $y=5$ does not check in the original equation, the solution is $y=8$.

35.   a.   Solving the equation:

$$x - 3 = 0$$
$$x = 3$$

b.   Solving the equation:

$$\sqrt{x} - 3 = 0$$
$$\sqrt{x} = 3$$
$$\left(\sqrt{x}\right)^2 = (3)^2$$
$$x = 9$$

This value checks in the original equation.

c.   Solving the equation:

$$\sqrt{x-3} = 0$$
$$\left(\sqrt{x-3}\right)^2 = (0)^2$$
$$x - 3 = 0$$
$$x = 3$$

This value checks in the original equation.

d.   Solving the equation:
$$\sqrt{x} + 3 = 0$$
$$\sqrt{x} = -3$$
$$\left(\sqrt{x}\right)^2 = (-3)^2$$
$$x = 9$$

This value does not check in the original equation, so there is no solution $(\emptyset)$.

e.   Solving the equation:
$$\sqrt{x} + 3 = 5$$
$$\sqrt{x} = 2$$
$$\left(\sqrt{x}\right)^2 = (2)^2$$
$$x = 4$$

This value checks in the original equation.

f.   Solving the equation:
$$\sqrt{x} + 3 = -5$$
$$\sqrt{x} = -8$$
$$\left(\sqrt{x}\right)^2 = (-8)^2$$
$$x = 64$$

This value does not check in the original equation, so there is no solution $(\emptyset)$.

g.   Solving the equation:
$$x - 3 = \sqrt{5 - x}$$
$$(x - 3)^2 = \left(\sqrt{5 - x}\right)^2$$
$$x^2 - 6x + 9 = 5 - x$$
$$x^2 - 5x + 4 = 0$$
$$(x - 1)(x - 4) = 0$$
$$x = 1, 4$$

Since $x = 1$ does not check in the original equation, the solution is $x = 4$.

37.   Drawing a line graph:

39. Complete the table:

x	y
0	0
1	1
2	1.4
3	1.7
4	2

Sketching the graph:

41. Complete the table:

x	y
0	0
1	2
4	4
9	6

Sketching the graph:

43. Complete the table:

x	y
0	2
1	3
2	3.4
4	4
9	5

Sketching the graph:

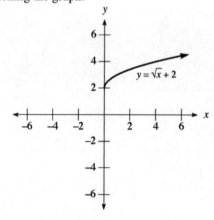

45. Let x represent the number. The equation is:

$$x + 2 = \sqrt{8x}$$
$$(x+2)^2 = \left(\sqrt{8x}\right)^2$$
$$x^2 + 4x + 4 = 8x$$
$$x^2 - 4x + 4 = 0$$
$$(x-2)^2 = 0$$
$$x - 2 = 0$$
$$x = 2$$

The number is 2.

47. Let x represent the number. The equation is:

$$x - 3 = 2\sqrt{x}$$
$$(x-3)^2 = \left(2\sqrt{x}\right)^2$$
$$x^2 - 6x + 9 = 4x$$
$$x^2 - 10x + 9 = 0$$
$$(x-9)(x-1) = 0$$
$$x = 9 \qquad \left(x = 1 \text{ does not check in the original equation}\right)$$

The number is 9.

49. Substituting $T = 2$:

$$2 = \frac{11}{7}\sqrt{\frac{L}{2}}$$
$$14 = 11\sqrt{\frac{L}{2}}$$
$$\frac{14}{11} = \sqrt{\frac{L}{2}}$$
$$\left(\frac{14}{11}\right)^2 = \left(\sqrt{\frac{L}{2}}\right)^2$$
$$\frac{196}{121} = \frac{L}{2}$$
$$L = \frac{392}{121} \approx 3.2 \text{ feet}$$

51. Reducing the rational expression: $\dfrac{x^2 - x - 6}{x^2 - 9} = \dfrac{(x-3)(x+2)}{(x+3)(x-3)} = \dfrac{x+2}{x+3}$

53. Performing the operations:

$$\frac{x^2 - 25}{x+4} \cdot \frac{2x+8}{x^2 - 9x + 20} = \frac{(x+5)(x-5)}{x+4} \cdot \frac{2(x+4)}{(x-4)(x-5)} = \frac{2(x+5)(x-5)(x+4)}{(x+4)(x-4)(x-5)} = \frac{2(x+5)}{x-4}$$

55. Performing the operations: $\dfrac{x}{x^2 - 16} + \dfrac{4}{x^2 - 16} = \dfrac{x+4}{x^2 - 16} = \dfrac{x+4}{(x+4)(x-4)} = \dfrac{1}{x-4}$

57. Simplifying the complex fraction: $\dfrac{1 - \dfrac{25}{x^2}}{1 - \dfrac{8}{x} + \dfrac{15}{x^2}} \cdot \dfrac{x^2}{x^2} = \dfrac{x^2 - 25}{x^2 - 8x + 15} = \dfrac{(x+5)(x-5)}{(x-5)(x-3)} = \dfrac{x+5}{x-3}$

59. Multiplying each side of the equation by $x^2 - 9 = (x+3)(x-3)$:

$$(x+3)(x-3)\left(\frac{x}{x^2-9} - \frac{3}{x-3}\right) = (x+3)(x-3) \cdot \frac{1}{x+3}$$
$$x - 3(x+3) = x - 3$$
$$x - 3x - 9 = x - 3$$
$$-2x - 9 = x - 3$$
$$-3x = 6$$
$$x = -2$$

Since $x = -2$ checks in the original equation, the solution is $x = -2$.

61. Let t represent the time to fill the pool with both pipes open. The equation is:

$$\frac{1}{8} - \frac{1}{12} = \frac{1}{t}$$
$$24t\left(\frac{1}{8} - \frac{1}{12}\right) = 24t \cdot \frac{1}{t}$$
$$3t - 2t = 24$$
$$t = 24$$

It will take 24 hours to fill the pool with both pipes left open.

63. The variation equation is $y = Kx$. Finding K:

$$8 = K \cdot 12$$
$$K = \frac{2}{3}$$

So $y = \frac{2}{3}x$. Substituting $x = 36$: $y = \frac{2}{3}(36) = 24$

Chapter 8 Review/Test

1. Finding the root: $\sqrt{25} = 5$

2. Finding the root: $\sqrt{169} = 13$

3. Finding the root: $\sqrt[3]{-1} = -1$

4. Finding the root: $\sqrt[4]{625} = 5$

5. Finding the root: $\sqrt{100x^2y^4} = 10xy^2$

6. Finding the root: $\sqrt[3]{8a^3} = 2a$

7. Simplifying the expression: $\sqrt{24} = \sqrt{4 \cdot 6} = 2\sqrt{6}$

8. Simplifying the expression: $\sqrt{60x^2} = \sqrt{4x^2 \cdot 15} = 2x\sqrt{15}$

9. Simplifying the expression: $\sqrt{90x^3y^4} = \sqrt{9x^2y^4 \cdot 10x} = 3xy^2\sqrt{10x}$

10. Simplifying the expression: $-\sqrt{32} = -\sqrt{16 \cdot 2} = -4\sqrt{2}$

11. Simplifying the expression: $3\sqrt{20x^3y} = 3\sqrt{4x^2 \cdot 5xy} = 3 \cdot 2x\sqrt{5xy} = 6x\sqrt{5xy}$

12. Simplifying the expression: $\sqrt{\dfrac{3}{49}} = \dfrac{\sqrt{3}}{\sqrt{49}} = \dfrac{\sqrt{3}}{7}$

13. Simplifying the expression: $\sqrt{\dfrac{8}{81}} = \dfrac{\sqrt{8}}{\sqrt{81}} = \dfrac{\sqrt{4 \cdot 2}}{9} = \dfrac{2\sqrt{2}}{9}$

14. Simplifying the expression: $\sqrt{\dfrac{49}{64}} = \dfrac{\sqrt{49}}{\sqrt{64}} = \dfrac{7}{8}$

15. Simplifying the expression: $\sqrt{\dfrac{49a^2b^2}{16}} = \dfrac{\sqrt{49a^2b^2}}{\sqrt{16}} = \dfrac{7ab}{4}$

16. Simplifying the expression: $\sqrt{\dfrac{80}{49}} = \dfrac{\sqrt{80}}{\sqrt{49}} = \dfrac{\sqrt{16 \cdot 5}}{7} = \dfrac{4\sqrt{5}}{7}$

17. Simplifying the expression: $\sqrt{\dfrac{40a^2}{121}} = \dfrac{\sqrt{40a^2}}{\sqrt{121}} = \dfrac{\sqrt{4a^2 \cdot 10}}{11} = \dfrac{2a\sqrt{10}}{11}$

18. Simplifying the expression: $\dfrac{5\sqrt{84}}{7} = \dfrac{5\sqrt{4 \cdot 21}}{7} = \dfrac{5 \cdot 2\sqrt{21}}{7} = \dfrac{10\sqrt{21}}{7}$

19. Simplifying the expression: $\dfrac{3\sqrt{120a^2b^2}}{\sqrt{25}} = \dfrac{3\sqrt{4a^2b^2 \cdot 30}}{5} = \dfrac{3 \cdot 2ab\sqrt{30}}{5} = \dfrac{6ab\sqrt{30}}{5}$

20. Simplifying the expression: $\dfrac{-5\sqrt{20x^3y^2}}{\sqrt{144}} = \dfrac{-5\sqrt{4x^2y^2 \cdot 5x}}{12} = \dfrac{-5 \cdot 2xy\sqrt{5x}}{12} = \dfrac{-10xy\sqrt{5x}}{12} = \dfrac{-5xy\sqrt{5x}}{6}$

21. Simplifying the radical expression: $\dfrac{2}{\sqrt{7}} \cdot \dfrac{\sqrt{7}}{\sqrt{7}} = \dfrac{2\sqrt{7}}{7}$

22. Simplifying the radical expression: $\sqrt{\dfrac{32}{5}} = \dfrac{\sqrt{32}}{\sqrt{5}} \cdot \dfrac{\sqrt{5}}{\sqrt{5}} = \dfrac{\sqrt{160}}{\sqrt{25}} = \dfrac{\sqrt{16 \cdot 10}}{5} = \dfrac{4\sqrt{10}}{5}$

23. Simplifying the radical expression: $\sqrt{\dfrac{5}{48}} = \dfrac{\sqrt{5}}{\sqrt{48}} \cdot \dfrac{\sqrt{3}}{\sqrt{3}} = \dfrac{\sqrt{15}}{\sqrt{144}} = \dfrac{\sqrt{15}}{12}$

24. Simplifying the radical expression: $\dfrac{-3\sqrt{60}}{\sqrt{5}} = -3\sqrt{\dfrac{60}{5}} = -3\sqrt{12} = -3\sqrt{4 \cdot 3} = -3 \cdot 2\sqrt{3} = -6\sqrt{3}$

25. Simplifying the radical expression: $\sqrt{\dfrac{32ab^2}{3}} = \dfrac{\sqrt{32ab^2}}{\sqrt{3}} \cdot \dfrac{\sqrt{3}}{\sqrt{3}} = \dfrac{\sqrt{96ab^2}}{\sqrt{9}} = \dfrac{\sqrt{16b^2 \cdot 6a}}{3} = \dfrac{4b\sqrt{6a}}{3}$

26. Simplifying the radical expression: $\sqrt[3]{\dfrac{3}{4}} = \dfrac{\sqrt[3]{3}}{\sqrt[3]{4}} \cdot \dfrac{\sqrt[3]{2}}{\sqrt[3]{2}} = \dfrac{\sqrt[3]{6}}{\sqrt[3]{8}} = \dfrac{\sqrt[3]{6}}{2}$

27. Rationalizing the denominator: $\dfrac{3}{\sqrt{3}-4} \cdot \dfrac{\sqrt{3}+4}{\sqrt{3}+4} = \dfrac{3\left(\sqrt{3}+4\right)}{3-16} = \dfrac{3\sqrt{3}+12}{-13} = \dfrac{-3\sqrt{3}-12}{13}$

28. Rationalizing the denominator: $\dfrac{2}{3+\sqrt{7}} \cdot \dfrac{3-\sqrt{7}}{3-\sqrt{7}} = \dfrac{2\left(3-\sqrt{7}\right)}{9-7} = \dfrac{2\left(3-\sqrt{7}\right)}{2} = 3-\sqrt{7}$

29. Rationalizing the denominator: $\dfrac{3}{\sqrt{5}-\sqrt{2}} \cdot \dfrac{\sqrt{5}+\sqrt{2}}{\sqrt{5}+\sqrt{2}} = \dfrac{3\left(\sqrt{5}+\sqrt{2}\right)}{5-2} = \dfrac{3\left(\sqrt{5}+\sqrt{2}\right)}{3} = \sqrt{5}+\sqrt{2}$

30. Rationalizing the denominator: $\dfrac{\sqrt{5}}{\sqrt{3}-\sqrt{5}} \cdot \dfrac{\sqrt{3}+\sqrt{5}}{\sqrt{3}+\sqrt{5}} = \dfrac{\sqrt{15}+\sqrt{25}}{3-5} = \dfrac{\sqrt{15}+5}{-2} = \dfrac{-\sqrt{15}-5}{2}$

31. Rationalizing the denominator: $\dfrac{\sqrt{5}-\sqrt{2}}{\sqrt{5}+\sqrt{2}} \cdot \dfrac{\sqrt{5}-\sqrt{2}}{\sqrt{5}-\sqrt{2}} = \dfrac{5-\sqrt{10}-\sqrt{10}+2}{5-2} = \dfrac{7-2\sqrt{10}}{3}$

32. Rationalizing the denominator: $\dfrac{\sqrt{x}+3}{\sqrt{x}-3} \cdot \dfrac{\sqrt{x}+3}{\sqrt{x}+3} = \dfrac{x+3\sqrt{x}+3\sqrt{x}+9}{x-9} = \dfrac{x+6\sqrt{x}+9}{x-9}$

33. Combining the expressions: $3\sqrt{5} - 7\sqrt{5} = -4\sqrt{5}$

34. Combining the expressions: $3\sqrt{27} - 5\sqrt{48} = 3\sqrt{9 \cdot 3} - 5\sqrt{16 \cdot 3} = 3 \cdot 3\sqrt{3} - 5 \cdot 4\sqrt{3} = 9\sqrt{3} - 20\sqrt{3} = -11\sqrt{3}$

35. Combining the expressions:

$$-2\sqrt{45} - 5\sqrt{80} + 2\sqrt{20} = -2\sqrt{9 \cdot 5} - 5\sqrt{16 \cdot 5} + 2\sqrt{4 \cdot 5}$$
$$= -2 \cdot 3\sqrt{5} - 5 \cdot 4\sqrt{5} + 2 \cdot 2\sqrt{5}$$
$$= -6\sqrt{5} - 20\sqrt{5} + 4\sqrt{5}$$
$$= -22\sqrt{5}$$

36. Combining the expressions:

$$3\sqrt{50x^2} - x\sqrt{200} = 3\sqrt{25x^2 \cdot 2} - x\sqrt{100 \cdot 2} = 3 \cdot 5x\sqrt{2} - x \cdot 10\sqrt{2} = 15x\sqrt{2} - 10x\sqrt{2} = 5x\sqrt{2}$$

37. Combining the expressions:

$$\sqrt{40a^3b^2} - a\sqrt{90ab^2} = \sqrt{4a^2b^2 \cdot 10a} - a\sqrt{9b^2 \cdot 10a} = 2ab\sqrt{10a} - 3ab\sqrt{10a} = -ab\sqrt{10a}$$

38.   Multiplying the expressions: $\sqrt{3}\left(\sqrt{3}+3\right)=\sqrt{9}+3\sqrt{3}=3+3\sqrt{3}$

39.   Multiplying the expressions: $4\sqrt{2}\left(\sqrt{3}+\sqrt{5}\right)=4\sqrt{6}+4\sqrt{10}$

40.   Multiplying the expressions: $\left(\sqrt{x}+7\right)\left(\sqrt{x}-7\right)=\left(\sqrt{x}\right)^2-\left(7\right)^2=x-49$

41.   Multiplying the expressions: $\left(2\sqrt{5}-4\right)\left(\sqrt{5}+3\right)=2\sqrt{25}-4\sqrt{5}+6\sqrt{5}-12=10+2\sqrt{5}-12=-2+2\sqrt{5}$

42.   Multiplying the expressions: $\left(\sqrt{x}+5\right)^2=\left(\sqrt{x}\right)^2+2\left(\sqrt{x}\right)(5)+(5)^2=x+10\sqrt{x}+25$

43.   Solving the equation by squaring:

$$\sqrt{x-3}=3$$
$$\left(\sqrt{x-3}\right)^2=(3)^2$$
$$x-3=9$$
$$x=12$$

This value checks in the original equation.

44.   Solving the equation by squaring:

$$\sqrt{3x-5}=4$$
$$\left(\sqrt{3x-5}\right)^2=(4)^2$$
$$3x-5=16$$
$$3x=21$$
$$x=7$$

This value checks in the original equation.

45.   Solving the equation by squaring:

$$5\sqrt{a}=20$$
$$\sqrt{a}=4$$
$$\left(\sqrt{a}\right)^2=(4)^2$$
$$a=16$$

This value checks in the original equation.

46.   Solving the equation by squaring:

$$\sqrt{3x-7}+6=2$$
$$\sqrt{3x-7}=-4$$
$$\left(\sqrt{3x-7}\right)^2=(-4)^2$$
$$3x-7=16$$
$$3x=23$$
$$x=\frac{23}{3}$$

Since $\sqrt{3\bullet\frac{23}{3}-7}+6=\sqrt{23-7}+6=\sqrt{16}+6=4+6=10\neq 2$, there is no solution to the equation.

47.   Solving the equation by squaring:

$$\sqrt{2x+1}+10=8$$
$$\sqrt{2x+1}=-2$$
$$\left(\sqrt{2x+1}\right)^2=(-2)^2$$
$$2x+1=4$$
$$2x=3$$
$$x=\frac{3}{2}$$

Since $\sqrt{2\bullet\frac{3}{2}+1}+10=\sqrt{3+1}+10=\sqrt{4}+10=2+10=12\neq 8$, there is no solution to the equation.

48. Solving the equation by squaring:

$$\sqrt{7x+1} = x+1$$
$$\left(\sqrt{7x+1}\right)^2 = (x+1)^2$$
$$7x+1 = x^2 + 2x + 1$$
$$0 = x^2 - 5x$$
$$0 = x(x-5)$$
$$x = 0, 5$$

Both values check in the original equation.

49. Using the Pythagorean Theorem:

$$x^2 = \left(\sqrt{2}\right)^2 + (1)^2 = 2 + 1 = 3$$
$$x = \sqrt{3}$$

50. Using the Pythagorean Theorem:

$$x^2 = (8)^2 + (8)^2 = 64 + 64 = 128$$
$$x = \sqrt{128} = \sqrt{64 \cdot 2} = 8\sqrt{2}$$

Chapter 9
Quadratic Equations

9.1 More Quadratic Equations

1. Solving the equation:
$$x^2 = 9$$
$$x = \pm\sqrt{9}$$
$$x = \pm 3$$

3. Solving the equation:
$$a^2 = 25$$
$$a = \pm\sqrt{25}$$
$$a = \pm 5$$

5. Solving the equation:
$$y^2 = 8$$
$$y = \pm\sqrt{8}$$
$$y = \pm 2\sqrt{2}$$

7. Solving the equation:
$$2x^2 = 100$$
$$x^2 = 50$$
$$x = \pm\sqrt{50}$$
$$x = \pm 5\sqrt{2}$$

9. Solving the equation:
$$3a^2 = 54$$
$$a^2 = 18$$
$$a = \pm\sqrt{18}$$
$$a = \pm 3\sqrt{2}$$

11. Solving the equation:
$$(x+2)^2 = 4$$
$$x+2 = \pm\sqrt{4}$$
$$x+2 = \pm 2$$
$$x = -2 \pm 2$$
$$x = -4, 0$$

13. Solving the equation:
$$(x+1)^2 = 25$$
$$x+1 = \pm\sqrt{25}$$
$$x+1 = \pm 5$$
$$x = -1 \pm 5$$
$$x = -6, 4$$

15. Solving the equation:
$$(a-5)^2 = 75$$
$$a-5 = \pm\sqrt{75}$$
$$a-5 = \pm 5\sqrt{3}$$
$$a = 5 \pm 5\sqrt{3}$$

17. Solving the equation:
$$(y+1)^2 = 50$$
$$y+1 = \pm\sqrt{50}$$
$$y+1 = \pm 5\sqrt{2}$$
$$y = -1 \pm 5\sqrt{2}$$

19. Solving the equation:
$$(2x+1)^2 = 25$$
$$2x+1 = \pm\sqrt{25}$$
$$2x+1 = \pm 5$$
$$2x = -1 \pm 5$$
$$2x = -6, 4$$
$$x = -3, 2$$

21. Solving the equation:
$$(4a-5)^2 = 36$$
$$4a-5 = \pm\sqrt{36}$$
$$4a-5 = \pm 6$$
$$4a = 5 \pm 6$$
$$4a = -1, 11$$
$$a = -\tfrac{1}{4}, \tfrac{11}{4}$$

23. Solving the equation:
$$(3y-1)^2 = 12$$
$$3y-1 = \pm\sqrt{12}$$
$$3y-1 = \pm 2\sqrt{3}$$
$$3y = 1 \pm 2\sqrt{3}$$
$$y = \frac{1 \pm 2\sqrt{3}}{3}$$

25.   Solving the equation:
$$(6x+2)^2 = 27$$
$$6x+2 = \pm\sqrt{27}$$
$$6x+2 = \pm3\sqrt{3}$$
$$6x = -2 \pm 3\sqrt{3}$$
$$x = \frac{-2 \pm 3\sqrt{3}}{6}$$

27.   Solving the equation:
$$(3x-9)^2 = 27$$
$$3x-9 = \pm\sqrt{27}$$
$$3x-9 = \pm3\sqrt{3}$$
$$3x = 9 \pm 3\sqrt{3}$$
$$x = 3 \pm \sqrt{3}$$

29.   Solving the equation:
$$(3x+6)^2 = 45$$
$$3x+6 = \pm\sqrt{45}$$
$$3x+6 = \pm3\sqrt{5}$$
$$3x = -6 \pm 3\sqrt{5}$$
$$x = -2 \pm \sqrt{5}$$

31.   Solving the equation:
$$(2y-4)^2 = 8$$
$$2y-4 = \pm\sqrt{8}$$
$$2y-4 = \pm2\sqrt{2}$$
$$2y = 4 \pm 2\sqrt{2}$$
$$y = 2 \pm \sqrt{2}$$

33.   Solving the equation:
$$\left(x-\tfrac{2}{3}\right)^2 = \tfrac{25}{9}$$
$$x-\tfrac{2}{3} = \pm\sqrt{\tfrac{25}{9}}$$
$$x-\tfrac{2}{3} = \pm\tfrac{5}{3}$$
$$x = \tfrac{2}{3} \pm \tfrac{5}{3}$$
$$x = -1, \tfrac{7}{3}$$

35.   Solving the equation:
$$\left(x+\tfrac{1}{2}\right)^2 = \tfrac{7}{4}$$
$$x+\tfrac{1}{2} = \pm\sqrt{\tfrac{7}{4}}$$
$$x+\tfrac{1}{2} = \pm\frac{\sqrt{7}}{2}$$
$$x = \frac{-1 \pm \sqrt{7}}{2}$$

37.   Solving the equation:
$$\left(a-\tfrac{4}{5}\right)^2 = \tfrac{12}{25}$$
$$a-\tfrac{4}{5} = \pm\sqrt{\tfrac{12}{25}}$$
$$a-\tfrac{4}{5} = \pm\frac{2\sqrt{3}}{5}$$
$$a = \frac{4 \pm 2\sqrt{3}}{5}$$

39.   Solving the equation:
$$x^2 + 10x + 25 = 7$$
$$(x+5)^2 = 7$$
$$x+5 = \pm\sqrt{7}$$
$$x = -5 \pm \sqrt{7}$$

41.   Solving the equation:
$$x^2 - 2x + 1 = 9$$
$$(x-1)^2 = 9$$
$$x-1 = \pm\sqrt{9}$$
$$x-1 = \pm3$$
$$x = 1 \pm 3$$
$$x = -2, 4$$

43.   Solving the equation:
$$x^2 + 12x + 36 = 8$$
$$(x+6)^2 = 8$$
$$x+6 = \pm\sqrt{8}$$
$$x+6 = \pm2\sqrt{2}$$
$$x = -6 \pm 2\sqrt{2}$$

45.   a.   No, as long as only rational numbers are used in the factoring.
   b.   Solving the equation:
$$x^2 = 3$$
$$x = \pm\sqrt{3}$$

47.   a.   Yes, the equation can be solved by factoring:
$$(x-3)^2 = 4$$
$$x^2 - 6x + 9 = 4$$
$$x^2 - 6x + 5 = 0$$
$$(x-1)(x-5) = 0$$
$$x = 1, 5$$

   b.   Solving the equation by taking square roots:
$$(x-3)^2 = 4$$
$$x-3 = \pm\sqrt{4}$$
$$x-3 = \pm2$$
$$x = 3 \pm 2$$
$$x = 1, 5$$

49. Checking the solution: $(2x-6)^2 = \left(2\left(3+\sqrt{3}\right)-6\right)^2 = \left(6+2\sqrt{3}-6\right)^2 = \left(2\sqrt{3}\right)^2 = 4 \cdot 3 = 12$

Yes, it is a solution to the equation.

51. **a.** Solving the equation:

$2x-1=0$

$2x=1$

$x=\frac{1}{2}$

b. Solving the equation:

$2x-1=4$

$2x=5$

$x=\frac{5}{2}$

c. Solving the equation:

$(2x-1)^2 = 4$

$2x-1=\pm\sqrt{4}$

$2x-1=\pm 2$

$2x=1\pm 2$

$2x=-1,3$

$x=-\frac{1}{2},\frac{3}{2}$

d. Solving the equation:

$\sqrt{2x}-1 \neq 4$

$\sqrt{2x}=5$

$\left(\sqrt{2x}\right)^2 = (5)^2$

$2x=25$

$x=\frac{25}{2}$

e. Solving the equation:

$\dfrac{1}{2x}-1=\dfrac{1}{4}$

$4x\left(\dfrac{1}{2x}-1\right)=4x\left(\dfrac{1}{4}\right)$

$2-4x=x$

$2=5x$

$x=\frac{2}{5}$

53. Checking the solution: $(x+1)^2 = \left(-1+5\sqrt{2}+1\right)^2 = \left(5\sqrt{2}\right)^2 = 25 \cdot 2 = 50$

55. **a.** Finding the sum: $\left(5+\sqrt{3}\right)+\left(5-\sqrt{3}\right)=5+\sqrt{3}+5-\sqrt{3}=10$

b. Finding the product: $\left(5+\sqrt{3}\right)\left(5-\sqrt{3}\right)=25+5\sqrt{3}-5\sqrt{3}-3=22$

57. Finding the radius r:

$\pi r^2 = 36\pi$

$r^2 = 36$

$r = \sqrt{36}$

$r = 6$

59. Solving for t:

$16t^2 = 100$

$t^2 = \frac{25}{4}$

$t = \sqrt{\frac{25}{4}}$

$t = \frac{5}{2}$ seconds

61. Let x represent the number. The equation is:

$(x+3)^2 = 16$

$x+3=\pm\sqrt{16}$

$x+3=\pm 4$

$x=-3\pm 4$

$x=-7,1$

The number is either –7 or 1.

63. Solving for r:

$100(1+r)^2 = A$

$(1+r)^2 = \dfrac{A}{100}$

$1+r = \sqrt{\dfrac{A}{100}}$ $\left(\text{since } 1+r>0\right)$

$1+r = \dfrac{\sqrt{A}}{10}$

$r = -1+\dfrac{\sqrt{A}}{10}$

65.   Let x represent the height of the triangle. Draw the figure:

Using the Pythagorean theorem:
$$5^2 + x^2 = 10^2$$
$$25 + x^2 = 100$$
$$x^2 = 75$$
$$x = \sqrt{75}$$
$$x = 5\sqrt{3} \approx 8.66$$

The height is $5\sqrt{3} \approx 8.66$ feet.

67.   Let x represent the height of the triangle. Draw the figure:

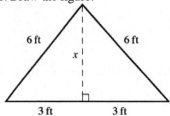

Using the Pythagorean theorem:
$$3^2 + x^2 = 6^2$$
$$9 + x^2 = 36$$
$$x^2 = 27$$
$$x = \sqrt{27}$$
$$x = 3\sqrt{3} \approx 5.2 \text{ feet}$$

No, a person 5 ft 8 in. tall cannot stand up inside the tent.

69.   Let x represent the height of the triangle. Draw the figure:

Using the Pythagorean theorem:
$$4^2 + x^2 = 5^2$$
$$16 + x^2 = 25$$
$$x^2 = 9$$
$$x = \sqrt{9} = 3$$

The height is 3 feet.

71. Substituting $c = 6$, $s = 3.53$, and $d = 192$ in the formula:

$$192 = (3.14)(3.53)(6)\left(\tfrac{1}{2}b\right)^2$$
$$\left(\tfrac{1}{2}b\right)^2 \approx 2.8870$$
$$\tfrac{1}{2}b \approx 1.6991$$
$$b \approx 3.4 \text{ inches}$$

73. Multiplying using the square of binomial formula: $(x-5)^2 = x^2 - 2(x)(5) + (5)^2 = x^2 - 10x + 25$

75. Factoring the polynomial: $x^2 - 12x + 36 = (x-6)(x-6) = (x-6)^2$

77. Factoring the polynomial: $x^2 + 4x + 4 = (x+2)(x+2) = (x+2)^2$

79. Finding the root: $\sqrt[3]{8} = 2$

81. Finding the root: $\sqrt[4]{16} = 2$

83. Simplifying: $\left(\tfrac{1}{2} \bullet 18\right)^2 = (9)^2 = 81$

85. Simplifying: $\left[\tfrac{1}{2}(-2)\right]^2 = (-1)^2 = 1$

87. Simplifying: $\left(\tfrac{1}{2} \bullet 3\right)^2 = \left(\tfrac{3}{2}\right)^2 = \tfrac{9}{4}$

89. Simplifying: $\dfrac{2x^2 + 16}{2} = \dfrac{2\left(x^2 + 8\right)}{2} = x^2 + 8$

91. Factoring: $x^2 + 6x + 9 = (x+3)^2$

93. Factoring: $y^2 - 3y + \tfrac{9}{4} = \left(y - \tfrac{3}{2}\right)^2$

9.2 Completing the Square

1. The correct term is 9, since: $x^2 + 6x + 9 = (x+3)^2$

3. The correct term is 1, since: $x^2 + 2x + 1 = (x+1)^2$

5. The correct term is 16, since: $y^2 - 8y + 16 = (y-4)^2$

7. The correct term is 1, since: $y^2 - 2y + 1 = (y-1)^2$

9. The correct term is 64, since: $x^2 + 16x + 64 = (x+8)^2$

11. The correct term is $\tfrac{9}{4}$, since: $a^2 - 3a + \tfrac{9}{4} = \left(a - \tfrac{3}{2}\right)^2$

13. The correct term is $\tfrac{49}{4}$, since: $x^2 - 7x + \tfrac{49}{4} = \left(x - \tfrac{7}{2}\right)^2$

15. The correct term is $\tfrac{1}{4}$, since: $y^2 + y + \tfrac{1}{4} = \left(y + \tfrac{1}{2}\right)^2$

17. The correct term is $\tfrac{9}{16}$, since: $x^2 - \tfrac{3}{2}x + \tfrac{9}{16} = \left(x - \tfrac{3}{4}\right)^2$

19. Solve by completing the square:
$$x^2 + 4x = 12$$
$$x^2 + 4x + 4 = 12 + 4$$
$$(x+2)^2 = 16$$
$$x + 2 = \pm 4$$
$$x = -2 \pm 4$$
$$x = -6, 2$$

21. Solve by completing the square:
$$x^2 - 6x = 16$$
$$x^2 - 6x + 9 = 16 + 9$$
$$(x-3)^2 = 25$$
$$x - 3 = \pm 5$$
$$x = 3 \pm 5$$
$$x = -2, 8$$

23. Solve by completing the square:
$$a^2 + 2a = 3$$
$$a^2 + 2a + 1 = 3 + 1$$
$$(a+1)^2 = 4$$
$$a + 1 = \pm 2$$
$$a = -1 \pm 2$$
$$a = -3, 1$$

25. Solve by completing the square:
$$x^2 - 10x = 0$$
$$x^2 - 10x + 25 = 0 + 25$$
$$(x-5)^2 = 25$$
$$x - 5 = \pm 5$$
$$x = 5 \pm 5$$
$$x = 0, 10$$

27. Solve by completing the square:
$$y^2 + 2y - 15 = 0$$
$$y^2 + 2y = 15$$
$$y^2 + 2y + 1 = 15 + 1$$
$$(y+1)^2 = 16$$
$$y + 1 = \pm 4$$
$$y = -1 \pm 4$$
$$y = -5, 3$$

29. Solve by completing the square:
$$x^2 + 4x - 3 = 0$$
$$x^2 + 4x = 3$$
$$x^2 + 4x + 4 = 3 + 4$$
$$(x+2)^2 = 7$$
$$x + 2 = \pm\sqrt{7}$$
$$x = -2 \pm \sqrt{7}$$

31. Solve by completing the square:
$$x^2 - 4x = 4$$
$$x^2 - 4x + 4 = 4 + 4$$
$$(x-2)^2 = 8$$
$$x - 2 = \pm\sqrt{8}$$
$$x - 2 = \pm 2\sqrt{2}$$
$$x = 2 \pm 2\sqrt{2}$$

33. Solve by completing the square:
$$a^2 = 7a + 8$$
$$a^2 - 7a = 8$$
$$a^2 - 7a + \frac{49}{4} = 8 + \frac{49}{4}$$
$$\left(a - \frac{7}{2}\right)^2 = \frac{81}{4}$$
$$a - \frac{7}{2} = \pm\sqrt{\frac{81}{4}}$$
$$a - \frac{7}{2} = \pm\frac{9}{2}$$
$$a = \frac{7}{2} \pm \frac{9}{2}$$
$$a = -1, 8$$

35. Solve by completing the square:
$$4x^2 + 8x - 4 = 0$$
$$x^2 + 2x - 1 = 0$$
$$x^2 + 2x = 1$$
$$x^2 + 2x + 1 = 1 + 1$$
$$(x+1)^2 = 2$$
$$x + 1 = \pm\sqrt{2}$$
$$x = -1 \pm \sqrt{2}$$

37. Solve by completing the square:
$$2x^2 + 2x - 4 = 0$$
$$x^2 + x - 2 = 0$$
$$x^2 + x = 2$$
$$x^2 + x + \frac{1}{4} = 2 + \frac{1}{4}$$
$$\left(x + \frac{1}{2}\right)^2 = \frac{9}{4}$$
$$x + \frac{1}{2} = \pm\sqrt{\frac{9}{4}}$$
$$x + \frac{1}{2} = \pm\frac{3}{2}$$
$$x = -\frac{1}{2} \pm \frac{3}{2}$$
$$x = -2, 1$$

39. Solve by completing the square:
$$4x^2 + 8x + 1 = 0$$
$$x^2 + 2x + \frac{1}{4} = 0$$
$$x^2 + 2x = -\frac{1}{4}$$
$$x^2 + 2x + 1 = -\frac{1}{4} + 1$$
$$(x+1)^2 = \frac{3}{4}$$
$$x + 1 = \pm\sqrt{\frac{3}{4}}$$
$$x + 1 = \pm\frac{\sqrt{3}}{2}$$
$$x = -1 \pm \frac{\sqrt{3}}{2}$$
$$x = \frac{-2 \pm \sqrt{3}}{2}$$

41. Solve by completing the square:
$$2x^2 - 2x = 1$$
$$x^2 - x = \frac{1}{2}$$
$$x^2 - x + \frac{1}{4} = \frac{1}{2} + \frac{1}{4}$$
$$\left(x - \frac{1}{2}\right)^2 = \frac{3}{4}$$
$$x - \frac{1}{2} = \pm\sqrt{\frac{3}{4}}$$
$$x - \frac{1}{2} = \pm\frac{\sqrt{3}}{2}$$
$$x = \frac{1}{2} \pm \frac{\sqrt{3}}{2}$$
$$x = \frac{1 \pm \sqrt{3}}{2}$$

43. Solve by completing the square:

$$4a^2 - 4a + 1 = 0$$
$$a^2 - a + \frac{1}{4} = 0$$
$$\left(a - \frac{1}{2}\right)^2 = 0$$
$$a - \frac{1}{2} = 0$$
$$a = \frac{1}{2}$$

45. Solve by completing the square:

$$3y^2 - 9y = 2$$
$$y^2 - 3y = \frac{2}{3}$$
$$y^2 - 3y + \frac{9}{4} = \frac{2}{3} + \frac{9}{4}$$
$$\left(y - \frac{3}{2}\right)^2 = \frac{8}{12} + \frac{27}{12} = \frac{35}{12}$$
$$y - \frac{3}{2} = \pm\sqrt{\frac{35}{12}} \cdot \frac{\sqrt{3}}{\sqrt{3}}$$
$$y - \frac{3}{2} = \pm\frac{\sqrt{105}}{6}$$
$$y = \frac{3}{2} \pm \frac{\sqrt{105}}{6}$$
$$y = \frac{9 \pm \sqrt{105}}{6}$$

47. a. Solving by factoring:

$$x^2 - 2x = 0$$
$$x(x - 2) = 0$$
$$x = 0, 2$$

b. Solving by completing the square:

$$x^2 - 2x = 0$$
$$x^2 - 2x + 1 = 1$$
$$(x - 1)^2 = 1$$
$$x - 1 = \pm\sqrt{1}$$
$$x - 1 = -1, 1$$
$$x = 0, 2$$

49. Substituting into the equation: $x^2 + 6x = \left(3 - \sqrt{2}\right)^2 + 6\left(3 - \sqrt{2}\right) = 9 - 6\sqrt{2} + 2 + 18 - 6\sqrt{2} = 29 - 12\sqrt{2}$

No, it is not a solution to the equation.

51. Yes, it is faster to solve the equation by factoring:

$$x^2 - 6x = 0$$
$$x(x - 6) = 0$$
$$x = 0, 6$$

53. Yes, another method is to solve the equation by factoring:

$$x^2 - 6x = 7$$
$$x^2 - 6x - 7 = 0$$
$$(x + 1)(x - 7) = 0$$
$$x = -1, 7$$

55. Solving by taking square roots:

$$(11a - 24)^2 = 144$$
$$11a - 24 = \pm\sqrt{144}$$
$$11a - 24 = -12, 12$$
$$11a = 12, 36$$
$$a = \frac{12}{11}, \frac{36}{11}$$

57. Solving by factoring:

$$t^2 - 6t = 0$$
$$t(t - 6) = 0$$
$$t = 0, 6$$

59. Solving by factoring:

$$t^2 - 6t = 7$$
$$t^2 - 6t - 7 = 0$$
$$(t + 1)(t - 7) = 0$$
$$t = -1, 7$$

61. Solving by completing the square:

$$t^2 - 6t = -7$$
$$t^2 - 6t + 9 = -7 + 9$$
$$(t - 3)^2 = 2$$
$$t - 3 = \pm\sqrt{2}$$
$$t = 3 \pm \sqrt{2}$$

63. Solving by taking square roots:

$$(6y-8)^2 = 20$$
$$6y-8 = \pm\sqrt{20}$$
$$6y-8 = \pm 2\sqrt{5}$$
$$6y = 8 \pm 2\sqrt{5}$$
$$y = \frac{8 \pm 2\sqrt{5}}{6} = \frac{4 \pm \sqrt{5}}{3}$$

65. Since the solutions are conjugate pairs, the other solution is $\dfrac{3+2\sqrt{5}}{2}$.

67. Solving the equation (note that $x > 0$):

$$(x+4)^2 = 100$$
$$x+4 = \sqrt{100}$$
$$x+4 = 10$$
$$x = 6$$

69. Using the Pythagorean theorem:

$$(3x)^2 + (4x)^2 = 14^2$$
$$9x^2 + 16x^2 = 196$$
$$25x^2 = 196$$
$$x^2 = \frac{196}{25}$$
$$x = \sqrt{\frac{196}{25}} = 2.8$$
$$3x = 8.4, 4x = 11.2$$

The length (height) of the screen is 8.4 inches and the width of the screen is 11.2 inches.

71. Evaluating when $a = 2$: $2a = 2(2) = 4$

73. Evaluating when $a = 2$ and $c = -3$: $4ac = 4(2)(-3) = -24$

75. Evaluating when $a = 2$, $b = 4$, and $c = -3$: $\sqrt{b^2 - 4ac} = \sqrt{(4)^2 - 4(2)(-3)} = \sqrt{16+24} = \sqrt{40} = 2\sqrt{10}$

77. Simplifying the radical: $\sqrt{12} = \sqrt{4 \cdot 3} = \sqrt{4}\sqrt{3} = 2\sqrt{3}$

79. Simplifying the radical: $\sqrt{20x^2y^3} = \sqrt{4x^2y^2 \cdot 5y} = \sqrt{4x^2y^2}\sqrt{5y} = 2xy\sqrt{5y}$

81. Simplifying the radical: $\sqrt{\dfrac{81}{25}} = \dfrac{\sqrt{81}}{\sqrt{25}} = \dfrac{9}{5}$

83. Writing in standard form: $2x^2 + 4x - 3 = 0$

85. Writing in standard form:

$$(x-2)(x+3) = 5$$
$$x^2 + 3x - 2x - 6 = 5$$
$$x^2 + x - 11 = 0$$

87. The coefficients and constant term are: $1, -5, -6$ **89.** The coefficients and constant term are: $2, 4, -3$

91. Evaluating $b^2 - 4ac$: $b^2 - 4ac = (-5)^2 - 4(1)(-6) = 25 + 24 = 49$

93. Evaluating $b^2 - 4ac$: $b^2 - 4ac = (4)^2 - 4(2)(-3) = 16 + 24 = 40$

95. Simplifying: $\dfrac{5+\sqrt{49}}{2} = \dfrac{5+7}{2} = \dfrac{12}{2} = 6$ **97.** Simplifying: $\dfrac{-4-\sqrt{40}}{4} = \dfrac{-4-2\sqrt{10}}{4} = \dfrac{-2-\sqrt{10}}{2}$

9.3 The Quadratic Formula

1. Using the quadratic formula with $a = 1$, $b = 3$, and $c = 2$:

$$x = \frac{-b \pm \sqrt{b^2 - 4ac}}{2a} = \frac{-3 \pm \sqrt{(3)^2 - 4(1)(2)}}{2(1)} = \frac{-3 \pm \sqrt{9 - 8}}{2} = \frac{-3 \pm 1}{2} = -2, -1$$

3. Using the quadratic formula with $a = 1$, $b = 5$, and $c = 6$:

$$x = \frac{-b \pm \sqrt{b^2 - 4ac}}{2a} = \frac{-5 \pm \sqrt{(5)^2 - 4(1)(6)}}{2(1)} = \frac{-5 \pm \sqrt{25 - 24}}{2} = \frac{-5 \pm 1}{2} = -3, -2$$

5. Using the quadratic formula with $a = 1$, $b = 6$, and $c = 9$:

$$x = \frac{-b \pm \sqrt{b^2 - 4ac}}{2a} = \frac{-6 \pm \sqrt{(6)^2 - 4(1)(9)}}{2(1)} = \frac{-6 \pm \sqrt{36 - 36}}{2} = \frac{-6 \pm 0}{2} = -3$$

7. Using the quadratic formula with $a = 1$, $b = 6$, and $c = 7$:

$$x = \frac{-b \pm \sqrt{b^2 - 4ac}}{2a} = \frac{-6 \pm \sqrt{(6)^2 - 4(1)(7)}}{2(1)} = \frac{-6 \pm \sqrt{36 - 28}}{2} = \frac{-6 \pm \sqrt{8}}{2} = \frac{-6 \pm 2\sqrt{2}}{2} = -3 \pm \sqrt{2}$$

9. Using the quadratic formula with $a = 2$, $b = 5$, and $c = 3$:

$$x = \frac{-b \pm \sqrt{b^2 - 4ac}}{2a} = \frac{-5 \pm \sqrt{(5)^2 - 4(2)(3)}}{2(2)} = \frac{-5 \pm \sqrt{25 - 24}}{4} = \frac{-5 \pm 1}{4} = -\frac{3}{2}, -1$$

11. Using the quadratic formula with $a = 4$, $b = 8$, and $c = 1$:

$$x = \frac{-b \pm \sqrt{b^2 - 4ac}}{2a} = \frac{-8 \pm \sqrt{(8)^2 - 4(4)(1)}}{2(4)} = \frac{-8 \pm \sqrt{64 - 16}}{8} = \frac{-8 \pm \sqrt{48}}{8} = \frac{-8 \pm 4\sqrt{3}}{8} = \frac{-2 \pm \sqrt{3}}{2}$$

13. Using the quadratic formula with $a = 1$, $b = -2$, and $c = 1$:

$$x = \frac{-b \pm \sqrt{b^2 - 4ac}}{2a} = \frac{-(-2) \pm \sqrt{(-2)^2 - 4(1)(1)}}{2(1)} = \frac{2 \pm \sqrt{4 - 4}}{2} = \frac{2 \pm 0}{2} = 1$$

15. First write the equation as $x^2 - 5x - 7 = 0$. Using $a = 1$, $b = -5$, and $c = -7$ in the quadratic formula:

$$x = \frac{-b \pm \sqrt{b^2 - 4ac}}{2a} = \frac{-(-5) \pm \sqrt{(-5)^2 - 4(1)(-7)}}{2(1)} = \frac{5 \pm \sqrt{25 + 28}}{2} = \frac{5 \pm \sqrt{53}}{2}$$

17. Using $a = 6$, $b = -1$, and $c = -2$ in the quadratic formula:

$$x = \frac{-b \pm \sqrt{b^2 - 4ac}}{2a} = \frac{-(-1) \pm \sqrt{(-1)^2 - 4(6)(-2)}}{2(6)} = \frac{1 \pm \sqrt{1 + 48}}{12} = \frac{1 \pm \sqrt{49}}{12} = \frac{1 \pm 7}{12} = -\frac{1}{2}, \frac{2}{3}$$

19. First simplify the equation:

$$(x - 2)(x + 1) = 3$$
$$x^2 - 2x + x - 2 = 3$$
$$x^2 - x - 5 = 0$$

Using $a = 1$, $b = -1$, and $c = -5$ in the quadratic formula:

$$x = \frac{-b \pm \sqrt{b^2 - 4ac}}{2a} = \frac{-(-1) \pm \sqrt{(-1)^2 - 4(1)(-5)}}{2(1)} = \frac{1 \pm \sqrt{1 + 20}}{2} = \frac{1 \pm \sqrt{21}}{2}$$

21. First simplify the equation:

$$(2x - 3)(x + 2) = 1$$
$$2x^2 - 3x + 4x - 6 = 1$$
$$2x^2 + x - 7 = 0$$

Using $a = 2$, $b = 1$, and $c = -7$ in the quadratic formula:

$$x = \frac{-b \pm \sqrt{b^2 - 4ac}}{2a} = \frac{-1 \pm \sqrt{(1)^2 - 4(2)(-7)}}{2(2)} = \frac{-1 \pm \sqrt{1 + 56}}{4} = \frac{-1 \pm \sqrt{57}}{4}$$

23. First write the equation as $2x^2 - 3x - 5 = 0$. Using $a = 2$, $b = -3$, and $c = -5$ in the quadratic formula:

$$x = \frac{-b \pm \sqrt{b^2 - 4ac}}{2a} = \frac{-(-3) \pm \sqrt{(-3)^2 - 4(2)(-5)}}{2(2)} = \frac{3 \pm \sqrt{9 + 40}}{4} = \frac{3 \pm \sqrt{49}}{4} = \frac{3 \pm 7}{4} = -1, \frac{5}{2}$$

25. First write the equation as $2x^2 + 6x - 7 = 0$. Using $a = 2$, $b = 6$, and $c = -7$ in the quadratic formula:

$$x = \frac{-b \pm \sqrt{b^2 - 4ac}}{2a} = \frac{-6 \pm \sqrt{(6)^2 - 4(2)(-7)}}{2(2)} = \frac{-6 \pm \sqrt{36 + 56}}{4} = \frac{-6 \pm \sqrt{92}}{4} = \frac{-6 \pm 2\sqrt{23}}{4} = \frac{-3 \pm \sqrt{23}}{2}$$

27. First write the equation as $3x^2 + 4x - 2 = 0$. Using $a = 3$, $b = 4$, and $c = -2$ in the quadratic formula:

$$x = \frac{-b \pm \sqrt{b^2 - 4ac}}{2a} = \frac{-4 \pm \sqrt{(4)^2 - 4(3)(-2)}}{2(3)} = \frac{-4 \pm \sqrt{16 + 24}}{6} = \frac{-4 \pm \sqrt{40}}{6} = \frac{-4 \pm 2\sqrt{10}}{6} = \frac{-2 \pm \sqrt{10}}{3}$$

29. First write the equation as $2x^2 - 2x - 5 = 0$. Using $a = 2$, $b = -2$, and $c = -5$ in the quadratic formula:

$$x = \frac{-b \pm \sqrt{b^2 - 4ac}}{2a} = \frac{-(-2) \pm \sqrt{(-2)^2 - 4(2)(-5)}}{2(2)} = \frac{2 \pm \sqrt{4 + 40}}{4} = \frac{2 \pm \sqrt{44}}{4} = \frac{2 \pm 2\sqrt{11}}{4} = \frac{1 \pm \sqrt{11}}{2}$$

31. Factoring out x results in the equation $x(2x^2 + 3x - 4) = 0$, so $x = 0$ is one solution. The other two solutions are found by using $a = 2$, $b = 3$, and $c = -4$ in the quadratic formula:

$$x = \frac{-b \pm \sqrt{b^2 - 4ac}}{2a} = \frac{-3 \pm \sqrt{(3)^2 - 4(2)(-4)}}{2(2)} = \frac{-3 \pm \sqrt{9 + 32}}{4} = \frac{-3 \pm \sqrt{41}}{4}$$

33. Using $a = 3$, $b = -4$, and $c = 0$ in the quadratic formula:

$$x = \frac{-b \pm \sqrt{b^2 - 4ac}}{2a} = \frac{-(-4) \pm \sqrt{(-4)^2 - 4(3)(0)}}{2(3)} = \frac{4 \pm \sqrt{16 - 0}}{6} = \frac{4 \pm 4}{6} = 0, \frac{4}{3}$$

35. Multiplying each side of the equation by 6 results in the equation $3x^2 - 3x - 1 = 0$. Using $a = 3$, $b = -3$, and $c = -1$ in the quadratic formula: $x = \dfrac{-b \pm \sqrt{b^2 - 4ac}}{2a} = \dfrac{-(-3) \pm \sqrt{(-3)^2 - 4(3)(-1)}}{2(3)} = \dfrac{3 \pm \sqrt{9 + 12}}{6} = \dfrac{3 \pm \sqrt{21}}{6}$

37. Yes. Solving by taking square roots:

$$(6x - 8)^2 = 32$$
$$6x - 8 = \pm\sqrt{32}$$
$$6x - 8 = \pm 4\sqrt{2}$$
$$6x = 8 \pm 4\sqrt{2}$$
$$x = \frac{8 \pm 4\sqrt{2}}{6} = \frac{4 \pm 2\sqrt{2}}{3}$$

39. No, since the roots are irrational.

41. **a.** No, it cannot be solved by factoring.

  b. Using $a = 1$, $b = -2$, and $c = -1$ in the quadratic formula:

$$x = \frac{-b \pm \sqrt{b^2 - 4ac}}{2a} = \frac{-(-2) \pm \sqrt{(-2)^2 - 4(1)(-1)}}{2(1)} = \frac{2 \pm \sqrt{4 + 4}}{2} = \frac{2 \pm \sqrt{8}}{2} = \frac{2 \pm 2\sqrt{2}}{2} = 1 \pm \sqrt{2}$$

43. Substituting: $(7 - \sqrt{5})^2 - 14(7 - \sqrt{5}) + 44 = 49 - 14\sqrt{5} + 5 - 98 + 14\sqrt{5} + 44 = 0$

Yes, it is a solution to the equation.

45. Solving by factoring:

$$2t^2 + 7t = 0$$
$$t(2t + 7) = 0$$
$$t = -\frac{7}{2}, 0$$

47. Solving by factoring:

$$2t^2 + 7t = -3$$
$$2t^2 + 7t + 3 = 0$$
$$(2t + 1)(t + 3) = 0$$
$$t = -3, -\frac{1}{2}$$

49. First rewrite the equation as $2t^2 + 7t - 3 = 0$. Using $a = 2$, $b = 7$, and $c = -3$ in the quadratic formula:

$$t = \frac{-b \pm \sqrt{b^2 - 4ac}}{2a} = \frac{-7 \pm \sqrt{7^2 - 4(2)(-3)}}{2(2)} = \frac{-7 \pm \sqrt{49 + 24}}{4} = \frac{-7 \pm \sqrt{73}}{4}$$

51. Solving by taking square roots:

$$(6x - 8)^2 = 32$$
$$6x - 8 = \pm\sqrt{32}$$
$$6x - 8 = \pm 4\sqrt{2}$$
$$6x = 8 \pm 4\sqrt{2}$$
$$x = \frac{8 \pm 4\sqrt{2}}{6} = \frac{4 \pm 2\sqrt{2}}{3}$$

53. Solving the equation:

$$56 = 8 + 64t - 16t^2$$
$$16t^2 - 64t + 48 = 0$$
$$t^2 - 4t + 3 = 0$$
$$(t - 1)(t - 3) = 0$$
$$t = 1, 3$$

The arrow is 56 feet above the ground after 1 second and after 3 seconds.

55. Simplifying: $\dfrac{3}{x - 2} + \dfrac{6}{2x - 4} = \dfrac{3}{x - 2} + \dfrac{6}{2(x - 2)} = \dfrac{3}{x - 2} + \dfrac{3}{x - 2} = \dfrac{6}{x - 2}$

57. Simplifying:

$$\frac{x}{x^2 - 9} + \frac{4}{4x - 12} = \frac{x}{(x + 3)(x - 3)} \cdot \frac{4}{4} + \frac{4}{4(x - 3)} \cdot \frac{x + 3}{x + 3}$$
$$= \frac{4x}{4(x + 3)(x - 3)} + \frac{4x + 12}{4(x + 3)(x - 3)}$$
$$= \frac{8x + 12}{4(x + 3)(x - 3)}$$
$$= \frac{4(2x + 3)}{4(x + 3)(x - 3)}$$
$$= \frac{2x + 3}{(x + 3)(x - 3)}$$

59. Simplifying:

$$\frac{2x}{x^2 - 1} + \frac{x}{x^2 - 3x + 2} = \frac{2x}{(x + 1)(x - 1)} \cdot \frac{x - 2}{x - 2} + \frac{x}{(x - 1)(x - 2)} \cdot \frac{x + 1}{x + 1}$$
$$= \frac{2x^2 - 4x}{(x + 1)(x - 1)(x - 2)} + \frac{x^2 + x}{(x + 1)(x - 1)(x - 2)}$$
$$= \frac{3x^2 - 3x}{(x + 1)(x - 1)(x - 2)}$$
$$= \frac{3x(x - 1)}{(x + 1)(x - 1)(x - 2)}$$
$$= \frac{3x}{(x + 1)(x - 2)}$$

61. Combining terms: $\left(3 + 4\sqrt{2}\right) + \left(2 - 6\sqrt{2}\right) = 5 - 2\sqrt{2}$

63. Combining terms: $\left(2 - 5\sqrt{2}\right) - \left(3 + 7\sqrt{2}\right) + \left(2 - \sqrt{2}\right) = 2 - 5\sqrt{2} - 3 - 7\sqrt{2} + 2 - \sqrt{2} = 1 - 13\sqrt{2}$

65. Multiplying: $4\sqrt{2}\left(3 + 5\sqrt{2}\right) = 12\sqrt{2} + 20\sqrt{4} = 12\sqrt{2} + 40$

67. Multiplying: $\left(3+2\sqrt{2}\right)\left(4-3\sqrt{2}\right)=12-9\sqrt{2}+8\sqrt{2}-12=-\sqrt{2}$

69. Rationalizing the denominator: $\dfrac{2}{3+\sqrt{5}}\cdot\dfrac{3-\sqrt{5}}{3-\sqrt{5}}=\dfrac{2\left(3-\sqrt{5}\right)}{9-5}=\dfrac{2\left(3-\sqrt{5}\right)}{4}=\dfrac{3-\sqrt{5}}{2}$

71. Rationalizing the denominator: $\dfrac{2+\sqrt{3}}{2-\sqrt{3}}\cdot\dfrac{2+\sqrt{3}}{2+\sqrt{3}}=\dfrac{4+2\sqrt{3}+2\sqrt{3}+3}{4-3}=\dfrac{7+4\sqrt{3}}{1}=7+4\sqrt{3}$

9.4 Complex Numbers

1. Combining the complex numbers: $(3-2i)+3i=3+(-2i+3i)=3+i$

3. Combining the complex numbers: $(6+2i)-10i=6+(2i-10i)=6-8i$

5. Combining the complex numbers: $(11+9i)-9i=11+(9i-9i)=11$

7. Combining the complex numbers: $(3+2i)+(6-i)=(3+6)+(2i-i)=9+i$

9. Combining the complex numbers: $(5+7i)-(6+8i)=5+7i-6-8i=-1-i$

11. Combining the complex numbers: $(9-i)+(2-i)=9-i+2-i=11-2i$

13. Combining the complex numbers: $(6+i)-4i-(2-i)=6+i-4i-2+i=4-2i$

15. Combining the complex numbers: $(6-11i)+3i+(2+i)=6-11i+3i+2+i=8-7i$

17. Combining the complex numbers: $(2+3i)-(6-2i)+(3-i)=2+3i-6+2i+3-i=-1+4i$

19. Multiplying the complex numbers: $3(2-i)=6-3i$

21. Multiplying the complex numbers: $2i(8-7i)=16i-14i^2=16i-14(-1)=14+16i$

23. Multiplying the complex numbers: $(2+i)(4-i)=8+4i-2i-i^2=8+2i-(-1)=8+2i+1=9+2i$

25. Multiplying the complex numbers: $(2+i)(3-5i)=6+3i-10i-5i^2=6-7i-5(-1)=6-7i+5=11-7i$

27. Multiplying the complex numbers: $(3+5i)(3-5i)=(3)^2-(5i)^2=9-25i^2=9-25(-1)=9+25=34$

29. Multiplying the complex numbers: $(2+i)(2-i)=(2)^2-(i)^2=4-i^2=4-(-1)=4+1=5$

31. Dividing the complex numbers: $\dfrac{2}{3-2i}\cdot\dfrac{3+2i}{3+2i}=\dfrac{2(3+2i)}{9-4i^2}=\dfrac{2(3+2i)}{9+4}=\dfrac{6+4i}{13}$

33. Dividing the complex numbers: $\dfrac{-3i}{2+3i}\cdot\dfrac{2-3i}{2-3i}=\dfrac{-6i+9i^2}{4-9i^2}=\dfrac{-6i-9}{4+9}=\dfrac{-9-6i}{13}$

35. Dividing the complex numbers: $\dfrac{2+i}{2-i}\cdot\dfrac{2+i}{2+i}=\dfrac{4+2i+2i+i^2}{4-i^2}=\dfrac{4+4i-1}{4+1}=\dfrac{3+4i}{5}$

37. **a.** Adding: $(2+3i)+(2-3i)=2+3i+2-3i=4$

b. Multiplying: $(2+3i)(2-3i)=(2)^2-(3i)^2=4-9i^2=4+9=13$

c. Squaring: $(2-3i)^2=(2-3i)(2-3i)=4-6i-6i+9i^2=4-12i-9=-5-12i$

d. Dividing: $\dfrac{2-3i}{2+3i}\cdot\dfrac{2-3i}{2-3i}=\dfrac{4-6i-6i+9i^2}{4-9i^2}=\dfrac{4-12i-9}{4+9}=\dfrac{-5-12i}{13}$

39. **a.** Adding: $(6i) + (3 - i) = 6i + 3 - i = 3 + 5i$

   b. Multiplying: $(6i)(3 - i) = 18i - 6i^2 = 18i + 6 = 6 + 18i$

   c. Dividing: $\dfrac{3 - i}{6i} \cdot \dfrac{i}{i} = \dfrac{3i - i^2}{6i^2} = \dfrac{3i + 1}{-6} = -\dfrac{1 + 3i}{6}$

   d. Dividing: $\dfrac{6i}{3 - i} \cdot \dfrac{3 + i}{3 + i} = \dfrac{18i + 6i^2}{9 - i^2} = \dfrac{18i - 6}{9 + 1} = \dfrac{-6 + 18i}{10} = \dfrac{-3 + 9i}{5}$

41. Simplifying: $(3 + 2i)^2 - 6(3 + 2i) = 9 + 6i + 6i + 4i^2 - 18 - 12i = 9 + 12i - 4 - 18 - 12i = -13$

43. Simplifying: $3(2 + i)^2 - 12(2 + i) = 3(4 + 4i + i^2) - 12(2 + i) = 3(3 + 4i) - 12(2 + i) = 9 + 12i - 24 - 12i = -15$

45. Simplifying: $(1 + i)^2 - 2(1 + i) + 2 = 1 + 2i + i^2 - 2 - 2i + 2 = 1 + 2i - 1 - 2 - 2i + 2 = 0$

47. Substituting the value: $(3i)^2 + 9 = 9i^2 + 9 = -9 + 9 = 0$

   Yes, it is a solution to the equation.

49. Solving the equation:
$$x^2 = -36$$
$$x = \pm\sqrt{-36}$$
$$x = \pm 6i$$

51. Substituting the value: $(1 + i)^2 - 2(1 + i) = 1 + 2i + i^2 - 2 - 2i = 1 + 2i - 1 - 2 - 2i = -2$

   Yes, it is a solution to the equation.

53. Multiplying the expressions: $(x + 3i)(x - 3i) = x^2 - (3i)^2 = x^2 - 9i^2 = x^2 - 9(-1) = x^2 + 9$

55. Simplifying: $\dfrac{1}{i} \cdot \dfrac{i}{i} = \dfrac{i}{i^2} = \dfrac{i}{-1} = -i$

57. Solving the equation:
$$(x - 3)^2 = 25$$
$$x - 3 = \pm\sqrt{25}$$
$$x - 3 = \pm 5$$
$$x = 3 \pm 5$$
$$x = -2, 8$$

59. Solving the equation:
$$(2x - 6)^2 = 16$$
$$2x - 6 = \pm\sqrt{16}$$
$$2x - 6 = \pm 4$$
$$2x = 6 \pm 4$$
$$2x = 2, 10$$
$$x = 1, 5$$

61. Solving the equation:
$$(x + 3)^2 = 12$$
$$x + 3 = \pm\sqrt{12}$$
$$x + 3 = \pm 2\sqrt{3}$$
$$x = -3 \pm 2\sqrt{3}$$

63. Simplifying the radical expression: $\sqrt{\dfrac{1}{2}} = \dfrac{\sqrt{1}}{\sqrt{2}} \cdot \dfrac{\sqrt{2}}{\sqrt{2}} = \dfrac{\sqrt{2}}{2}$

65. Simplifying the radical expression: $\sqrt{\dfrac{8x^2y^3}{3}} = \dfrac{\sqrt{8x^2y^3}}{\sqrt{3}} \cdot \dfrac{\sqrt{3}}{\sqrt{3}} = \dfrac{\sqrt{24x^2y^3}}{3} = \dfrac{\sqrt{4x^2y^2 \cdot 6y}}{3} = \dfrac{2xy\sqrt{6y}}{3}$

67. Simplifying the radical expression: $\sqrt[3]{\dfrac{1}{4}} = \dfrac{\sqrt[3]{1}}{\sqrt[3]{4}} \cdot \dfrac{\sqrt[3]{2}}{\sqrt[3]{2}} = \dfrac{\sqrt[3]{2}}{\sqrt[3]{8}} = \dfrac{\sqrt[3]{2}}{2}$

69. Simplifying: $\sqrt{36} = 6$

71. Simplifying: $\sqrt{75} = \sqrt{25 \cdot 3} = 5\sqrt{3}$

73. Solving the equation:

$$(x+2)^2 = 9$$
$$x+2 = \pm\sqrt{9} = \pm 3$$
$$x = -5, 1$$

75. Solving the equation:

$$\frac{1}{10}x^2 - \frac{1}{5}x = \frac{1}{2}$$
$$10\left(\frac{1}{10}x^2 - \frac{1}{5}x\right) = 10\left(\frac{1}{2}\right)$$
$$x^2 - 2x = 5$$
$$x^2 - 2x + 1 = 5 + 1$$
$$(x-1)^2 = 6$$
$$x-1 = \pm\sqrt{6}$$
$$x = 1 \pm \sqrt{6}$$

77. Solving the equation:

$$(2x-3)(2x-1) = 4$$
$$4x^2 - 2x - 6x + 3 = 4$$
$$4x^2 - 8x = 1$$
$$x^2 - 2x = \frac{1}{4}$$
$$x^2 - 2x + 1 = \frac{1}{4} + 1$$
$$(x-1)^2 = \frac{5}{4}$$
$$x - 1 = \pm\frac{\sqrt{5}}{2}$$
$$x = 1 \pm \frac{\sqrt{5}}{2} = \frac{2 \pm \sqrt{5}}{2}$$

9.5 Complex Solutions to Quadratic Equations

1. Writing as a complex number: $\sqrt{-16} = \sqrt{16(-1)} = \sqrt{16}\sqrt{-1} = 4i$

3. Writing as a complex number: $\sqrt{-49} = \sqrt{49(-1)} = \sqrt{49}\sqrt{-1} = 7i$

5. Writing as a complex number: $\sqrt{-6} = \sqrt{6(-1)} = \sqrt{6}\sqrt{-1} = i\sqrt{6}$

7. Writing as a complex number: $\sqrt{-11} = \sqrt{11(-1)} = \sqrt{11}\sqrt{-1} = i\sqrt{11}$

9. Writing as a complex number: $\sqrt{-32} = \sqrt{-16 \cdot 2} = \sqrt{-16}\sqrt{2} = 4i\sqrt{2}$

11. Writing as a complex number: $\sqrt{-50} = \sqrt{-25 \cdot 2} = \sqrt{-25}\sqrt{2} = 5i\sqrt{2}$

13. Writing as a complex number: $\sqrt{-8} = \sqrt{-4 \cdot 2} = \sqrt{-4}\sqrt{2} = 2i\sqrt{2}$

15. Writing as a complex number: $\sqrt{-48} = \sqrt{-16 \cdot 3} = \sqrt{-16}\sqrt{3} = 4i\sqrt{3}$

17. First write the equation as $x^2 - 2x + 2 = 0$. Using $a = 1$, $b = -2$, and $c = 2$ in the quadratic formula:

$$x = \frac{-b \pm \sqrt{b^2 - 4ac}}{2a} = \frac{-(-2) \pm \sqrt{(-2)^2 - 4(1)(2)}}{2(1)} = \frac{2 \pm \sqrt{4 - 8}}{2} = \frac{2 \pm \sqrt{-4}}{2} = \frac{2 \pm 2i}{2} = 1 \pm i$$

19. Solving the equation:

$$x^2 - 4x = -4$$
$$x^2 - 4x + 4 = 0$$
$$(x-2)^2 = 0$$
$$x - 2 = 0$$
$$x = 2$$

21. Solving the equation:
$$2x^2 + 5x = 12$$
$$2x^2 + 5x - 12 = 0$$
$$(2x - 3)(x + 4) = 0$$
$$x = \tfrac{3}{2}, -4$$

23. Solving the equation:
$$(x - 2)^2 = -4$$
$$x - 2 = \pm\sqrt{-4}$$
$$x - 2 = \pm 2i$$
$$x = 2 \pm 2i$$

25. Solving the equation:
$$\left(x + \tfrac{1}{2}\right)^2 = -\tfrac{9}{4}$$
$$x + \tfrac{1}{2} = \pm\sqrt{-\tfrac{9}{4}}$$
$$x + \tfrac{1}{2} = \pm\tfrac{3}{2}i$$
$$x = \frac{-1 \pm 3i}{2}$$

27. Solving the equation:
$$\left(x - \tfrac{1}{2}\right)^2 = -\tfrac{27}{36}$$
$$x - \tfrac{1}{2} = \pm\sqrt{-\tfrac{27}{36}}$$
$$x - \tfrac{1}{2} = \pm\frac{3i\sqrt{3}}{6}$$
$$x - \tfrac{1}{2} = \pm\frac{i\sqrt{3}}{2}$$
$$x = \frac{1 \pm i\sqrt{3}}{2}$$

29. Using $a = 1$, $b = 1$, and $c = 1$ in the quadratic formula:
$$x = \frac{-b \pm \sqrt{b^2 - 4ac}}{2a} = \frac{-1 \pm \sqrt{(1)^2 - 4(1)(1)}}{2(1)} = \frac{-1 \pm \sqrt{1 - 4}}{2} = \frac{-1 \pm \sqrt{-3}}{2} = \frac{-1 \pm i\sqrt{3}}{2}$$

31. Solving the equation:
$$x^2 - 5x + 6 = 0$$
$$(x - 2)(x - 3) = 0$$
$$x = 2, 3$$

33. First multiply by 6 to clear the equation of fractions:
$$6\left(\tfrac{1}{2}x^2 + \tfrac{1}{3}x + \tfrac{1}{6}\right) = 6(0)$$
$$3x^2 + 2x + 1 = 0$$
Using $a = 3$, $b = 2$, and $c = 1$ in the quadratic formula:
$$x = \frac{-b \pm \sqrt{b^2 - 4ac}}{2a} = \frac{-2 \pm \sqrt{(2)^2 - 4(3)(1)}}{2(3)} = \frac{-2 \pm \sqrt{4 - 12}}{6} = \frac{-2 \pm \sqrt{-8}}{6} = \frac{-2 \pm 2i\sqrt{2}}{6} = \frac{-1 \pm i\sqrt{2}}{3}$$

35. First multiply by 6 to clear the equation of fractions:
$$6\left(\tfrac{1}{3}x^2\right) = 6\left(-\tfrac{1}{2}x + \tfrac{1}{3}\right)$$
$$2x^2 = -3x + 2$$
$$2x^2 + 3x - 2 = 0$$
$$(2x - 1)(x + 2) = 0$$
$$x = \tfrac{1}{2}, -2$$

37. Solving the equation:
$$(x + 2)(x - 3) = 5$$
$$x^2 - x - 6 = 5$$
$$x^2 - x - 11 = 0$$
Using $a = 1$, $b = -1$, and $c = -11$ in the quadratic formula:
$$x = \frac{-b \pm \sqrt{b^2 - 4ac}}{2a} = \frac{-(-1) \pm \sqrt{(-1)^2 - 4(1)(-11)}}{2(1)} = \frac{1 \pm \sqrt{1 + 44}}{2} = \frac{1 \pm \sqrt{45}}{2} = \frac{1 \pm 3\sqrt{5}}{2}$$

39.   Solving the equation:
$$(x-5)(x-3)=-10$$
$$x^2-8x+15=-10$$
$$x^2-8x+25=0$$
Using $a=1$, $b=-8$, and $c=25$ in the quadratic formula:
$$x=\frac{-b\pm\sqrt{b^2-4ac}}{2a}=\frac{-(-8)\pm\sqrt{(-8)^2-4(1)(25)}}{2(1)}=\frac{8\pm\sqrt{64-100}}{2}=\frac{8\pm\sqrt{-36}}{2}=\frac{8\pm6i}{2}=4\pm3i$$

41.   Solving the equation:
$$(2x-2)(x-3)=9$$
$$2x^2-8x+6=9$$
$$2x^2-8x-3=0$$
Using $a=2$, $b=-8$, and $c=-3$ in the quadratic formula:
$$x=\frac{-b\pm\sqrt{b^2-4ac}}{2a}=\frac{-(-8)\pm\sqrt{(-8)^2-4(2)(-3)}}{2(2)}=\frac{8\pm\sqrt{64+24}}{4}=\frac{8\pm\sqrt{88}}{4}=\frac{8\pm2\sqrt{22}}{4}=\frac{4\pm\sqrt{22}}{2}$$

43.   The fastest method is taking square roots.

45.   No, since the solutions are complex numbers.

47.   Solving the equation by taking square roots:
$$(9x-11)^2=-16$$
$$9x-11=\pm\sqrt{-16}$$
$$9x-11=\pm4i$$
$$9x=11\pm4i$$
$$x=\frac{11\pm4i}{9}$$

49.   Solving the equation by taking square roots:
$$2y^2-6y=0$$
$$2y(y-3)=0$$
$$y=0,3$$

51.   First rewrite the equation as $2y^2-6y+5=0$. Solving the equation using the quadratic formula:
$$y=\frac{-b\pm\sqrt{b^2-4ac}}{2a}=\frac{-(-6)\pm\sqrt{(-6)^2-4(2)(5)}}{2(2)}=\frac{6\pm\sqrt{36-40}}{4}=\frac{6\pm\sqrt{-4}}{4}=\frac{6\pm2i}{4}=\frac{3\pm i}{2}$$

53.   Solving the equation by factoring:
$$2y^2-6y=20$$
$$2y^2-6y-20=0$$
$$2(y^2-3y-10)=0$$
$$2(y+2)(y-5)=0$$
$$y=-2,5$$

55.   Solving the equation by taking square roots:
$$(8x-12)^2=-36$$
$$8x-12=\pm\sqrt{-36}$$
$$8x-12=\pm6i$$
$$8x=12\pm6i$$
$$x=\frac{12\pm6i}{8}=\frac{6\pm3i}{4}$$

57.   Solving the equation by factoring:
$$1-\frac{3}{y}=\frac{10}{y^2}$$
$$y^2\left(1-\frac{3}{y}\right)=y^2\left(\frac{10}{y^2}\right)$$
$$y^2-3y=10$$
$$y^2-3y-10=0$$
$$(y+2)(y-5)=0$$
$$y=-2,5$$

59.   Using $a=1$, $b=-4$, and $c=5$ in the quadratic formula:
$$x=\frac{-b\pm\sqrt{b^2-4ac}}{2a}=\frac{-(-4)\pm\sqrt{(-4)^2-4(1)(5)}}{2(1)}=\frac{4\pm\sqrt{16-20}}{2}=\frac{4\pm\sqrt{-4}}{2}=\frac{4\pm2i}{2}=2\pm i$$
Since the solutions are non-real complex numbers, the graph cannot cross the x-axis.

61.   Substituting $x = 2 + 2i$ into the equation:

$$x^2 - 4x + 8 = (2 + 2i)^2 - 4(2 + 2i) + 8 = 4 + 8i + 4i^2 - 8 - 8i + 8 = 4 + 8i - 4 - 8 - 8i + 8 = 0$$

Yes, $x = 2 + 2i$ is a solution to the equation.

63.   The other solution is $3 - 7i$.

65.   Graphing the line:

67.   Graphing the line:

69.   Simplifying the radical expression: $3\sqrt{50} + 2\sqrt{32} = 3\sqrt{25 \cdot 2} + 2\sqrt{16 \cdot 2} = 3 \cdot 5\sqrt{2} + 2 \cdot 4\sqrt{2} = 15\sqrt{2} + 8\sqrt{2} = 23\sqrt{2}$

71.   Simplifying the radical expression: $\sqrt{24} - \sqrt{54} - \sqrt{150} = \sqrt{4 \cdot 6} - \sqrt{9 \cdot 6} - \sqrt{25 \cdot 6} = 2\sqrt{6} - 3\sqrt{6} - 5\sqrt{6} = -6\sqrt{6}$

73.   Simplifying the radical expression:

$$2\sqrt{27x^2} - x\sqrt{48} = 2\sqrt{9x^2 \cdot 3} - x\sqrt{16 \cdot 3} = 2 \cdot 3x\sqrt{3} - x \cdot 4\sqrt{3} = 6x\sqrt{3} - 4x\sqrt{3} = 2x\sqrt{3}$$

75.   Completing the square: $x^2 - 6x + 9 = (x - 3)^2$

77.   Evaluating when $x = -4$: $y = (-4 + 1)^2 - 3 = (-3)^2 - 3 = 9 - 3 = 6$

79.   Evaluating when $x = -2$: $y = (-2 + 1)^2 - 3 = (-1)^2 - 3 = 1 - 3 = -2$

81.   Evaluating when $x = 1$: $y = (1 + 1)^2 - 3 = (2)^2 - 3 = 4 - 3 = 1$

83.   Graphing the ordered pair:

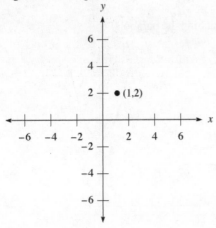

85.   Graphing the ordered pair:

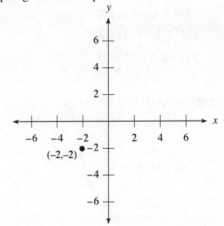

87.   Graphing the ordered pair:

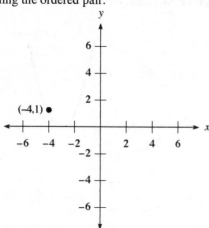

89.   Graphing the ordered pair:

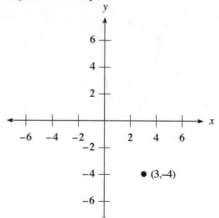

91.   Graphing the ordered pair:

9.6 Graphing Parabolas

1.   Graphing the parabola:

3.   Graphing the parabola:

5.   Graphing the parabola:

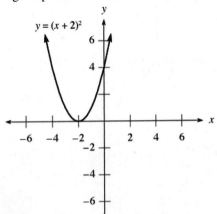

$y = (x + 2)^2$

7.   Graphing the parabola:

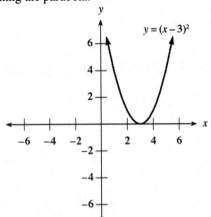

$y = (x - 3)^2$

9.   Graphing the parabola:

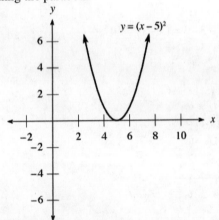

$y = (x - 5)^2$

11.   Graphing the parabola:

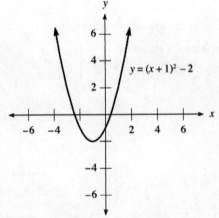

$y = (x + 1)^2 - 2$

13.   Graphing the parabola:

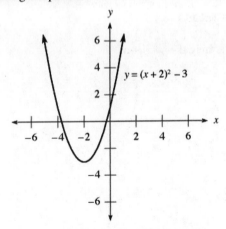

$y = (x + 2)^2 - 3$

15.   Graphing the parabola:

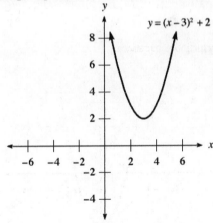

$y = (x - 3)^2 + 2$

17.   Begin by completing the square: $y = x^2 + 6x + 5 = \left(x^2 + 6x + 9\right) + 5 - 9 = (x+3)^2 - 4$

Graphing the parabola:

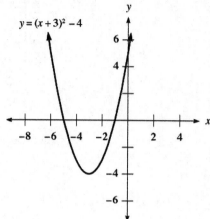

19.   Begin by completing the square: $y = x^2 - 2x - 3 = \left(x^2 - 2x + 1\right) - 3 - 1 = (x-1)^2 - 4$

Graphing the parabola:

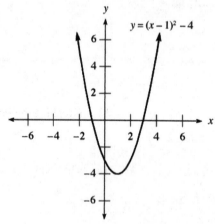

21. Begin by making a table of values:

x	y
-3	-5
-2	0
-1	3
0	4
1	3
2	0
3	-5

Graphing the parabola:

23. Begin by making a table of values:

x	y
-2	-5
-1	-2
0	-1
1	-2
2	-5

Graphing the parabola:

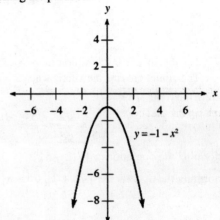

25. Graphing the line and the parabola:

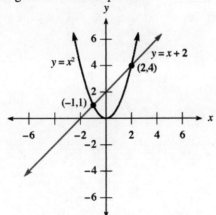

The intersection points are $(-1,1)$ and $(2,4)$.

27. Graphing the line and the parabola:

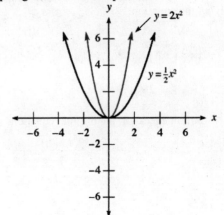

29.   **a.**   Graphing the orbit:

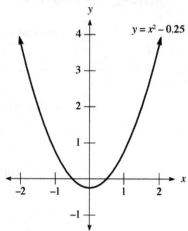

$y = x^2 - 0.25$

b.   The distance is 0.25 million miles, or 250,000 miles.

c.   The comet crosses the x-axis when $x = 0.5$, which corresponds to 0.5 million miles, or 500,000 miles.

31.   Finding the root: $\sqrt{49} = 7$ **33.**   Simplifying the radical: $\sqrt{50} = \sqrt{25 \cdot 2} = 5\sqrt{2}$

35.   Simplifying the radical: $\sqrt{\dfrac{2}{5}} = \dfrac{\sqrt{2}}{\sqrt{5}} \cdot \dfrac{\sqrt{5}}{\sqrt{5}} = \dfrac{\sqrt{10}}{5}$

37.   Performing the operations: $3\sqrt{12} + 5\sqrt{27} = 3\sqrt{4 \cdot 3} + 5\sqrt{9 \cdot 3} = 3 \cdot 2\sqrt{3} + 5 \cdot 3\sqrt{3} = 6\sqrt{3} + 15\sqrt{3} = 21\sqrt{3}$

39.   Performing the operations: $\left(\sqrt{6} + 2\right)\left(\sqrt{6} - 5\right) = \sqrt{36} + 2\sqrt{6} - 5\sqrt{6} - 10 = 6 - 3\sqrt{6} - 10 = -4 - 3\sqrt{6}$

41.   Rationalizing the denominator: $\dfrac{8}{\sqrt{5} - \sqrt{3}} \cdot \dfrac{\sqrt{5} + \sqrt{3}}{\sqrt{5} + \sqrt{3}} = \dfrac{8\left(\sqrt{5} + \sqrt{3}\right)}{5 - 3} = \dfrac{8\left(\sqrt{5} + \sqrt{3}\right)}{2} = 4\left(\sqrt{5} + \sqrt{3}\right) = 4\sqrt{5} + 4\sqrt{3}$

43.   Solving the equation:
$$\sqrt{2x - 5} = 3$$
$$\left(\sqrt{2x - 5}\right)^2 = 3^2$$
$$2x - 5 = 9$$
$$2x = 14$$
$$x = 7$$
This solution checks in the original equation.

Chapter 9 Review/Test

1. Solving the quadratic equation:

$$a^2 = 32$$
$$a = \pm\sqrt{32}$$
$$a = \pm 4\sqrt{2}$$

2. Solving the quadratic equation:

$$a^2 = 60$$
$$a = \pm\sqrt{60}$$
$$a = \pm 2\sqrt{15}$$

3. Solving the quadratic equation:

$$2x^2 = 32$$
$$x^2 = 16$$
$$x = \pm\sqrt{16}$$
$$x = \pm 4$$

4. Solving the quadratic equation:

$$(x+3)^2 = 36$$
$$x+3 = \pm\sqrt{36}$$
$$x+3 = \pm 6$$
$$x = -3 \pm 6$$
$$x = -9, 3$$

5. Solving the quadratic equation:

$$(x-2)^2 = 81$$
$$x-2 = \pm\sqrt{81}$$
$$x-2 = \pm 9$$
$$x = 2 \pm 9$$
$$x = -7, 11$$

6. Solving the quadratic equation:

$$(3x+2)^2 = 16$$
$$3x+2 = \pm\sqrt{16}$$
$$3x+2 = \pm 4$$
$$3x = -2 \pm 4$$
$$3x = -6, 2$$
$$x = -2, \frac{2}{3}$$

7. Solving the quadratic equation:

$$(2x+5)^2 = 32$$
$$2x+5 = \pm\sqrt{32}$$
$$2x+5 = \pm 4\sqrt{2}$$
$$2x = -5 \pm 4\sqrt{2}$$
$$x = \frac{-5 \pm 4\sqrt{2}}{2}$$

8. Solving the quadratic equation:

$$(3x-4)^2 = 27$$
$$3x-4 = \pm\sqrt{27}$$
$$3x-4 = \pm 3\sqrt{3}$$
$$3x = 4 \pm 3\sqrt{3}$$
$$x = \frac{4 \pm 3\sqrt{3}}{3}$$

9. Solving the quadratic equation:

$$\left(x-\frac{2}{3}\right)^2 = -\frac{25}{9}$$
$$x-\frac{2}{3} = \pm\sqrt{-\frac{25}{9}}$$
$$x-\frac{2}{3} = \pm\frac{5i}{3}$$
$$x = \frac{2}{3} \pm \frac{5}{3}i$$

10. Solving by completing the square:

$$x^2 + 8x = 4$$
$$x^2 + 8x + 16 = 4 + 16$$
$$(x+4)^2 = 20$$
$$x+4 = \pm\sqrt{20}$$
$$x+4 = \pm 2\sqrt{5}$$
$$x = -4 \pm 2\sqrt{5}$$

11. Solving by completing the square:

$$x^2 - 8x = 4$$
$$x^2 - 8x + 16 = 4 + 16$$
$$(x-4)^2 = 20$$
$$x-4 = \pm\sqrt{20}$$
$$x-4 = \pm 2\sqrt{5}$$
$$x = 4 \pm 2\sqrt{5}$$

12. Solving by completing the square:

$$x^2 - 4x - 7 = 0$$
$$x^2 - 4x = 7$$
$$x^2 - 4x + 4 = 7 + 4$$
$$(x-2)^2 = 11$$
$$x-2 = \pm\sqrt{11}$$
$$x = 2 \pm \sqrt{11}$$

13. Solving by completing the square:

$$x^2 + 4x + 3 = 0$$
$$x^2 + 4x = -3$$
$$x^2 + 4x + 4 = -3 + 4$$
$$(x+2)^2 = 1$$
$$x + 2 = \pm\sqrt{1}$$
$$x + 2 = \pm 1$$
$$x = -2 \pm 1$$
$$x = -3, -1$$

14. Solving by completing the square:

$$a^2 = 9a + 3$$
$$a^2 - 9a = 3$$
$$a^2 - 9a + \tfrac{81}{4} = 3 + \tfrac{81}{4}$$
$$\left(a - \tfrac{9}{2}\right)^2 = \tfrac{93}{4}$$
$$a - \tfrac{9}{2} = \pm\sqrt{\tfrac{93}{4}}$$
$$a - \tfrac{9}{2} = \pm\frac{\sqrt{93}}{2}$$
$$a = \frac{9 \pm \sqrt{93}}{2}$$

15. Solving by completing the square:

$$a^2 = 5a + 6$$
$$a^2 - 5a = 6$$
$$a^2 - 5a + \tfrac{25}{4} = 6 + \tfrac{25}{4}$$
$$\left(a - \tfrac{5}{2}\right)^2 = \tfrac{49}{4}$$
$$a - \tfrac{5}{2} = \pm\sqrt{\tfrac{49}{4}}$$
$$a - \tfrac{5}{2} = \pm\tfrac{7}{2}$$
$$a = \tfrac{5}{2} \pm \tfrac{7}{2}$$
$$a = -1, 6$$

16. Solving by completing the square:

$$2x^2 + 4x - 6 = 0$$
$$x^2 + 2x - 3 = 0$$
$$x^2 + 2x = 3$$
$$x^2 + 2x + 1 = 3 + 1$$
$$(x+1)^2 = 4$$
$$x + 1 = \pm\sqrt{4}$$
$$x + 1 = \pm 2$$
$$x = -1 \pm 2$$
$$x = -3, 1$$

17. Solving by completing the square:

$$3x^2 - 6x - 2 = 0$$
$$x^2 - 2x - \tfrac{2}{3} = 0$$
$$x^2 - 2x = \tfrac{2}{3}$$
$$x^2 - 2x + 1 = \tfrac{2}{3} + 1$$
$$(x-1)^2 = \tfrac{5}{3}$$
$$x - 1 = \pm\sqrt{\tfrac{5}{3}}$$
$$x - 1 = \pm\frac{\sqrt{5}}{\sqrt{3}} \cdot \frac{\sqrt{3}}{\sqrt{3}}$$
$$x - 1 = \pm\frac{\sqrt{15}}{3}$$
$$x = 1 \pm \frac{\sqrt{15}}{3}$$
$$x = \frac{3 \pm \sqrt{15}}{3}$$

18. Using $a = 1$, $b = 7$, and $c = 12$ in the quadratic formula:

$$x = \frac{-b \pm \sqrt{b^2 - 4ac}}{2a} = \frac{-7 \pm \sqrt{(7)^2 - 4(1)(12)}}{2(1)} = \frac{-7 \pm \sqrt{49 - 48}}{2} = \frac{-7 \pm 1}{2} = -4, -3$$

19. Using $a = 1$, $b = -8$, and $c = 16$ in the quadratic formula:

$$x = \frac{-b \pm \sqrt{b^2 - 4ac}}{2a} = \frac{-(-8) \pm \sqrt{(-8)^2 - 4(1)(16)}}{2(1)} = \frac{8 \pm \sqrt{64 - 64}}{2} = \frac{8 \pm 0}{2} = 4$$

20. Using $a = 1$, $b = 5$, and $c = 7$ in the quadratic formula:

$$x = \frac{-b \pm \sqrt{b^2 - 4ac}}{2a} = \frac{-5 \pm \sqrt{(5)^2 - 4(1)(7)}}{2(1)} = \frac{-5 \pm \sqrt{25 - 28}}{2} = \frac{-5 \pm \sqrt{-3}}{2} = \frac{-5 \pm i\sqrt{3}}{2}$$

21. First write the equation as $2x^2 + 8x - 5 = 0$. Using $a = 2$, $b = 8$, and $c = -5$ in the quadratic formula:

$$x = \frac{-b \pm \sqrt{b^2 - 4ac}}{2a} = \frac{-8 \pm \sqrt{(8)^2 - 4(2)(-5)}}{2(2)} = \frac{-8 \pm \sqrt{64 + 40}}{4} = \frac{-8 \pm \sqrt{104}}{4} = \frac{-8 \pm 2\sqrt{26}}{4} = \frac{-4 \pm \sqrt{26}}{2}$$

22. First write the equation as $3x^2 + 3x + 4 = 0$. Using $a = 3$, $b = 3$, and $c = 4$ in the quadratic formula:

$$x = \frac{-b \pm \sqrt{b^2 - 4ac}}{2a} = \frac{-3 \pm \sqrt{(3)^2 - 4(3)(4)}}{2(3)} = \frac{-3 \pm \sqrt{9 - 48}}{6} = \frac{-3 \pm \sqrt{-39}}{6} = \frac{-3 \pm i\sqrt{39}}{6}$$

23. First multiply by 10 to clear the equation of fractions:

$$10\left(\tfrac{1}{5}x^2 - \tfrac{1}{2}x\right) = 10\left(\tfrac{3}{10}\right)$$
$$2x^2 - 5x = 3$$
$$2x^2 - 5x - 3 = 0$$

Using $a = 2$, $b = -5$, and $c = -3$ in the quadratic formula:

$$x = \frac{-b \pm \sqrt{b^2 - 4ac}}{2a} = \frac{-(-5) \pm \sqrt{(-5)^2 - 4(2)(-3)}}{2(2)} = \frac{5 \pm \sqrt{25 + 24}}{4} = \frac{5 \pm \sqrt{49}}{4} = \frac{5 \pm 7}{4} = -\tfrac{1}{2}, 3$$

24. Simplifying the equation:

$$(2x+1)(2x-3) = -6$$
$$4x^2 - 4x - 3 = -6$$
$$4x^2 - 4x + 3 = 0$$

Using $a = 4$, $b = -4$, and $c = 3$ in the quadratic formula:

$$x = \frac{-b \pm \sqrt{b^2 - 4ac}}{2a} = \frac{-(-4) \pm \sqrt{(-4)^2 - 4(4)(3)}}{2(4)} = \frac{4 \pm \sqrt{16 - 48}}{8} = \frac{4 \pm \sqrt{-32}}{8} = \frac{4 \pm 4i\sqrt{2}}{8} = \frac{1 \pm i\sqrt{2}}{2}$$

25. Combining the complex numbers: $(4 - 3i) + 5i = 4 + 2i$

26. Combining the complex numbers: $(2 - 5i) - 3i = 2 - 8i$

27. Combining the complex numbers: $(5 + 6i) + (5 - i) = 10 + 5i$

28. Combining the complex numbers: $(2 + 5i) + (3 - 7i) = 5 - 2i$

29. Combining the complex numbers: $(3 - 2i) - (3 - i) = 3 - 2i - 3 + i = -i$

30. Combining the complex numbers: $(5 - 7i) - (6 - 2i) = 5 - 7i - 6 + 2i = -1 - 5i$

31. Combining the complex numbers: $(3 + i) - 5i - (4 - i) = 3 + i - 5i - 4 + i = -1 - 3i$

32. Combining the complex numbers: $(2 - 3i) - (5 - 2i) + (5 - i) = 2 - 3i - 5 + 2i + 5 - i = 2 - 2i$

33. Multiplying the complex numbers: $2(3 - i) = 6 - 2i$

34. Multiplying the complex numbers: $-6(4 + 2i) = -24 - 12i$

35. Multiplying the complex numbers: $4i(6 - 5i) = 24i - 20i^2 = 24i - 20(-1) = 20 + 24i$

36. Multiplying the complex numbers: $(2 - i)(3 + i) = 6 - 3i + 2i - i^2 = 6 - i - (-1) = 6 - i + 1 = 7 - i$

37. Multiplying the complex numbers: $(3 - 4i)(5 + i) = 15 - 20i + 3i - 4i^2 = 15 - 17i - 4(-1) = 15 - 17i + 4 = 19 - 17i$

38. Multiplying the complex numbers: $(3 - i)^2 = 9 - 3i - 3i + i^2 = 9 - 6i - 1 = 8 - 6i$

39. Multiplying the complex numbers: $(4 + i)(4 - i) = 16 - i^2 = 16 - (-1) = 16 + 1 = 17$

40. Multiplying the complex numbers: $(3 + 2i)(3 - 2i) = 9 - 4i^2 = 9 - 4(-1) = 9 + 4 = 13$

41. Dividing the complex numbers: $\dfrac{i}{3 + i} \cdot \dfrac{3 - i}{3 - i} = \dfrac{3i - i^2}{9 - i^2} = \dfrac{3i + 1}{9 + 1} = \dfrac{1 + 3i}{10}$

42.   Dividing the complex numbers: $\dfrac{i}{2-i} \cdot \dfrac{2+i}{2+i} = \dfrac{2i+i^2}{4-i^2} = \dfrac{2i-1}{4+1} = \dfrac{-1+2i}{5}$

43.   Dividing the complex numbers: $\dfrac{5}{2+5i} \cdot \dfrac{2-5i}{2-5i} = \dfrac{10-25i}{4-25i^2} = \dfrac{10-25i}{4+25} = \dfrac{10-25i}{29}$

44.   Dividing the complex numbers: $\dfrac{2i}{4-i} \cdot \dfrac{4+i}{4+i} = \dfrac{8i+2i^2}{16-i^2} = \dfrac{8i-2}{16+1} = \dfrac{-2+8i}{17}$

45.   Dividing the complex numbers: $\dfrac{-3i}{3-2i} \cdot \dfrac{3+2i}{3+2i} = \dfrac{-9i-6i^2}{9-4i^2} = \dfrac{-9i+6}{9+4} = \dfrac{6-9i}{13}$

46.   Dividing the complex numbers: $\dfrac{3+i}{3-i} \cdot \dfrac{3+i}{3+i} = \dfrac{9+3i+3i+i^2}{9-i^2} = \dfrac{9+6i-1}{9+1} = \dfrac{8+6i}{10} = \dfrac{4+3i}{5}$

47.   Dividing the complex numbers: $\dfrac{4-5i}{4+5i} \cdot \dfrac{4-5i}{4-5i} = \dfrac{16-20i-20i+25i^2}{16-25i^2} = \dfrac{16-40i-25}{16+25} = \dfrac{-9-40i}{41}$

48.   Dividing the complex numbers: $\dfrac{2+3i}{3-5i} \cdot \dfrac{3+5i}{3+5i} = \dfrac{6+9i+10i+15i^2}{9-25i^2} = \dfrac{6+19i-15}{9+25} = \dfrac{-9+19i}{34}$

49.   Writing as a complex number: $\sqrt{-36} = 6i$ **50.**   Writing as a complex number: $\sqrt{-144} = 12i$

51.   Writing as a complex number: $\sqrt{-17} = i\sqrt{17}$ **52.**   Writing as a complex number: $\sqrt{-31} = i\sqrt{31}$

53.   Writing as a complex number: $\sqrt{-40} = \sqrt{-4 \cdot 10} = 2i\sqrt{10}$

54.   Writing as a complex number: $\sqrt{-72} = \sqrt{-36 \cdot 2} = 6i\sqrt{2}$

55.   Writing as a complex number: $\sqrt{-200} = \sqrt{-100 \cdot 2} = 10i\sqrt{2}$

56.   Writing as a complex number: $\sqrt{-242} = \sqrt{-121 \cdot 2} = 11i\sqrt{2}$

57.   Graphing the parabola: **58.**   Graphing the parabola:

59.   Graphing the parabola:

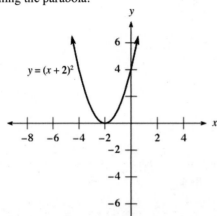

$y = (x + 2)^2$

60.   Graphing the parabola:

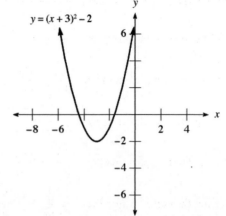

$y = (x + 3)^2 - 2$

61.   Begin by completing the square: $y = x^2 + 4x + 7 = \left(x^2 + 4x + 4\right) + 7 - 4 = (x + 2)^2 + 3$

Graphing the parabola:

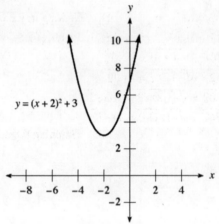

$y = (x + 2)^2 + 3$

62.   Begin by completing the square: $y = x^2 - 4x + 7 = \left(x^2 - 4x + 4\right) + 7 - 4 = (x - 2)^2 + 3$

Graphing the parabola:

$y = (x - 2)^2 + 3$